s a volume in
IPUTER SCIENCE AND APPLIED MATHEMATICS
ries of Monographs and Textbooks

pr: WERNER RHEINBOLDT

mplete list of titles in this series appears at the end of the volume.

MATF

This
COM
A Se

Edit

A co

INTRODUCTION TO MATRIX COMPUTATIONS

G. W. Stewart
The University of Texas at Austin

ACADEMIC PRESS
New York San Francisco London
A Subsidiary of Harcourt Brace Jovanovich, Publishers

ACADEMIC PRESS, INC.
111 Fifth Avenue, New York, New York 10003

United Kingdom Edition published by
ACADEMIC PRESS, INC. (LONDON) LTD.
24/28 Oval Road, London NW1

LIBRARY OF CONGRESS CATALOG CARD NUMBER: 72-82636

AMS (MOS) 1970 Subject Classifications: 65F05, 65F10, 65F15, 65F20, 65F25, 65F30, 65F35

PRINTED IN THE UNITED STATES OF AMERICA

To A. S. Householder

CONTENTS

PREFACE

Speaking in 1966 before the Society of Industrial and Applied Mathematics, the late George E. Forsythe, then Chairman of the Computer Science Department at Stanford University, stated:

> It is safe to say that matrix computation has passed well beyond the stage where an amateur is likely to think of computing methods which can compete with the better-known methods. Certainly one cannot learn theoretical linear algebra and an algebraic programming language, and nothing else, and start writing programs which will perform acceptably by today's standards. There is simply too much hard-earned experience behind the better algorithms, and yet this experience is hardly mentioned in mathematical textbooks of linear algebra.

Professor Forsythe went on to point out that most of this hard-earned experience has been accumulated since 1953. The impetus for this extraordinary development of numerical linear algebra has been provided by the demand for efficient, self-contained matrix algorithms suitable for use on a high-speed digital computer.

Today linear algebra textbooks still do not mention this hard-earned experience. The gains of the past 20 years are, for the most part, contained in advanced treatises, in journal articles, and even in unpublished technical reports. The consequence of this is that many people whose daily business involves computations with matrices are unacquainted with the best algorithms and their properties. Moreover, specialists in other areas of numerical analysis are frequently unaware of how the techniques of numerical linear algebra may be applied to their problems.

The purpose of this book is to provide a reasonably elementary introduction to some of the more important algorithms for matrix computations and to the techniques by which these algorithms may be analyzed. It is addressed to the beginner in numerical analysis as well as to the advanced student in the sciences who wishes to know more about the art of matrix computations. While the book is intended primarily as a text, it is hoped that its supplementary notes and bibliography will also make it a useful reference.

Numerical linear algebra is far too broad a subject to treat in a single introductory volume. I have chosen to treat algorithms for solving linear systems, linear least squares problems, and eigenvalue problems involving matrices whose elements can all be contained in the high-speed storage of a computer. By way of theory, I have chosen to discuss the theory of norms and perturbation theory for linear systems and for the algebraic eigenvalue problem. These choices exclude, among other things, the solution of large sparse linear systems by direct and iterative methods, linear programming, and the useful Perron–Frobenious theory and its extensions. However, a person who has fully mastered the material in this book should be well prepared for independent study in other areas of numerical linear algebra.

Since most of the algorithms discussed in this book have been published as ALGOL or FORTRAN programs, no program listings are given. However, it is a long step from a mathematical description of an algorithm to its efficient implementation on a computer. To illustrate the common techniques for conserving computer storage and operations, I have presented many of the algorithms in an informal algorithmic language, which is described in Chapter 2.

Some of the most useful results in numerical analysis consist of observations that cannot be proved in general but are nonetheless true most of the time. For example, the stability of Gaussian elimination with partial pivoting depends on the elements of the reduced matrices remaining of moderate size. Since matrices are known for which the elements become quite large, one cannot prove unconditionally that Gaussian elimination is stable. However, it has been observed that this growth does not occur with the matrices one usually encounters in practice. Similarly, error results phrased in terms of a vector norm give an imprecise idea of what is happening to the individual components of the vector; but in many applications such an imprecise idea is sufficient. It would be wrong to exclude such observations from the book because they are not mathematically rigorous. However, it is also important for the beginner to have a clear idea of what, on one hand, can be proved about an algorithm and what, on the other, is true of it most of the time. Accordingly, I have segregated the rigorous results into theorems and left discussions and empirical observations in the body of the text.

One of the major advances in numerical linear algebra has been the development of techniques for analyzing the effects of rounding error on matrix algorithms. However, even though the analyses are usually conceptually straightforward, they are often very tedious to present. Moreover, concentration on the details of a rounding-error analysis often obscures real purpose of the analysis, which is to demonstrate the stability of an algorithm or to expose the conditions under which it may become unstable. For this reason, although results from rounding-error analyses are frequently quoted in the text, only two representative analyses are given in Appendix 3.

The first chapter of the book contains a fairly complete review of elementary linear algebra, and a bright student might find it sufficient. Ordinarily, though, a course based on this book should require a year of calculus and a sophomore course in linear algebra or matrix theory, in which case the material in the first chapter can be used selectively to fill gaps in the background of the students. Particular attention should be paid to Sections 1.3 and 1.4, which treat material on matrix structure and matrix operations that is not usually stressed in linear algebra courses.

An instructor should be able to use the book at several levels, since really difficult material is introduced only in the later chapters. In particular, the content of the first three chapters supplemented by material from Sections 4.4 and 4.5 would comprise a respectable elementary course in the direct solution of linear systems. I have taught the contents of Chapters 1–5 in a one-semester introductory undergraduate course and the entire book in a one-semester graduate course. By emphasizing the derivations of the algorithms and the meanings of the theorems (as opposed to their proofs), it should be possible to present a large part of this book to a relatively unsophisticated audience. However, this can be carried too far. For example, a good grasp of the notion of orthogonality is required for the section on the linear least squares problem.

The nature of the material does not lend itself well to routine exercises, and the problems at the end of each section are relatively difficult. However, many of them have concise solutions that exploit the power of matrix methods, and the student will find it profitable to hunt for them. The instructor can of course supplement the problems with programming assignments.

ACKNOWLEDGMENTS

No book on numerical linear algebra can avoid drawing heavily on the pioneering work of J. H. Wilkinson, who, more than any one person, is responsible for the present high state of the art. I am happy to acknowledge my debt to him. I must also thank the many friends and colleagues who have helped and encouraged me in writing this book, especially those in the Center for Numerical Analysis at the University of Texas. I am specifically indebted to R. E. Funderlic, D. R. Kinkaid, W. R. Rheinboldt, and R. A. Tapia, who made detailed comments on the manuscript; to Linda Brothers who typed it; and to Margie Blevins, who has helped me at every stage and corrected my spelling errors.

More personally, I should like to thank Mrs. Ival Aslinger for introducing me to mathematics. I hope she will find this book by one of her grateful students a small reward for many years of inspired high school teaching.

I owe most to my friend and teacher A. S. Householder, to whom this book is dedicated.

PRELIMINARIES

CHAPTER 1

The subject of this book is the description and analysis of computational methods involving vectors and matrices. In this chapter we shall develop the elementary theory which underlies our subject. This development has two aspects: first the definition of vectors, matrices, and their operations; second the abstract relationships between various concepts that grow out of the idea of a vector or a matrix, such as linear dependence, column spaces, and so on. Facility with matrix operations is required to understand the description of the algorithms to be presented later; insight into matrix theory is required to understand their analyses.

We shall be concerned with real n-space. When $n = 2$ this is, in effect, the Euclidean plane, and when $n = 3$, the three-dimensional space of our everyday experience. It follows that many general theorems can be visualized as geometric facts about the plane or three-dimensional space. Conversely, our geometric knowledge of two- or three-dimensional space can often be directly extended to general theorems about n-space. We shall develop this geometric point of view informally in this chapter.

Most of the algorithms dealing with the rectangular arrays of numbers called matrices proceed by a succession of transformations that introduce new zeros into the array, finally arriving at a conveniently simple form. It is not surprising then that there is an extensive terminology associated with the distribution of zeros in a matrix. Moreover, it is important to know what distributions are preserved by the standard matrix operations. These points are also treated in this chapter.

1. THE SPACE $\mathbb{R}^n$

The idea of a vector in real n-dimensional space is a natural generalization of the representation of points in a plane by Cartesian coordinates. In this representation, a special point is distinguished as an origin and two perpendicular lines called coordinate axes are constructed through this point. Then each point p in the plane can be represented as an ordered pair (ξ_1, ξ_2) whose first and second elements are obtained by projecting p on the first and second coordinate axes, respectively (Fig. 1). Of course the point p and the ordered pair or vector (ξ_1, ξ_2) are different objects; however, their relation is so intimate that one often speaks of the point (ξ_1, ξ_2) and proves geometric theorems about the plane by manipulating ordered pairs of numbers rather than points.

Following this lead, we shall speak of ordered n-tuples as n-vectors. However, for reasons that will become clear later, it is convenient to arrange these numbers in a column rather than in a row.

DEFINITION 1.1. An *n-vector* x is a collection of n real numbers $\xi_1, \xi_2, \ldots, \xi_n$ arranged in order in a column:

$$x = \begin{pmatrix} \xi_1 \\ \xi_2 \\ \cdot \\ \cdot \\ \cdot \\ \xi_n \end{pmatrix}.$$

The numbers $\xi_1, \xi_2, \ldots, \xi_n$ are called the *components* of x.

When n is fixed or known from context, it is customary to refer to the vector x rather than the n-vector x. We shall denote the set of all n-vectors by $\mathbb{R}^n$, which will be called the *vector space* $\mathbb{R}^n$, or simply the *space* $\mathbb{R}^n$. The field of real numbers will be denoted by $\mathbb{R}$ and will be referred to as *scalars*.

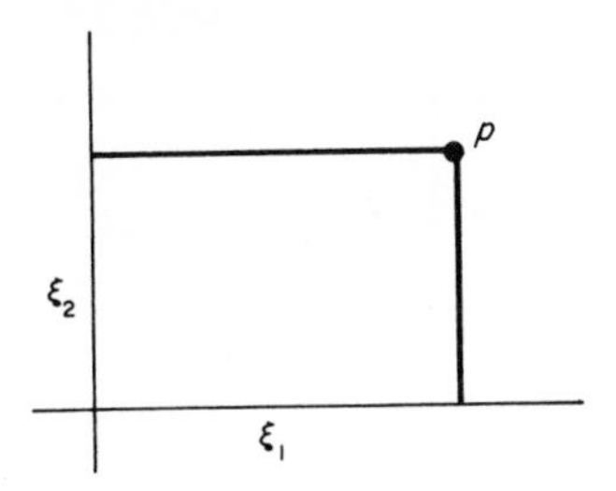

Fig. 1

EXAMPLE 1.2. The elements of the vector space $\mathbb{R}^1$ may be placed in a natural correspondence with the elements of the field $\mathbb{R}$ as follows. Each vector $(\alpha) \in \mathbb{R}^1$ is associated with the scalar $\alpha \in \mathbb{R}$ and vice versa. Strictly speaking, the space $\mathbb{R}^1$ and the field $\mathbb{R}$ are distinct mathematical objects. However, we shall often identify the two, using members of $\mathbb{R}^1$ as scalars and conversely regarding scalars as members of $\mathbb{R}^1$.

The following notational conventions will be used throughout the book. Lower case Greek letters will denote scalars; lower case Latin letters will denote vectors. Wherever possible we shall attempt to represent the components of a vector by the corresponding Greek letter. Thus, unless otherwise stated, the scalar α_i is to be taken as the ith component of the vector a. Since the correspondence between the Greek and Latin alphabets is not perfect, some of the associations, which are listed in Appendix 1, are artificial. Particular note should be made of the association of x with ξ and y with η. As an exception to the above conventions we shall often use lower case Latin letters as subscripts and summation indices.

As was noted above, a vector x in $\mathbb{R}^2$ is associated with a point in the plane whose coordinates are ξ_1 and ξ_2. Similarly, a vector x in $\mathbb{R}^3$ is associated with a point in three-dimensional space whose coordinates are ξ_1, ξ_2, and ξ_3. It is customary to represent a vector graphically by drawing an arrow from the origin to the point associated with the vector (Fig. 2).

Two n-vectors are equal if and only if their corresponding components are equal. Thus the vector equality

$$a = b$$

is equivalent to the set of scalar equalities

$$\alpha_i = \beta_i \qquad (i = 1, 2, \ldots, n).$$

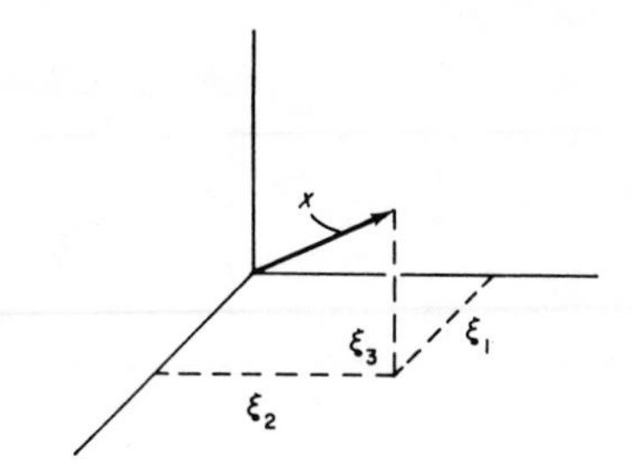

Fig. 2

In particular, to prove that two vectors are equal it is only necessary to show that their components are equal, and to define a new vector it is only necessary to specify how its components are formed.

We now turn to the first of the vector operations, the sum of two vectors.

DEFINITION 1.3. Let $a, b \in \mathbb{R}^n$. The *sum* of a and b, written $a + b$, is the n-vector c whose components are given by

$$\gamma_i = \alpha_i + \beta_i.$$

The sum of two vectors has the following geometric interpretation. The vectors a and b form the sides of a parallelogram with one corner at the origin. The sum of a and b is then the diagonal of the parallelogram that proceeds from the origin (Fig. 3).

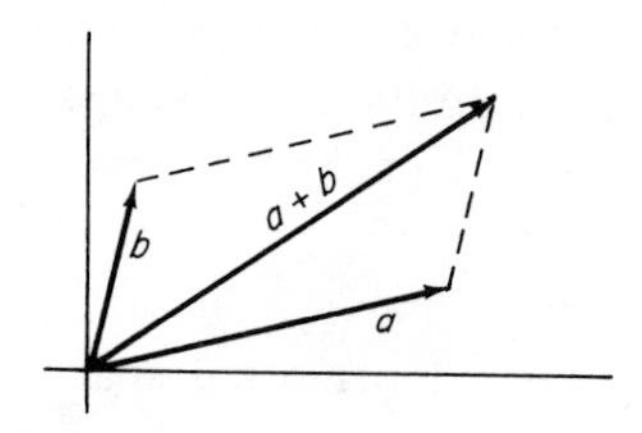

Fig. 3

The sum of two vectors has some of the properties of the usual sum of two scalars, as is shown in the following theorem.

THEOREM 1.4. Let $a, b, c \in \mathbb{R}^n$. Then

1. $a + b = b + a$,
2. $(a + b) + c = a + (b + c)$.

PROOF. To prove part 1, let $x = a + b$ and $y = b + a$. Then by the properties of $\mathbb{R}$,

$$\xi_i = \alpha_i + \beta_i = \beta_i + \alpha_i = \eta_i.$$

Hence by the above observations on the equality of vectors, $x = y$. To prove part 2, let $x = (a + b) + c$ and $y = a + (b + c)$. Then

$$\xi_i = (\alpha_i + \beta_i) + \gamma_i = \alpha_i + (\beta_i + \gamma_i) = \eta_i.$$

Hence $x = y$. ■

Property 1.4.2 says that the vector sum is an associative operation. More generally, if $a_1, a_2, \ldots, \alpha_n \in \mathbb{R}^n$, the sum $a_1 + a_2 + \cdots + a_n$ is the same, irrespective of the order in which sums are grouped. Likewise it follows from 1.4.1 that the sum $a_1 + a_2 + \cdots + a_n$ is unaltered when the order of the a_i's is changed. For example,

$$a_1 + a_2 + \cdots + a_k = a_k + a_{k-1} + \cdots + a_1.$$

DEFINITION 1.5. The *zero vector* in $\mathbb{R}^n$ is the vector whose n components are zero.

For all n we shall denote the zero vector in $\mathbb{R}^n$ by the same symbol "0", which is also used to denote the scalar zero. Where the meaning of the symbol "0" is not specified, it will be clear from the context what is meant.

The zero vector has some of the properties of the number zero.

THEOREM 1.6. Let $a \in \mathbb{R}^n$. Then

1. $a + 0 = a$,
2. there is a vector, written $-a$, in $\mathbb{R}^n$ such that $a + (-a) = 0$.

PROOF. Let $b = a + 0$. Then

$$\beta_i = \alpha_i + 0 = \alpha_i$$

and $b = a$, which establishes part 1. For part 2, let

$$b = \begin{pmatrix} -\alpha_1 \\ -\alpha_2 \\ \vdots \\ -\alpha_n \end{pmatrix},$$

and $c = a + b$. Then

$$\gamma_i = \alpha_i + \beta_i = \alpha_i + (-\alpha_i) = 0.$$

Thus $c = 0$ and b is the vector $-a$ sought in part 2. ∎

The vector $-a$ of the second part of Theorem 1.6 is simply the vector obtained by changing the signs of the components of a. Geometrically, this means that the vector a is reflected through the origin (Fig. 4).

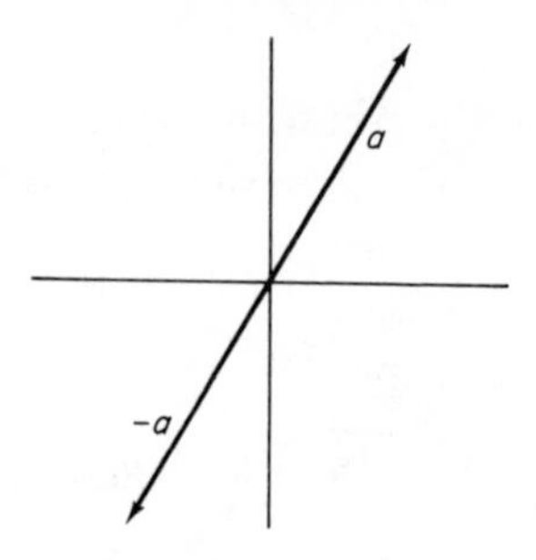

Fig. 4

The symbol "$-$" has been used for the unitary operation that, given the vector a, produces the vector $-a$ of Theorem 1.6. We shall also use the symbol to denoted the binary operation of subtraction. Specifically, given the vectors a and b, we defined their *difference* $a - b$ by

$$a - b = a + (-b).$$

It is easily verified that the difference operation satisfies the usual laws of subtraction among scalars; for example,

$$a - b = -(b - a).$$

The second important operation with vectors is the operation of scalar multiplication.

DEFINITION 1.7. Let $\lambda \in \mathbb{R}$ and $a \in \mathbb{R}^n$. The product of λ and a, written $\lambda \cdot a$ or λa, is the n-vector b whose components are given by

$$\beta_i = \lambda \alpha_i.$$

Geometrically, the operation of multiplying the vector a by the scalar λ

changes the length of a by a factor of $|\lambda|$. If λ is negative, a is also reflected through the origin.

THEOREM 1.8. Let $\lambda, \mu \in \mathbb{R}$ and $a, b \in \mathbb{R}^n$. Then

1. $(\lambda\mu)a = \lambda(\mu a)$,
2. $(\lambda + \mu)a = \lambda a + \mu a$,
3. $\lambda(a + b) = \lambda a + \lambda b$,
4. $1 \cdot a = a$.

PROOF. We shall establish property 3, leaving the rest for Exercise 1.3. In stating the theorem we have followed the usual convention of allowing multiplication to take precedence over addition, so that $\lambda a + \mu b$ means $(\lambda a) + (\mu b)$. Let $x = \lambda(a + b)$ and $y = \lambda a + \lambda b$. Then

$$\xi_i = \lambda(\alpha_i + \beta_i) = (\lambda\alpha_i) + (\lambda\beta_i) = \eta_i,$$

which establishes part 3. ■

In the above development we have defined vectors and their operations in terms of the real numbers. The properties listed in Theorems 1.4, 1.6, 1.8 are then immediate consequences of the properties of real numbers. Alternatively, we could take the properties of the theorems as axioms describing the properties of a sum and product over some set of objects $\mathcal{V}$. Specifically we call a set $\mathcal{V}$ an *abstract vector space* (over the real numbers) if

1. there is a sum "+" defined among the elements of $\mathcal{V}$ that satisfies the properties listed in Theorem 1.4,
2. there is an element "0" of $\mathcal{V}$ that satisfies the properties listed in Theorem 1.6,
3. there is a product "·" defined between the real numbers and the elements of $\mathcal{V}$ that satisfies the properties listed in Theorem 1.8.

The elements of an abstract vector space need not be n-tuples of real numbers, as the following example shows.

EXAMPLE 1.9. Let $\mathcal{V}$ be the set of all real-valued functions defined on $[0, 1]$. If $f, g \in \mathcal{V}$, define $h = f + g$ as the function whose values are $h(\xi) = f(\xi) + g(\xi)$, $\xi \in [0, 1]$. If $f \in \mathcal{V}$ and $\lambda \in \mathbb{R}$, define $h = \lambda f$ as the function whose values are $h(\xi) = \lambda f(\xi)$. Let "0" be the function that is identically zero on $[0, 1]$. Then with these definitions, $\mathcal{V}$ is a vector space.

The properties of an abstract vector space allow us to establish theorems that are not immediately obvious. However, the proofs may be tedious. For example, the following is a proof of the theorem that if $\mathcal{V}$ is an abstract vector space and $a \in \mathcal{V}$, then $0 \cdot a = 0$. The properties that ensure each equality are listed to the side. (Incidentally, note that in the equation $0 \cdot a = 0$, the symbol "0" is used in two ways; on the left it is the scalar zero, on the right the zero vector.)

$$\begin{aligned} 0 &= a + (-a), && 1.6.2, \\ &= 1 \cdot a + (-a), && 1.8.4, \\ &= (0 + 1) \cdot a + (-a), && 1 + 0 = 1, \\ &= (0 \cdot a + 1 \cdot a) + (-a), && 1.8.2, \\ &= 0 \cdot a + [1 \cdot a + (-a)], && 1.4.2, \\ &= 0 \cdot a + [a + (-a)], && 1.8.4, \\ &= 0 \cdot a + 0, && 1.6.2, \\ &= 0 \cdot a, && 1.6.1. \end{aligned}$$

On the other hand, to verify this fact about $\mathbb{R}^n$ is easy. Let $b = 0 \cdot a$. Then $\beta_i = 0 \cdot \alpha_i = 0$; hence $b = 0$. It is often the case that theorems concerning abstract vector spaces are trivialities when stated about $\mathbb{R}^n$. Since in this chapter we shall be concerned exclusively with $\mathbb{R}^n$, we shall not develop the theory of abstract vector spaces. Whenever a fact can be easily demonstrated by appealing to the properties of real numbers, we will use it, leaving its verification as an exercise.

EXERCISES

1. Perform the indicated calculations.

(a) $\begin{pmatrix} 1 \\ -3 \end{pmatrix} + 2\begin{pmatrix} -1 \\ 1 \end{pmatrix}$, (b) $\begin{pmatrix} \xi_1 \\ \xi_2 \end{pmatrix} - \begin{pmatrix} \eta_1 \\ \eta_2 \end{pmatrix}$,

(c) $\beta\begin{pmatrix} 1 \\ \alpha \\ \alpha^2 \end{pmatrix} + \begin{pmatrix} 1 \\ 1 \\ 1 \end{pmatrix}$, (d) $\begin{pmatrix} 0 \\ 1 \\ 0 \end{pmatrix} + \tau\begin{pmatrix} 2 \\ 10 \\ 1 \end{pmatrix}$,

(e) $\xi_1\begin{pmatrix} 1 \\ 0 \\ 0 \end{pmatrix} + \xi_2\begin{pmatrix} 0 \\ 1 \\ 0 \end{pmatrix} + \xi_3\begin{pmatrix} 0 \\ 0 \\ 1 \end{pmatrix}$.

2. In the plane sketch the locus of the following vectors.

(a) $\begin{pmatrix} \xi \\ |\xi| \end{pmatrix}, \quad -\infty < \xi < \infty,$

(b) $\begin{pmatrix} \xi \\ \xi^2 \end{pmatrix}, \quad -\infty < \xi < \infty,$

(c) $\begin{pmatrix} 1 \\ -1 \end{pmatrix} + \begin{pmatrix} \cos\theta \\ \sin\theta \end{pmatrix}, \quad 0 \le \theta \le 2\pi,$

(d) $\begin{pmatrix} \cosh\tau \\ \cosh\tau \end{pmatrix}, \quad -\infty < \tau < \infty.$

3. Complete the proof of Theorem 1.8.

4. Establish the following identities in $\mathbb{R}^n$.

$$a + (b + (c + d)) = ((a + b) + c) + d,$$
$$(-1) \cdot a = -a,$$
$$a + b + c + d = d + c + b + a.$$

Show that the same identities hold in any abstract vector space.

5. Prove that in $\mathbb{R}^n$

$$\lambda x = 0 \quad \Rightarrow \quad \lambda = 0 \text{ or } x = 0.$$

6. Verify in detail that the set $\mathcal{V}$ of Example 1.9 is an abstract vector space.

7. The function $\phi : [0, 1] \to \mathbb{R}$ is a *step function* if there are points $0 = \xi_0 < \xi_1 < \xi_2 < \cdots < \xi_n = 1$ and constants $\eta_1, \eta_2, \ldots, \eta_n$ such that $\phi(\xi) = \eta_i$ whenever $\xi \in (\xi_{i-1}, \xi_i)$. Define a sum and scalar product for step functions as in Example 1.9, and let "0" be represented by the function that is identically zero. Prove that with these definitions the set of all step functions on [0, 1] is an abstract vector space.

2. LINEAR INDEPENDENCE, SUBSPACES, AND BASES

Given a collection of vectors $\mathcal{X}$, it is possible to obtain new vectors by repeatedly applying the operations of addition and scalar multiplication.

Any vector obtained in this way can be written in the form

$$\alpha_1 x_1 + \alpha_2 x_2 + \cdots + \alpha_m x_m, \tag{2.1}$$

where $x_1, x_2, \ldots, x_m \in \mathcal{X}$. It is useful to have a name for the form (2.1).

DEFINITION 2.1. Let $x_1, x_2, \ldots, x_m \in R^n$ and $\alpha_1, \alpha_2, \ldots, \alpha_m \in R$. Then the vector (2.1) is called a *linear combination* of $x_1, x_2, \ldots, x_m$. The numbers $\alpha_1, \alpha_2, \ldots, \alpha_m$ are called the *coefficients* of the linear combination. If $\alpha_1 = \alpha_2 = \cdots = \alpha_m = 0$, the linear combination is said to be *trivial*; otherwise it is said to be *nontrivial*.

EXAMPLE 2.2. Let e_i $(i = 1, 2, \ldots, n)$ denote the n-vector whose ith component is unity and whose other components are zero. Then for any n-vector x

$$x = \xi_1 e_1 + \xi_2 e_2 + \cdots + \xi_n e_n. \tag{2.2}$$

In other words *any n-vector is a linear combination of* $e_1, e_2, \ldots, e_n$. These vectors will continue to appear, and we shall reserve the symbols e_i $(i = 1, 2, \ldots, n)$ to refer to them. When it is not clear from the context, the dimension n will be specified explicitly.

EXAMPLE 2.3. Let x and y be members of R^3 that do not lie along the same line. Then the set of all linear combinations of x and y is the plane passing through the points x, y, and the origin (Fig. 1). This plane is called the plane *spanned* by x and y.

In Example 2.3 no nontrivial linear combination of x and y can be the zero vector; for if $\alpha x + \beta y = 0$ with, say $\alpha \neq 0$, then $x = (\beta/\alpha)y$ and x

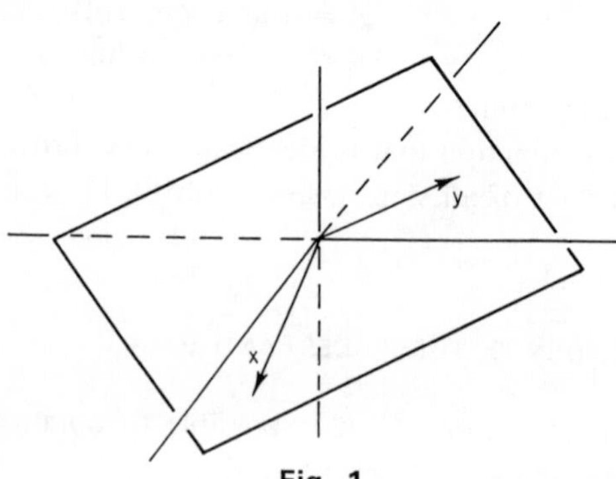

Fig. 1

and y lie along the same line. On the other hand if z lies in the plane spanned by x and y, then $\alpha x + \beta y - 1 \cdot z = 0$ for some scalars α and β. It is important to have names for sets like $\{x, y\}$ whose elements are independent and sets like $\{x, y, z\}$ whose elements depend on one another.

DEFINITION 2.4. Let $\mathcal{X} \subset \mathbb{R}^n$. The elements of $\mathcal{X}$ are said to be *linearly independent* if no nontrivial linear combination of elements of $\mathcal{X}$ is zero. Otherwise the elements of $\mathcal{X}$ are said to be *linearly dependent.*

An equivalent way of saying that the elements of $\mathcal{X}$ are linearly independent is to say that if $x_1, x_2, \ldots, x_m \in \mathcal{X}$ and $\alpha_1 x_1 + \alpha_2 x_2 + \cdots + \alpha_m x_m = 0$, then $\alpha_1 = \alpha_2 = \cdots = \alpha_m = 0$. Thus to prove that $x_1, x_2, \ldots, x_m$ are linearly independent it is sufficient to assume that $\alpha_1 x_1 + \alpha_2 x_2 + \cdots + \alpha_m x_m = 0$ and then to show that all the α_i are zero. On the other hand, to prove that $x_1, x_2, \ldots, x_m$ are linearly dependent one must find scalars $\alpha_1, \alpha_2, \ldots, \alpha_m$, not all zero, such that $\alpha_1 x_1 + \alpha_2 x_2 + \cdots + \alpha_m x_m = 0$.

EXAMPLE 2.5. In $\mathbb{R}^n$ suppose that

$$a = \alpha_1 e_1 + \alpha_2 e_2 + \cdots + \alpha_n e_n = 0.$$

Then since α_i is the ith component of a, it must be zero. Thus *the vectors* $e_1, e_2, \ldots, e_n$ *are linearly independent.*

EXAMPLE 2.6. Any set containing the zero vector has linearly dependent elements, for the linear combination $1 \cdot 0$ is nontrivial and is certainly zero.

Suppose that the elements of $\mathcal{X}$ are linearly dependent. Then there are vectors $x_1, x_2, \ldots, x_m \in \mathcal{X}$ and scalars $\alpha_1, \alpha_2, \ldots, \alpha_m$, not all zero, such that

$$\alpha_1 x_1 + \alpha_2 x_2 + \cdots + \alpha_m x_m = 0. \tag{2.3}$$

In particular, suppose that $\alpha_k \neq 0$. Then (2.3) may be solved for x_k

$$x_k = -\frac{\alpha_1}{\alpha_k} x_1 - \cdots - \frac{\alpha_{k-1}}{\alpha_k} x_{k-1} - \frac{\alpha_{k+1}}{\alpha_k} x_{k+1} - \cdots - \frac{\alpha_m}{\alpha_k} x_m. \tag{2.4}$$

Thus any linear combination of vectors in $\mathcal{X}$ can be written without the vector x_k. So far as linear combinations are concerned x_k contributes no additional information.

A set of linearly independent vectors has the important property that different linear combinations of its vectors give different results. More precisely, we have the following theorem.

THEOREM 2.7. Let $x_1, x_2, \ldots, x_m \in \mathbb{R}^n$ be linearly independent. Suppose that both

$$y = \alpha_1 x_1 + \alpha_2 x_2 + \cdots + \alpha_m x_m \tag{2.5}$$

and

$$y = \beta_1 x_1 + \beta_2 x_2 + \cdots + \beta_m x_m. \tag{2.6}$$

Then

$$\alpha_i = \beta_i \qquad (i = 1, 2, \ldots, m). \tag{2.7}$$

PROOF. From (2.5) and (2.6) it follows that

$$0 = y - y = (\alpha_1 - \beta_1)x_1 + (\alpha_2 - \beta_2)x_2 + \cdots + (\alpha_m - \beta_m)x_m.$$

Since the vectors x_k are linearly independent, we must have

$$\alpha_i - \beta_i = 0 \qquad (i = 1, 2, \ldots, m),$$

which is just (2.7). ∎

Let z_1 and z_2 be members of the plane described in Example 2.3. Then $z_1 = \alpha_1 x + \beta_1 y$ and $z_2 = \alpha_2 x + \beta_2 y$ for some scalars $\alpha_1, \alpha_2, \beta_1$, and β_2. The linear combination $\gamma_1 z_1 + \gamma_2 z_2$ can be written in the form

$$\gamma_1 z_1 + \gamma_2 z_2 = (\gamma_1 \alpha_1 + \gamma_2 \alpha_2)x + (\gamma_1 \beta_1 + \gamma_2 \beta_2)y,$$

which is a linear combination of x and y and hence lies in the plane. Thus a linear combination of any two vectors in the plane is again in the plane, which for this reason is called a subspace of $\mathbb{R}^3$. This notion can be generalized to arbitrary subsets of $\mathbb{R}^n$.

DEFINITION 2.8. Let $\mathcal{S}$ be a nonempty subset of $\mathbb{R}^n$. Then $\mathcal{S}$ is a *subspace* of $\mathbb{R}^n$ if

1. $x, y \in \mathcal{S} \;\Rightarrow\; x + y \in \mathcal{S}$,
2. $x \in \mathcal{S}$ and $\alpha \in \mathbb{R} \;\Rightarrow\; \alpha x \in \mathcal{S}$.

It follows immediately from the definition that if $\mathcal{S}$ is a subspace of $\mathbb{R}^n$, linear combinations of elements in $\mathcal{S}$ are again in $\mathcal{S}$. Conversely, any nonempty subset of $\mathbb{R}^n$ that is closed under linear combinations is a subspace.

EXAMPLE 2.9. The space $\mathbb{R}^n$ is itself a subspace of $\mathbb{R}^n$. The set $\{0\}$ consisting of only the zero vector is the *trivial* subspace of $\mathbb{R}^n$.

The plane of Example 2.3 was formed by taking all linear combinations of the vectors x and y. This suggests the following technique for generating subspaces.

THEOREM 2.10. Let $\mathcal{A} \subset \mathbb{R}^n$ be nonempty and let $\mathcal{S}$ be the set of all linear combinations of elements of $\mathcal{A}$. Then $\mathcal{S}$ is a subspace of $\mathbb{R}^n$.

PROOF. Let $x, y \in \mathcal{S}$. Then, for some vectors $a_1, a_2, \ldots, a_m \in \mathcal{A}$ and scalars $\sigma_1, \ldots, \sigma_m, \tau_1, \ldots, \tau_m$, we have $x = \sum_{i=1}^m \sigma_i a_i$ and $y = \sum_{i=1}^m \tau_i a_i$. Then $x + y = \sum_{i=1}^m (\sigma_i + \tau_i) a_i \in \mathcal{S}$. Likewise, if $x = \sum_{i=1}^m \sigma_i a_i \in \mathcal{S}$, then $\alpha x = \sum_{i=1}^m (\alpha\sigma_i) a_i \in \mathcal{S}$. ■

The subspace $\mathcal{S}$ of Theorem 2.10 is called the subspace *generated* by $\mathcal{A}$. The elements of $\mathcal{A}$ are said to *span* $\mathcal{S}$. Thus the plane of Example 2.3 is generated by the set $\{x, y\}$ and is spanned by the vectors x and y.

In many situations it is more convenient to work with a set that generates a subspace than to work with the subspace itself, since the former may contain only a finite number of vectors while the latter, if it is nontrivial, must contain an infinite number of vectors. If the elements of a generating set are linearly dependent, it follows from the observation after Example 2.6 that at least one of the vectors in the set is superfluous. Thus there is good reason for studying generating sets whose elements are linearly independent. Such a set is called a basis for the subspace.

DEFINITION 2.11. Let $\mathcal{S} \subset \mathbb{R}^n$ be a subspace, and let $\mathcal{B} \subset \mathcal{S}$. Then $\mathcal{B}$ is a *basis* for $\mathcal{S}$ if

1. the elements of $\mathcal{B}$ are linearly independent,
2. $\mathcal{B}$ generates $\mathcal{S}$.

If $\mathcal{B} = \{b_1, b_2, \ldots, b_m\}$ is a basis for $\mathcal{S} \subset \mathbb{R}^n$, and $x \in \mathcal{S}$, then by 2.11.2

$$x = \gamma_1 b_1 + \gamma_2 b_2 + \cdots + \gamma_m b_m,$$

for some numbers $\gamma_1, \gamma_2, \ldots, \gamma_m$. By Theorem 2.7, the numbers $\gamma_1, \gamma_2, \ldots, \gamma_m$ are uniquely determined by the vector x. They are called the *components* of x with respect to the basis $\mathcal{B}$.

EXAMPLE 2.12. The vectors

$$\begin{pmatrix} 1 \\ 1 \end{pmatrix}, \quad \begin{pmatrix} 1 \\ -1 \end{pmatrix}$$

form a basis for $\mathbb{R}^2$. The components of the vector x with respect to this basis are $(\xi_1 + \xi_2)/2$ and $(\xi_1 - \xi_2)/2$.

EXAMPLE 2.13. By Example 2.5, the vectors $e_1, e_2, \ldots, e_n \in \mathbb{R}^n$ are linearly independent. By Example 2.2, they span $\mathbb{R}^n$. Hence the set $\{e_1, e_2, \ldots, e_n\}$ is a basis for $\mathbb{R}^n$. From the equality (2.2), we see that the components of the vector x with respect to the basis $\{e_1, e_2, \ldots, e_n\}$ are the numbers $\xi_1, \xi_2, \ldots, \xi_n$, which are just the usual components of the vector x.

So far we have left open the question of whether an arbitrary subspace of $\mathbb{R}^n$ has a basis. In answering this question, we shall also establish that any two bases for a given subspace of $\mathbb{R}^n$ must have the same finite number of elements. The development begins with the following useful result.

THEOREM 2.14. Let $\mathcal{A} = \{a_1, a_2, \ldots, a_l\}$ be a basis for the subspace $\mathcal{S} \subset \mathbb{R}^n$. Let the vectors $b_1, b_2, \ldots, b_m \in \mathcal{S}$ be linearly independent. Then there is an integer k such that the vectors $a_k, b_2, b_3, \ldots, b_m$ are linearly independent.

PROOF. Suppose on the contrary that for each $i = 1, 2, \ldots, l$ the vectors $a_i, b_2, \ldots, b_m$ are linearly dependent. Then for each $i = 1, 2, \ldots, l$, there are constants $\sigma_{i1}, \sigma_{i2}, \ldots, \sigma_{im}$, not all zero, such that

$$\sigma_{i1}a_i + \sigma_{i2}b_2 + \cdots + \sigma_{im}b_m = 0. \tag{2.8}$$

Now σ_{i1} cannot be zero, for then the vectors $b_2, b_3, \ldots, b_m$ would be linearly dependent. Hence

$$\begin{aligned} a_i &= -\sigma_{i1}^{-1}(\sigma_{i2}b_2 + \sigma_{i3}b_3 + \cdots + \sigma_{im}b_m) \\ &\equiv \eta_{i2}b_2 + \eta_{i3}b_3 + \cdots + \eta_{im}b_m. \end{aligned} \tag{2.9}$$

Since $\mathcal{A}$ is a basis for $\mathcal{S}$ and $b_1 \in \mathcal{S}$, there are constants $\gamma_1, \gamma_2, \ldots, \gamma_l$ such that

$$b_1 = \sum_{i=1}^{l} \gamma_i a_i. \tag{2.10}$$

If (2.9) and (2.10) are combined, the result is

$$b_1 = \sum_{j=2}^{m} \left(\sum_{i=1}^{l} \gamma_i \eta_{ij} \right) b_j. \tag{2.11}$$

Equation (2.11) expresses b_1 as a linear combination of $b_2, b_3, \ldots, b_m$, which is impossible since the vectors $b_1, b_2, \ldots, b_m$ are linearly independent. The contradiction establishes the theorem. ■

The main results follow from Theorem 2.14.

THEOREM 2.15. Any basis for $\mathbb{R}^n$ must have exactly n elements.

PROOF. We first show that no basis can have more than n elements. Suppose to the contrary that $\{b_1, b_2, \ldots\}$ is a basis with more than n elements. Then the vectors $b_1, b_2, \ldots, b_n, b_{n+1}$ are linearly independent. Since the set $\{e_1, e_2, \ldots, e_n\}$ is a basis for $\mathbb{R}^n$, we may apply Theorem 2.14 to find an integer m_1 such that the vectors $e_{m_1}, b_2, b_3, \ldots, b_{n+1}$ are linearly independent. A second application of Theorem 2.14 yields an integer m_2 such that the vectors $e_{m_1}, e_{m_2}, b_3, \ldots, b_{n+1}$ are linearly independent. Of course $m_1 \neq m_2$. Repeated applications of this process finally gives n distinct integers $m_1, m_2, \ldots, m_n$ such that the vectors $e_{m_1}, e_{m_2}, \ldots, e_{m_n}, b_{n+1}$ are linearly independent. However, $\{e_{m_1}, e_{m_2}, \ldots, e_{m_n}\} = \{e_1, e_2, \ldots, e_n\}$; hence b_{n+1} can be written as a linear combination of $e_{m_1}, e_{m_2}, \ldots, e_{m_n}$, a contradiction.

A similar argument shows that a basis for $\mathbb{R}^n$ can have no fewer than n elements. ■

In proving Theorem 2.15, we also proved the following corollary.

COROLLARY 2.16. No set of linearly independent vectors in $\mathbb{R}^n$ has more than n elements.

The existence of bases for subspaces of $\mathbb{R}^n$ is a corollary of the following theorem, which enables us to construct a basis that includes a given set of linearly independent vectors.

THEOREM 2.17. Let $\mathcal{S}$ be a subspace of $\mathbb{R}^n$ and let $b_1, b_2, \ldots, b_k \in \mathcal{S}$ be linearly independent. Then there is a basis for $\mathcal{S}$ containing the vectors $b_1, b_2, \ldots, b_k$.

PROOF. Let $\mathcal{B}_k = \{b_1, b_2, \ldots, b_k\}$. If there is a vector b_{k+1} in $\mathcal{S}$ such that the vectors $b_1, b_2, \ldots, b_k, b_{k+1}$ are linearly independent, set $\mathcal{B}_{k+1} = \{b_1, b_2, \ldots, b_{k+1}\}$. Continue in this way to construct sets

$$\mathcal{B}_k \subset \mathcal{B}_{k+1} \subset \mathcal{B}_{k+2} \subset \cdots$$

such that $\mathcal{B}_l$ consists of l linearly independent vectors of $\mathcal{S}$. Since no set of linearly independent vectors in $\mathbb{R}^n$ can have more than n elements, the above construction must stop with some $\mathcal{B}_m$, where $m \leq n$.

We claim that $\mathcal{B}_m$ is a basis for $\mathcal{S}$. By construction the members of $\mathcal{B}_m$ are linearly independent. It remains to show that $\mathcal{B}_m$ generates $\mathcal{S}$. Let $x \in \mathcal{S}$. Then the vectors $x, b_1, b_2, \ldots, b_m$ must be linearly dependent, since otherwise we could construct the set $\mathcal{B}_{m+1} = \{b_1, b_2, \ldots, b_m, x\}$ of linearly independent vectors of $\mathcal{S}$, contradicting the maximality of m. Since $b_1, b_2, \ldots, b_m$ are linearly independent it follows (compare the proof of Theorem 2.14) that x is a linear combination of $b_1, b_2, \ldots, b_m$.

Since $b_1, b_2, \ldots, b_k \in \mathcal{B}_m$, the set $\mathcal{B}_m$ is the required basis. ∎

COROLLARY 2.18. Every nontrivial subspace of $\mathbb{R}^n$ has a basis.

PROOF. If $\mathcal{S} \subset \mathbb{R}^n$ is a nontrivial subspace, there is a nonzero vector $b_1 \in \mathcal{S}$. The vector b_1 is linearly independent; hence Theorem 2.17 applies to give a basis for $\mathcal{S}$, incidentally containing b_1. ∎

In proving Theorem 2.15, we started with the basis $\{e_1, e_2, \ldots, e_n\}$ for $\mathbb{R}^n$ and showed first that any other basis must have fewer elements and then that any other basis must have more elements. Starting with a fixed basis for a subspace $\mathcal{S} \subset \mathbb{R}^n$, we may use the same arguments to show that any other basis must have the same number of elements as $\mathcal{S}$. Of course this number must be less than or equal to n. This justifies the following definition.

DEFINITION 2.19. If $\mathcal{S}$ is a nontrivial subspace of $\mathbb{R}^n$, the *dimension* of $\mathcal{S}$, written $\dim(\mathcal{S})$, is the number of elements in a basis for $\mathcal{S}$. The dimension of the trivial subspace is zero.

The above definition agrees with our intuitive notion of dimension in $\mathbb{R}^3$. If $\dim(\mathcal{S}) = 1$, then a basis for $\mathcal{S}$ consists of a single vector and every other element of $\mathcal{S}$ must be a multiple of this vector. In other words, the elements of $\mathcal{S}$ lie along a line passing through the origin. If $\dim(\mathcal{S}) = 2$,

are linearly independent. [*Hint*: If $\sum_{i=1}^{n} \alpha_i v_i = 0$, then the polynomial $\pi(\xi) = \alpha_n \xi^{n-1} + \alpha_{n-1}\xi^{n-2} + \cdots + \alpha_1$ has the n distinct zeros $\xi_1, \xi_2, \ldots, \xi_n$.]

3. Show that a subspace of $\mathbb{R}^n$ is an abstract vector space.

4. Why does the trivial subspace $\{0\}$ not have a basis?

5. Let $\{\mathcal{S}_\alpha : \alpha \in I\}$ be a set of subspaces of $\mathbb{R}^n$. Show that $\bigcap_{\alpha \in I} \mathcal{S}_\alpha$ is a subspace.

6. Let $\mathcal{A} \subset \mathbb{R}^n$. Show that the subspace generated by $\mathcal{A}$ is $\bigcap \{\mathcal{S} : \mathcal{S}$ is a subspace and $\mathcal{A} \subset \mathcal{S}\}$.

7. Let $\mathcal{A} \subset \mathbb{R}^n$ have a nonzero element. Show that there is a basis for the subspace generated by $\mathcal{A}$ consisting entirely of elements of $\mathcal{A}$.

8. In Theorem 2.14, show that if $\{b_1, b_2, \ldots, b_m\}$ is also a basis for $\mathcal{S}$, then $\{a_k, b_2, \ldots, b_m\}$ is a basis for $\mathcal{S}$.

9. Complete the proof of Theorem 2.15.

10. Give the details of the proof that any two bases for a given subspace of $\mathbb{R}^n$ must have the same number of elements.

11. Let $\mathcal{S}$ and $\mathcal{T}$ be subspaces of $\mathbb{R}^n$. Prove that if $\mathcal{S} \subset \mathcal{T}$ and $\dim(\mathcal{S}) = \dim(\mathcal{T})$, then $\mathcal{S} = \mathcal{T}$.

12. Let $\mathcal{S}, \mathcal{T} \subset \mathbb{R}^n$. The *sum* of $\mathcal{S}$ and $\mathcal{T}$ is the set $\mathcal{S} + \mathcal{T} = \{s + t : s \in \mathcal{S}$ and $t \in \mathcal{T}\}$. Show that for all $\mathcal{S}, \mathcal{T}, \mathcal{U} \subset \mathbb{R}^n$

    1. $\mathcal{S} + \mathcal{T} = \mathcal{T} + \mathcal{S}$,
    2. $\mathcal{S} + (\mathcal{T} + \mathcal{U}) = (\mathcal{S} + \mathcal{T}) + \mathcal{U}$.

13. Prove that if $\mathcal{S}$ and $\mathcal{T}$ are subspaces, then $\mathcal{S} + \mathcal{T}$ is a subspace.

14. Let $\mathcal{S}$ and $\mathcal{T}$ be subspaces of $\mathbb{R}^n$ generated by $\mathcal{A}$ and $\mathcal{B}$, respectively. Show that $\mathcal{S} + \mathcal{T}$ is the subspace generated by $\mathcal{A} \cup \mathcal{B}$.

15. Let $\mathcal{A}$ and $\mathcal{B}$ be subspaces of $\mathbb{R}^n$. Let $a \in \mathcal{A}$, $b \in \mathcal{B}$, $c \in \mathcal{A} \cap \mathcal{B}$. Show that if $a, b \notin \mathcal{A} \cap \mathcal{B}$, then a and b are linearly independent. Show that if, in addition, $c \neq 0$, then a, b, and c are linearly independent.

16. Let $\mathcal{A}$ and $\mathcal{B}$ be subspaces of $\mathbb{R}^n$. Show that

$$\dim(\mathcal{A} + \mathcal{B}) = \dim(\mathcal{A}) + \dim(\mathcal{B}) - \dim(\mathcal{A} \cap \mathcal{B}).$$

[*Hint*: Start with a basis for $\mathcal{A} \cap \mathcal{B}$ and extend it first to a basis for $\mathcal{A}$ and then to a basis for $\mathcal{B}$. Prove that the resulting collection of vectors is a basis for $\mathcal{A} + \mathcal{B}$.]

17. Prove that in $\mathbb{R}^3$ any two planes passing through the origin must have a line in common.

3. MATRICES

The greater part of this book will be concerned with manipulating rectangular arrays of numbers called matrices. Matrices arise in many different connections. One of the most important is their use to represent linear transformations. Accordingly we shall introduce the idea of a linear transformation from $\mathbb{R}^2$ into $\mathbb{R}^2$ to motivate and illustrate the definitions of the next two sections. Linear transformations will be considered in full generality in Section 5.

A function $f: \mathbb{R}^2 \to \mathbb{R}^2$ is said to be a linear transformation if

$$f(\alpha x + \beta y) = \alpha f(x) + \beta f(y) \tag{3.1}$$

for any vectors x and y and any scalars α and β.

The linear transformation f is uniquely determined by its values at the basis vectors e_1 and e_2. In fact let

$$f(e_1) = a_1 = \begin{pmatrix} \alpha_{11} \\ \alpha_{21} \end{pmatrix} \quad \text{and} \quad f(e_2) = a_2 = \begin{pmatrix} \alpha_{12} \\ \alpha_{22} \end{pmatrix}.$$

Then, since $x = \xi_1 e_1 + \xi_2 e_2$, it follows from (3.1) that

$$\begin{aligned} y = f(x) &= f(\xi_1 e_1 + \xi_2 e_2) \\ &= \xi_1 f(e_1) + \xi_2 f(e_2) \\ &= \xi_1 a_1 + \xi_2 a_2. \end{aligned}$$

In terms of individual components, if $y = f(x)$, then

$$\eta_1 = \alpha_{11}\xi_1 + \alpha_{12}\xi_2, \qquad \eta_2 = \alpha_{21}\xi_1 + \alpha_{22}\xi_2. \tag{3.2}$$

Equations (3.2) show that the linear transformation f is completely determined by the array of numbers

$$\begin{pmatrix} \alpha_{11} & \alpha_{12} \\ \alpha_{21} & \alpha_{22} \end{pmatrix}. \tag{3.3}$$

The array (3.3) is called the matrix representing the linear transformation f. The following definition generalizes this idea of a matrix.

DEFINITION 3.1. An $m \times n$ *matrix* is a rectangular array of numbers having m rows and n columns. The numbers comprising the array are called the *elements* of the matrix. The numbers m and n are called the *dimensions* of the matrix. The set of all $m \times n$ matrices is denoted by $\mathbb{R}^{m \times n}$.

We shall ordinarily denote a matrix by an upper case Latin or Greek letter. Whenever possible, an element of a matrix will be denoted by the corresponding lower case Greek letter with two subscripts, the first specifying the row that contains the element and the second the column. Thus the 3×4 matrix A has the form

$$A = \begin{pmatrix} \alpha_{11} & \alpha_{12} & \alpha_{13} & \alpha_{14} \\ \alpha_{21} & \alpha_{22} & \alpha_{23} & \alpha_{24} \\ \alpha_{31} & \alpha_{32} & \alpha_{33} & \alpha_{34} \end{pmatrix}.$$

The scalar α_{ij} is also called the (i, j)-element of A.

In some applications, notably those involving partitioned matrices, considerable notational simplification can be achieved by permitting matrices with one or both its dimensions zero. Such matrices will be said to be *void*.

Note that two matrices are equal if and only if they have the same dimensions and their corresponding elements are equal.

EXAMPLE 3.2. The $n \times 1$ matrix A has the form

$$A = \begin{pmatrix} \alpha_{11} \\ \alpha_{21} \\ \vdots \\ \alpha_{n1} \end{pmatrix}$$

which looks exactly like a member of $\mathbb{R}^n$. In this book we shall not distinguish between $n \times 1$ matrices and n-vectors; they will be denoted by upper or lower case Latin letters as convenience dictates.

EXAMPLE 3.3. The $1 \times n$ matrix R has the form

$$R = (\varrho_{11}, \varrho_{12}, \ldots, \varrho_{1n}).$$

Such a matrix will be called a *row vector*.

The remainder of this section will be devoted to describing the structure of specific kinds of matrices that will prove important later.

DEFINITION 3.4. An $m \times n$ matrix A is *square* if $m = n$. In this case A is said to be of *order* n.

In the sequel the dimensions and properties of a matrix will often be determined by context. As an example of this, the statement that A is of order n carries the implication that A is square.

DEFINITION 3.5. Let A be an $m \times n$ matrix and let $k = \min\{m, n\}$. The elements α_{ii} ($i = 1, 2, \ldots, k$) are said to lie on the *diagonal* of A, and α_{ii} is called the ith diagonal element of A. The elements $\alpha_{i,i+1}$ are said to lie on the *superdiagonal* of A, the elements $\alpha_{i,i-1}$ on the *subdiagonal* of A. If A is square, the elements $\alpha_{n-i+1,i}$ ($i = 1, 2, \ldots, n$) are said to lie on the *secondary* diagonal of A.

In the array below, the diagonal elements are labeled by δ, the superdiagonal elements by μ, and the subdiagonal elements by λ:

$$\begin{pmatrix} \delta & \mu & x & x & x & x \\ \lambda & \delta & \mu & x & x & x \\ x & \lambda & \delta & \mu & x & x \\ x & x & \lambda & \delta & \mu & x \end{pmatrix}.$$

The secondary diagonal of a matrix is the diagonal proceeding from the lower left to the upper right. (Here the x's indicate elements of the matrix which may be, but are not required to be, nonzero.)

The presence of zero elements in a matrix often simplifies calculations involving that matrix. In fact matrices with certain special distributions of zero elements are so important that they have been given names.

DEFINITION 3.6. A square matrix is *diagonal* if its only nonzero elements lie on the diagonal. We write

$$\mathrm{diag}(\delta_1, \delta_2, \ldots, \delta_n)$$

to denote the diagonal matrix whose diagonal elements are $\delta_1, \delta_2, \ldots, \delta_n$.

DEFINITION 3.7. A $m \times n$ matrix A is *upper trapezoidal* if

$$i > j \quad \Rightarrow \quad \alpha_{ij} = 0.$$

It is *lower trapezoidal* if

$$i < j \quad \Rightarrow \quad \alpha_{ij} = 0.$$

A square upper (lower) trapezoidal matrix is said to be *upper* (*lower*) *triangular*.

A 3×5 upper trapezoidal matrix has the form

$$\begin{pmatrix} x & x & x & x & x \\ 0 & x & x & x & x \\ 0 & 0 & x & x & x \end{pmatrix}.$$

The name comes from the fact that the nonzero elements form a trapezoid in the upper part of the matrix.

Some special kinds of triangular matrices are important. If T is upper (lower) triangular with zero diagonal elements, then T is said to be *strictly upper* (*lower*) *triangular*. If the diagonal elements of T are unity, T is said to be *unit upper* (*lower*) *triangular*. The arrays below are strictly lower triangular and unit upper triangular, respectively:

$$\begin{pmatrix} 0 & 0 & 0 & 0 \\ x & 0 & 0 & 0 \\ x & x & 0 & 0 \\ x & x & x & 0 \end{pmatrix}, \qquad \begin{pmatrix} 1 & x & x & x \\ 0 & 1 & x & x \\ 0 & 0 & 1 & x \\ 0 & 0 & 0 & 1 \end{pmatrix}.$$

EXAMPLE 3.8. A matrix is diagonal if and only if it is both upper and lower triangular.

DEFINITION 3.9. A square matrix A is *upper Hessenberg* if

$$i > j + 1 \quad \Rightarrow \quad \alpha_{ij} = 0.$$

It is *lower Hessenberg* if

$$i < j - 1 \quad \Rightarrow \quad \alpha_{ij} = 0.$$

It is *tridiagonal* if it is both upper and lower Hessenberg.

The form of a 4×4 upper Hessenberg matrix is

$$\begin{pmatrix} x & x & x & x \\ x & x & x & x \\ 0 & x & x & x \\ 0 & 0 & x & x \end{pmatrix}.$$

An upper Hessenberg matrix is zero below its subdiagonal. A lower Hessenberg matrix is zero above its superdiagonal. A tridiagonal matrix has its nonzero elements arranged in a band along its diagonals:

$$\begin{pmatrix} x & x & 0 & 0 & 0 \\ x & x & x & 0 & 0 \\ 0 & x & x & x & 0 \\ 0 & 0 & x & x & x \\ 0 & 0 & 0 & x & x \end{pmatrix}.$$

One important class of matrices consists of those whose elements above the diagonal are equal to the corresponding elements below the diagonal, as illustrated by the array

$$\begin{pmatrix} 1 & 2 & 3 & 4 \\ 2 & 4 & 6 & 8 \\ 3 & 6 & 9 & 12 \\ 4 & 8 & 12 & 16 \end{pmatrix}.$$

DEFINITION 3.10. A matrix A of order n is *symmetric* if

$$\alpha_{ij} = \alpha_{ji} \qquad (i, j = 1, 2, \ldots, n).$$

Let A be a given matrix, and suppose that certain rows and columns of A have been distinguished. The rectangular array of elements of A lying in the intersection of these rows and columns is again a matrix and is called a submatrix of A. For example, the array

$$S = \begin{pmatrix} \alpha_{22} & \alpha_{23} & \alpha_{25} \\ \alpha_{42} & \alpha_{43} & \alpha_{45} \end{pmatrix}$$

is a submatrix obtained from A by distinguishing rows 2 and 4 and columns 2, 3, and 5 as follows:

$$A = \begin{pmatrix} \alpha_{11} & \alpha_{12} & \alpha_{13} & \alpha_{14} & \alpha_{15} \\ \alpha_{21} & \alpha_{22} & \alpha_{23} & \alpha_{24} & \alpha_{25} \\ \alpha_{31} & \alpha_{32} & \alpha_{33} & \alpha_{34} & \alpha_{35} \\ \alpha_{41} & \alpha_{42} & \alpha_{43} & \alpha_{44} & \alpha_{45} \end{pmatrix}.$$

The elements of A lying at the intersection of the lines comprise the elements of S. The following is a more precise definition of the idea of a submatrix.

DEFINITION 3.11. Let A be an $m \times n$ matrix and let $1 \leq i_1 < i_2 < \cdots < i_k \leq m$ and $1 \leq j_1 < j_2 < \cdots < j_l \leq n$. The $k \times l$ matrix S whose (μ, ν)-element is

$$\sigma_{\mu\nu} = \alpha_{i_\mu, j_\nu}$$

is called a *submatrix* of A. If $k = l$ and $i_1 = j_1, i_2 = j_2, \ldots, i_k = j_k$, then S is called a *principal submatrix* of A. If $i_1 = 1, i_2 = 2, \ldots, i_k = k$ and $j_1 = 1, j_2 = 2, \ldots, j_l = l$, then S is called a *leading submatrix* of A.

According to the definition a principal submatrix is a square submatrix in which the distinguished rows and columns are the same. For example, the matrix

$$\begin{pmatrix} \alpha_{22} & \alpha_{24} \\ \alpha_{42} & \alpha_{44} \end{pmatrix}$$

is a 2×2 principal submatrix of A. A leading submatrix of A is a submatrix that lies in the upper left-hand corner of A; e.g.,

$$\begin{pmatrix} \alpha_{11} & \alpha_{12} & \alpha_{13} \\ \alpha_{21} & \alpha_{22} & \alpha_{23} \end{pmatrix}.$$

The leading principal submatrices of a matrix occur frequently enough to require their own notation. We shall denote the leading principal submatrix of order k of A by $A^{[k]}$. For example,

$$A^{[3]} = \begin{pmatrix} \alpha_{11} & \alpha_{12} & \alpha_{13} \\ \alpha_{21} & \alpha_{22} & \alpha_{23} \\ \alpha_{31} & \alpha_{32} & \alpha_{33} \end{pmatrix}.$$

In the next section we shall define a number of operations with matrices. We shall be particularly concerned with determining conditions under which a given distribution of zero elements is preserved. Implicit in the definition of submatrix is the operation of forming a submatrix. It turns out that the operation of forming a *principal* submatrix preserves all of the distributions defined above.

THEOREM 3.12. A principal submatrix of a diagonal (tridiagonal, triangular, strictly triangular, unit triangular, Hessenberg, symmetric) matrix is diagonal (tridiagonal, triangular, strictly triangular, unit triangular, Hessenberg, symmetric).

PROOF. We prove the theorem for an upper Hessenberg matrix A of order n, leaving the other forms for Exercise 3.4. Let S be a principal submatrix of A of order k. Then there are integers $1 \leq i_1 < i_2 < \cdots < i_k \leq n$ such that

$$\sigma_{\mu\nu} = \alpha_{i_\mu, i_\nu}.$$

We must show that if $\mu > \nu + 1$, then $\sigma_{\mu\nu} = 0$. However, if $\mu > \nu + 1$, then $i_\mu > i_\nu + 1$, and since A is upper Hessenberg $0 = \alpha_{i_\mu, i_\nu} = \sigma_{\mu\nu}$. ■

The 4×4 matrix A can be written as a matrix of 2×2 submatrices in the form

$$A = \begin{pmatrix} A_{11} & A_{12} \\ A_{21} & A_{22} \end{pmatrix},$$

where

$$A_{11} = \begin{pmatrix} \alpha_{11} & \alpha_{12} \\ \alpha_{21} & \alpha_{22} \end{pmatrix}, \qquad A_{12} = \begin{pmatrix} \alpha_{13} & \alpha_{14} \\ \alpha_{23} & \alpha_{24} \end{pmatrix},$$

$$A_{21} = \begin{pmatrix} \alpha_{31} & \alpha_{32} \\ \alpha_{41} & \alpha_{42} \end{pmatrix}, \qquad A_{22} = \begin{pmatrix} \alpha_{33} & \alpha_{34} \\ \alpha_{43} & \alpha_{44} \end{pmatrix}.$$

We say that A has been partitioned into submatrices. A partitioning can be accomplished in different ways. For example,

$$A = (A_1, A_2),$$

where

$$A_1 = \begin{pmatrix} \alpha_{11} & \alpha_{12} \\ \alpha_{21} & \alpha_{22} \\ \alpha_{31} & \alpha_{32} \\ \alpha_{41} & \alpha_{42} \end{pmatrix}, \qquad A_2 = \begin{pmatrix} \alpha_{13} & \alpha_{14} \\ \alpha_{23} & \alpha_{24} \\ \alpha_{33} & \alpha_{34} \\ \alpha_{43} & \alpha_{44} \end{pmatrix}.$$

DEFINITION 3.13. An $m \times n$ matrix A is said to be *partitioned into submatrices* when it is written in the form

$$A = \begin{pmatrix} A_{11} & A_{12} & \cdots & A_{1l} \\ A_{21} & A_{22} & \cdots & A_{2l} \\ \vdots & \vdots & \cdots & \vdots \\ A_{k1} & A_{k2} & \cdots & A_{kl} \end{pmatrix}, \tag{3.4}$$

where each A_{ij} is an $m_i \times n_j$ submatrix of A.

Strictly speaking, Definition 3.13 entails an abuse of notation. The right-hand side of (3.4) is a rectangular array of matrices, not scalars, and is therefore itself not a matrix, but a matrix of matrices. However, it can be identified with a matrix by the simple device of arranging the elements of the matrices of the partition in their natural order. In the definition we assume that such an identification has been made.

The partition in the definition has k rows and l columns. Each submatrix in the ith row of the partition has m_i rows, and of course $m_1 + m_2 + \cdots + m_k = m$, the number of rows in A. Similarly, the submatrices of the jth column of the partition have n_j columns and $n_1 + n_2 + \cdots + n_l = n$. The matrix A_{11} of the partition is a leading submatrix of A.

Some of the elements of a partition may be $n \times 1$ matrices. Whenever it is convenient, we shall regard such matrices as vectors and denote them by lower case Latin letters. Similarly 1×1 matrices in a partition may be denoted by lower case Greek letters.

EXAMPLE 3.14. Let $A \in \mathbb{R}^{m \times n}$. The vector

$$a_j = \begin{pmatrix} \alpha_{1j} \\ \alpha_{2j} \\ \vdots \\ \alpha_{mj} \end{pmatrix}$$

is called the jth *column* of A. When A is written in the form

$$A = (a_1, a_2, \ldots, a_n),$$

it is said to be *partitioned by columns.* Similarly the row vector

$$A_i = (\alpha_{i1}, \alpha_{i2}, \ldots, \alpha_{in})$$

is called the ith *row* of A. When A is written in the form

$$A = \begin{pmatrix} A_1 \\ A_2 \\ \vdots \\ A_m \end{pmatrix},$$

it is said to be *partitioned by rows*.

EXERCISES

1. Verify Example 3.8.

2. Prove that a symmetrix triangular matrix is diagonal.

3. Prove that a symmetric Hessenberg matrix is tridiagonal.

4. Complete the proof of Theorem 3.12. [*Hint*: Do not prove the theorem for diagonal matrices directly. Instead prove it for triangular matrices and apply Example 3.8.]

5. Prove that A is tridiagonal if and only if

$$|i - j| > 1 \quad \Rightarrow \quad \alpha_{ij} = 0.$$

6. The square matrix A is a *band matrix* of *bandwidth* $2k + 1$ if

$$|i - j| > k \quad \Rightarrow \quad \alpha_{ij} = 0.$$

(For $k = 0, 1, 2$, A is called diagonal, tridiagonal, and pentadiagonal, respectively.) Show that a principal submatrix of a band matrix is a band matrix of the same bandwidth.

7. In Definition 3.11, if $k = l$, the submatrix S is sometimes written

$$A\begin{pmatrix} i_1, & i_2, & \ldots, & i_k \\ j_1, & j_2, & \ldots, & j_k \end{pmatrix}.$$

Prove that if A is upper triangular (Hessenberg) and $i_1 \geq j_1$, $i_2 \geq j_2$, $\ldots, i_k \geq j_k$, then

$$A\begin{pmatrix} i_1, & i_2, & \ldots, & i_k \\ j_1, & j_2, & \ldots, & j_k \end{pmatrix}$$

is upper triangular (Hessenberg). State and prove a similar theorem for lower triangular (Hessenberg) matrices. What about unit triangular matrices? strictly triangular matrices?

4. OPERATIONS WITH MATRICES

Let f and g be linear transformations of $\mathbb{R}^2$ into $\mathbb{R}^2$. With f and g there is associated a new function $h : \mathbb{R}^2 \to \mathbb{R}^2$, called the "sum" of f and g, that is defined by

$$h(x) = f(x) + g(x). \tag{4.1}$$

Since

$$\begin{aligned} h(\alpha x + \beta y) &= f(\alpha x + \beta y) + g(\alpha x + \beta y) \\ &= \alpha f(x) + \beta f(y) + \alpha g(x) + \beta g(y) \\ &= \alpha[f(x) + g(x)] + \beta[f(y) + g(y)] \\ &= \alpha h(x) + \beta h(y), \end{aligned}$$

the function h is a linear transformation. We write $h = f + g$.

The transformations f, g, and h may be represented by matrices A, B, and C, and it is natural to ask what relations exist between A, B, and C. Let the matrices be partitioned by columns

$$A = (a_1, a_2), \qquad B = (b_1, b_2), \qquad C = (c_1, c_2). \tag{4.2}$$

Then from the comments at the beginning of Section 3,

$$a_i = f(e_i), \qquad b_i = g(e_i), \qquad c_i = h(e_i) \qquad (i = 1, 2).$$

Hence from (4.1)

$$c_i = h(e_i) = f(e_i) + g(e_i) = a_i + b_i \qquad (i = 1, 2).$$

In terms of the elements of A, B, and C

$$\gamma_{ij} = \alpha_{ij} + \beta_{ij} \qquad (i, j = 1, 2). \tag{4.3}$$

Since h is the sum of f and g, it is natural to call the matrix C defined by (4.3) the sum of A and B and to write $C = A + B$. This suggests the following definition.

DEFINITION 4.1. Let A and B be $m \times n$ matrices. The *sum* of A and B is the matrix C whose elements are given by

$$\gamma_{ij} = \alpha_{ij} + \beta_{ij} \qquad (i = 1, 2, \ldots, m; \quad j = 1, 2, \ldots, n).$$

We write

$$C = A + B.$$

The sum of two matrices is defined only when their dimensions are the same. Whenever we write $A + B$ without giving the dimensions of A or B, it will be understood that their dimensions are equal.

A theorem analogous to Theorem 1.4 holds for the matrix sum.

THEOREM 4.2. Let A, B, and C be $m \times n$ matrices. Then

1. $A + B = B + A$,
2. $(A + B) + C = A + (B + C)$.

PROOF. As in the proof of Theorem 1.4, the assertions follow from the scalar relations

$$\alpha_{ij} + \beta_{ij} = \beta_{ij} + \alpha_{ij} \qquad \text{and} \qquad (\alpha_{ij} + \beta_{ij}) + \gamma_{ij} = \alpha_{ij} + (\beta_{ij} + \gamma_{ij})$$
$$(i = 1, 2, \ldots, m; \; j = 1, 2, \ldots, n). \quad \blacksquare$$

The following theorem, whose proof is left as Exercise 4.3, shows that the sum preserves most of the special distributions of zeros defined in Section 3.

THEOREM 4.3. The sum of two diagonal (upper trapezoidal, lower trapezoidal, upper Hessenberg, lower Hessenberg, tridiagonal, symmetric) matrices is diagonal (upper trapezoidal, lower trapezoidal, upper Hessenberg, lower Hessenberg, tridiagonal, symmetric).

In line with Definition 1.5 we can define zero matrices whose properties are analogous to those of the zero vector.

DEFINITION 4.4. Any matrix whose elements are all zero will be called a *zero matrix* and will be denoted by the symbol 0.

This definition adds one more meaning to the overworked symbol 0. However, it will still be clear from context just what is meant. For example, in the expression $A + 0$, the symbol 0 must denote a zero matrix whose dimensions are the same as those of A.

THEOREM 4.5. For any matrix A:

1. $A + 0 = A$,
2. there is a matrix $-A$ such that $A + (-A) = 0$.

PROOF. The proof of part 1 is left as an exercise. For the proof of part 2, let $-A$ be the matrix whose elements are $-\alpha_{ij}$. Then it is easily verified that $A + (-A) = 0$. ■

As with scalars and vectors, we write $A - B$ for $A + (-B)$.

A second matrix operation is suggested by the following considerations. Let $f : \mathbb{R}^2 \to \mathbb{R}^2$ be a linear transformation and $\lambda \in \mathbb{R}$. Define the function $g = \lambda f$ by the equation

$$g(x) = \lambda f(x).$$

It is easy to verify that g is a linear transformation. Moreover, if A and B are the matrices representing f and g, respectively, then

$$\beta_{ij} = \lambda \alpha_{ij} \qquad (i, j = 1, 2).$$

This suggests the following definition of the product between a matrix and a scalar.

DEFINITION 4.6. Let A be an $m \times n$ matrix and $\lambda \in \mathbb{R}$. The product of λ and A, written $\lambda \cdot A$ or λA, is the $m \times n$ matrix whose elements are given by

$$\lambda \alpha_{ij} \qquad (i = 1, 2, \ldots, m;\ j = 1, 2, \ldots, n).$$

If $\lambda \neq 0$, then A and λA have the same nonzero elements. Thus diagonal matrices remain diagonal after being multiplied by a scalar. Similarly, the properties of being trapezoidal or Hessenberg are preserved under scalar multiplication. Symmetry is also preserved.

The proof of the following theorem is left for Exercise 4.6.

THEOREM 4.7. Let A and B be matrices and $\lambda, \mu \in \mathbb{R}$. Then

1. $(\lambda\mu)A = \lambda(\mu A)$,
2. $(\lambda + \mu)A = \lambda A + \mu A$,
3. $\lambda(A + B) = \lambda A + \lambda B$,
4. $1 \cdot A = A$.

Theorems 4.2, 4.5, and 4.7 together show that the set $\mathbb{R}^{m\times n}$ of all $m\times n$ matrices with addition and scalar multiplication as defined here is an abstract vector space. All theorems applying to abstract vector spaces are also true of matrices, although their truth for matrices is often obvious (e.g., $(-1)\cdot A = -A$). A basis for $\mathbb{R}^{m\times n}$ is given by the matrices E_{ij} whose elements are zero except the (i, j)th, which is unity. In fact the E_{ij} are obviously linearly independent, and if $A \in \mathbb{R}^{m\times n}$, then

$$A = \sum_{i=1}^{m} \sum_{j=1}^{n} \alpha_{ij} E_{ij}.$$

There are exactly mn distinct E_{ij}. Hence the dimension of the space of all $m\times n$ matrices is mn.

EXAMPLE 4.8. The set $\mathbb{R}^{1\times n}$ of all row vectors with n components is a vector space. Except for the fact that its components are arranged in a row rather than a column, $\mathbb{R}^{1\times n}$ is indistinguishable from the vector space $\mathbb{R}^n$. In particular, all the definitions and theorems about $\mathbb{R}^n$ hold, with suitable modifications, for row vectors. This justifies, for example, speaking of matrices with linearly independent rows.

The composition of linear transformations gives rise to yet another matrix operation. Let $f, g : \mathbb{R}^2 \to \mathbb{R}^2$ be linear transformations. The composition of f and g is the function $h = g \circ f$ defined by

$$h(x) = g[f(x)].$$

Since

$$\begin{aligned} h(\alpha x + \beta y) = g[f(\alpha x + \beta y)] &= g[\alpha f(x) + \beta f(y)] \\ &= \alpha g[f(x)] + \beta g[f(y)] \\ &= \alpha h(x) + \beta h(y), \end{aligned}$$

h is a linear transformation.

Let the matrices representing f, g, and h be A, B, and C, partitioned by columns as in (4.2). Then

$$c_i = h(e_i) = g[f(e_i)] = g(a_i) \qquad (i = 1, 2).$$

In terms of elements

$$\gamma_{ij} = \beta_{i1}\alpha_{1j} + \beta_{i2}\alpha_{2j} \qquad (i, j = 1, 2). \tag{4.4}$$

We call the matrix C defined by (4.4) the product of B and A.

DEFINITION 4.9. Let A be an $l \times m$ matrix and B be an $m \times n$ matrix. The *product* of A and B is the $l \times n$ matrix C whose elements are given by

$$\gamma_{ij} = \sum_{k=1}^{m} \alpha_{ik}\beta_{kj} \qquad (i = 1, 2, \ldots, l; \;\; j = 1, 2, \ldots, n). \tag{4.5}$$

We write

$$C = AB.$$

Note that the product AB is defined only when the number of columns of A is equal to the number of rows of B. The product has the same number of rows as A and the same number of columns as B. Whenever AB is written, it will be implicitly assumed that the product is defined.

When forming matrix products, it is easier to think in terms of the elements as they appear in the array rather than to apply (4.5) formally. Specifically to form γ_{ij}, pick out the ith row of A and the jth column B, multiply their corresponding elements, and add the multiples. The process for forming γ_{23} is illustrated below:

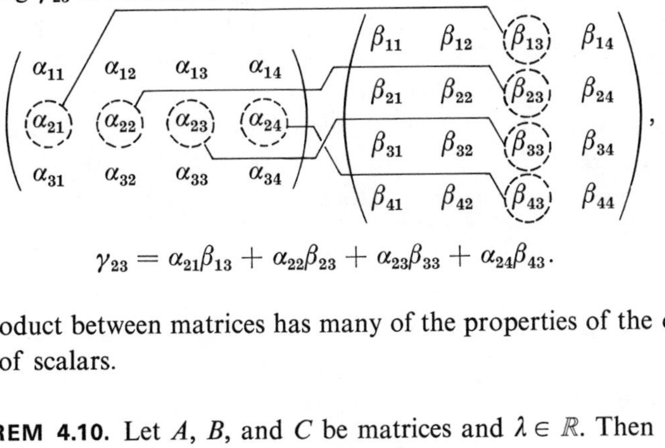

$$\begin{pmatrix} \alpha_{11} & \alpha_{12} & \alpha_{13} & \alpha_{14} \\ \alpha_{21} & \alpha_{22} & \alpha_{23} & \alpha_{24} \\ \alpha_{31} & \alpha_{32} & \alpha_{33} & \alpha_{34} \end{pmatrix} \begin{pmatrix} \beta_{11} & \beta_{12} & \beta_{13} & \beta_{14} \\ \beta_{21} & \beta_{22} & \beta_{23} & \beta_{24} \\ \beta_{31} & \beta_{32} & \beta_{33} & \beta_{34} \\ \beta_{41} & \beta_{42} & \beta_{43} & \beta_{44} \end{pmatrix},$$

$$\gamma_{23} = \alpha_{21}\beta_{13} + \alpha_{22}\beta_{23} + \alpha_{23}\beta_{33} + \alpha_{24}\beta_{43}.$$

The product between matrices has many of the properties of the ordinary product of scalars.

THEOREM 4.10. Let A, B, and C be matrices and $\lambda \in \mathbb{R}$. Then

1. $(AB)C = A(BC)$,
2. $A(B + C) = AB + AC$,
3. $(A + B)C = AC + BC$,
4. $\lambda(AB) = (\lambda A)B = A(\lambda B)$.

PROOF. We prove only part 1, leaving the rest for Exercise 4.9. First we must show that if one side of part 1 is defined, then so is the other. Suppose $(AB)C$ is defined, and let A be $k \times l$. Then since AB is defined B must have l rows. Let B be $l \times m$. Then AB is $k \times m$ and C must have m rows, say, C

is $m \times n$. However, with these dimensions $A(BC)$ is also defined. A similar argument shows that if $A(BC)$ is defined, then $(AB)C$ is defined.

We must now show that $(AB)C = A(BC)$. The (i, μ)-element of AB is $\sum_{\lambda=1}^{l} \alpha_{i\lambda}\beta_{\lambda\mu}$. Thus the (i, j)-element of $(AB)C$ is

$$\sum_{\mu=1}^{m} \left(\sum_{\lambda=1}^{l} \alpha_{i\lambda}\beta_{\lambda\mu} \right) \gamma_{\mu j}. \tag{4.6}$$

On the other hand, the (λ, j)-element of BC is $\sum_{\mu=1}^{m} \beta_{\lambda\mu}\gamma_{\mu j}$. Hence the (i, j)-element of $A(BC)$ is

$$\sum_{\lambda=1}^{l} \alpha_{i\lambda} \left(\sum_{\mu=1}^{m} \beta_{\lambda\mu}\gamma_{\mu j} \right). \tag{4.7}$$

The quantity (4.6) is obviously equal to (4.7), which establishes part 1. ■

EXAMPLE 4.11. Theorem 4.10 shows that the matrix product satisfies associative and distributive laws. However, it is not commutative, and it is instructive to see how this property may fail. Let A be $l \times m$ and B be $m \times n$ so that AB is defined. Then if $l \neq n$, BA is not defined so that AB cannot be equal to BA. On the other hand, if $l = n \neq m$, then both AB and BA are defined; but AB is square of order n while BA is of order $m \neq n$, so again $AB \neq BA$. The only remaining case is $l = m = n$, for which AB and BA are of the same order. However, commutativity may still fail, as the following example shows:

$$\begin{pmatrix} 1 & 0 \\ 0 & -1 \end{pmatrix} \begin{pmatrix} 0 & 1 \\ 1 & 0 \end{pmatrix} = \begin{pmatrix} 0 & 1 \\ -1 & 0 \end{pmatrix} \neq \begin{pmatrix} 0 & -1 \\ 1 & 0 \end{pmatrix} = \begin{pmatrix} 0 & 1 \\ 1 & 0 \end{pmatrix} \begin{pmatrix} 1 & 0 \\ 0 & -1 \end{pmatrix}.$$

EXAMPLE 4.12. The matrix equation

$$\begin{pmatrix} 1 & 1 \\ 1 & 1 \end{pmatrix} \begin{pmatrix} 1 & 1 \\ -1 & -1 \end{pmatrix} = \begin{pmatrix} 0 & 0 \\ 0 & 0 \end{pmatrix}$$

shows that it is possible for the product of two nonzero matrices to be zero. In particular, this means that from

$$AX = AY, \qquad A \neq 0$$

we cannot conclude that $X = Y$; that is, *the cancellation law for real numbers does not extend to matrices.* This point cannot be stressed too strongly.

The illegitimate use of the cancellation law is one of the greatest sources of errors among beginners.

Some of the special matrices defined in Section 3 are unaltered by multiplication.

THEOREM 4.13. The product of a diagonal matrix and a diagonal (trapezoidal, Hessenberg, tridiagonal) matrix is diagonal (trapezoidal, Hessenberg, tridiagonal). The product of two upper (lower) trapezoidal matrices is upper (lower) trapezoidal.

PROOF. We prove the last assertion and leave the rest for Exercise 4.9. Let A and B be upper trapezoidal of dimensions $l\times m$ and $m\times n$, respectively. Let $C = AB$. We must show that if $i > j$, then $\gamma_{ij} = 0$. Now

$$\gamma_{ij} = \sum_{k=1}^{m} \alpha_{ik}\beta_{kj} = \sum_{k=1}^{i-1} \alpha_{ik}\beta_{kj} + \sum_{k=i}^{m} \alpha_{ik}\beta_{kj}. \tag{4.8}$$

Because A is upper trapezoidal, $\alpha_{ik} = 0$ for $k = 1, 2, \ldots, i-1$. Because B is upper trapezoidal and $i > j$, $\beta_{kj} = 0$ for $k = i, i+1, \ldots, m$. Hence both sums on the right-hand side of (4.8) are zero and $\gamma_{ij} = 0$. ■

The above proof works formally with the definition of the matrix product. However, it is instructive to write down two upper trapezoidal matrices, form their product by the row–column process described in the discussion after Definition 4.9, and observe how the zeros in each matrix conspire to make the product upper trapezoidal. The other assertions of Theorem 4.13 may be verified similarly, and the verification should in turn suggest the details of the formal proof.

There is a matrix analog of the number one.

DEFINITION 4.14. The matrix of order n

$$I_n = \operatorname{diag}(1, 1, \ldots, 1)$$

is called the *identity matrix of order n.*

Usually the order of a particular identity matrix will be clear from context. In such cases we shall drop the subscript n and refer to $I_n = I$ simply as the identity matrix.

Note that

$$I_n = (e_1, e_2, \ldots, e_n),$$

where the e_i are the distinguished basis for $\mathbb{R}^n$ introduced in Example 2.2. Otherwise put, e_i is the ith column of the identity matrix.

THEOREM 4.15. Let A be an $m \times n$ matrix. Then

$$I_m A = A I_n = A.$$

This theorem, whose proof is left as an exercise, shows that the identity matrices do not alter the matrices they multiply. The analogous equation for scalars is $1 \cdot \alpha = \alpha \cdot 1 = \alpha$.

The final matrix operation, transposition, cannot be introduced simply in terms of linear transformations.

DEFINITION 4.16. Let A be an $m \times n$ matrix. The *transpose* of A is the $n \times m$ matrix B whose elements are given by

$$\beta_{ij} = \alpha_{ji} \qquad (i = 1, 2, \ldots, n; \quad j = 1, 2, \ldots, m).$$

We write

$$B = A^{\mathrm{T}}.$$

Thus the transpose of A is the matrix obtained by taking the rows of A and standing them on end so that they become columns.

EXAMPLE 4.17. The transpose of the vector x is the row vector

$$x^{\mathrm{T}} = (\xi_1, \xi_2, \ldots, \xi_n).$$

Conversely

$$x = (\xi_1, \xi_2, \ldots, \xi_n)^{\mathrm{T}}. \tag{4.9}$$

Equation (4.9) represents a particularly economical way of writing out vectors. For example the 3-vector

$$\begin{pmatrix} 1 \\ -2 \\ 4 \end{pmatrix}$$

can be written on a single line as $(1, -2, 4)^{\mathrm{T}}$.

EXAMPLE 4.18. Let $x, y \in \mathbb{R}^n$. Then the product $x^{\mathrm{T}}y$ is a scalar whose value is

$$x^{\mathrm{T}}y = \sum_{i=1}^{n} \xi_i \eta_i.$$

The scalar $x^{\mathrm{T}}y$ is called the *inner product* of x and y.

EXAMPLE 4.19. A matrix A is symmetric if and only if $A^{\mathrm{T}} = A$.

THEOREM 4.20. Let A and B be matrices and $\lambda \in \mathbb{R}$. Then

1. $(A^{\mathrm{T}})^{\mathrm{T}} = A$,
2. $(A + B)^{\mathrm{T}} = A^{\mathrm{T}} + B^{\mathrm{T}}$,
3. $(\lambda A)^{\mathrm{T}} = \lambda A^{\mathrm{T}}$,
4. $(AB)^{\mathrm{T}} = B^{\mathrm{T}}A^{\mathrm{T}}$.

PROOF. We give only the proof of part 4 and leave the rest for Exercise 4.9. Let A be $l \times m$ and B be $m \times n$ so that AB is defined. Then A^{T} is $m \times l$ and B^{T} is $n \times m$ so that $B^{\mathrm{T}}A^{\mathrm{T}}$ is also defined and has the same dimensions as $(AB)^{\mathrm{T}}$. Conversely if $B^{\mathrm{T}}A^{\mathrm{T}}$ is defined, so is AB, and $B^{\mathrm{T}}A^{\mathrm{T}}$ and $(AB)^{\mathrm{T}}$ have the same dimensions.

Let $C = (AB)^{\mathrm{T}}$. Since the (j, i)-element of AB is $\sum_{k=1}^{m} \alpha_{jk}\beta_{ki}$, it follows that

$$\gamma_{ij} = \sum_{k=1}^{m} \alpha_{jk}\beta_{ki}.$$

Let $A' = A^{\mathrm{T}}$ and $B' = B^{\mathrm{T}}$ and $D = B^{\mathrm{T}}A^{\mathrm{T}}$.

Then

$$\delta_{ij} = \sum_{k=1}^{m} \beta'_{ik}\alpha'_{kj} = \sum_{k=1}^{m} \beta_{ki}\alpha_{jk} = \gamma_{ij},$$

which shows that $D = C$, the required result. ■

EXAMPLE 4.21. For any matrix A, the matrices $A^{\mathrm{T}}A$ and AA^{T} are defined and symmetric. To show that, say, $A^{\mathrm{T}}A$ is symmetric note that, by Theorem 4.20,

$$(A^{\mathrm{T}}A)^{\mathrm{T}} = A^{\mathrm{T}}(A^{\mathrm{T}})^{\mathrm{T}} = A^{\mathrm{T}}A.$$

Hence by Example 4.19, $A^{\mathrm{T}}A$ is symmetric.

The matrix operations introduced in this section have been defined in terms of scalar addition and multiplication. This suggests the possibility of defining operations among partitioned matrices by substituting the appropriate elements from the partitions into the definitions, but using matrix addition and multiplication instead of the scalar operations. For example, let the 4×4 matrices A and B be partitioned into 2×2 submatrices

$$A=\begin{pmatrix} A_{11} & A_{12} \\ A_{21} & A_{22} \end{pmatrix}, \qquad B=\begin{pmatrix} B_{11} & B_{12} \\ B_{21} & B_{22} \end{pmatrix}. \tag{4.10}$$

Then it is possible to define a partitioned sum $\oplus$ by

$$A \oplus B=\begin{pmatrix} A_{11}+B_{11} & A_{12}+B_{12} \\ A_{21}+B_{21} & A_{22}+B_{22} \end{pmatrix} \tag{4.11}$$

and a partitioned product $\odot$ by

$$A \odot B=\begin{pmatrix} A_{11}B_{11}+A_{12}B_{21} & A_{11}B_{12}+A_{12}B_{22} \\ A_{21}B_{11}+A_{22}B_{21} & A_{21}B_{12}+A_{22}B_{22} \end{pmatrix}.$$

The above examples raise two questions. First, when can a partitioned operation be defined? For example, if A_{11} is taken to be a 3×3 submatrix (hence A_{12}, A_{21}, and A_{22} are 3×1, 1×3, and 1×1, respectively), $A_{11}+B_{11}$ is not defined and (4.11) makes no sense. Second, does the partitioned operation correspond to the usual matrix operation; that is, under what circumstances do we have

$$A \oplus B = A + B \qquad \text{and} \qquad A \odot B = AB?$$

The answer to these questions is given in the following theorem, which also defines the operations among partitioned matrices in full generality.

THEOREM 4.22. Let

$$A=\begin{pmatrix} A_{11} & \cdots & A_{1l} \\ \vdots & & \vdots \\ A_{k1} & \cdots & A_{kl} \end{pmatrix} \qquad \text{and} \qquad B=\begin{pmatrix} B_{11} & \cdots & B_{1n} \\ \vdots & & \vdots \\ B_{m1} & \cdots & B_{mn} \end{pmatrix},$$

where A_{ij} is $\varkappa_i \times \lambda_j$ and B_{ij} is $\mu_i \times \nu_j$. Then we have the following equalities.

1. $\varrho A = \begin{pmatrix} \varrho A_{11} & \cdots & \varrho A_{1l} \\ \vdots & & \vdots \\ \varrho A_{k1} & \cdots & \varrho A_{kl} \end{pmatrix}.$

2. $A^{\mathrm{T}} = \begin{pmatrix} A_{11}^{\mathrm{T}} & \cdots & A_{k1}^{\mathrm{T}} \\ \vdots & & \vdots \\ A_{1l}^{\mathrm{T}} & \cdots & A_{kl}^{\mathrm{T}} \end{pmatrix}.$

3. If $k = m$, $l = n$, $\varkappa_i = \mu_i$, and $\lambda_j = \nu_j$, then

$$A + B = \begin{pmatrix} C_{11} & \cdots & C_{1l} \\ \vdots & & \vdots \\ C_{k1} & \cdots & C_{kl} \end{pmatrix},$$

where

$$C_{ij} = A_{ij} + B_{ij}.$$

4. If $l = m$ and $\lambda_i = \mu_i$, then

$$AB = \begin{pmatrix} C_{11} & \cdots & C_{1n} \\ \vdots & & \vdots \\ C_{k1} & \cdots & C_{kn} \end{pmatrix},$$

where

$$C_{ij} = \sum_{s=1}^{l} A_{is} B_{sj}.$$

PROOF. It is easy to verify that, with the restrictions on the dimensions, all the operations can be performed. It is quite tedious to verify that the equalities hold. We shall give the proof for part 4, which is the most difficult and leave the rest for Exercise 4.9.

Let $k_1 = 0$ and for $i = 2, 3, \ldots, k+1$ let $k_i = \varkappa_1 + \varkappa_2 + \cdots + \varkappa_{i-1}$. Define l_i, m_i, and n_i similarly. Then the (p, r)-element of A_{is} is

$$\alpha_{pr}^{(i,s)} = \alpha_{k_i+p,\, l_s+r}$$

and the (r, q)-element of B_{sj} is

$$\beta_{rq}^{(s,j)} = \beta_{l_s+r, n_j+q}.$$

Hence the (p, q)-element of $A_{is}B_{sj}$ is

$$\sum_{r=1}^{\lambda_s} \alpha_{k_i+p, l_s+r} \beta_{l_s+r, n_j+q},$$

and the (p, q)-element of C_{ij} is

$$\sum_{s=1}^{l} \sum_{r=1}^{\lambda_s} \alpha_{k_i+p, l_s+r} \beta_{l_s+r, n_j+q} = \sum_{t=1}^{l_{l+1}} \alpha_{k_i+p, t} \beta_{t, n_j+q},$$

(n.b., l_{l+1} is the number of columns of A and the number of rows of B). However, this is just the $(k_i + p, n_j + q)$-element of AB, which establishes the result. ■

In the operations of addition and multiplication, the dimensions of the submatrices in the partitions must be restricted so that the defining operations can be performed. Two matrices satisfying such restriction are said to be partitioned *conformally* with respect to the operation in question. For addition, conformity requires that both matrices be partitioned in the same way.

In forming the product AB, there may be several partitions of A that conform with a fixed partition of B. For example, if the matrix A of Equation (4.10) is partitioned into 4×2 submatrices

$$A = (A_1, A_2),$$

the partition is still conformal with that of B. The product is

$$AB = (A_1B_{11} + A_2B_{21}, A_1B_{12} + A_2B_{22}).$$

Since an n-vector x can be regarded as an $n \times 1$ matrix, the product Ax of x with any $m \times n$ matrix is well defined. In fact the product is an m-vector y whose components are

$$\eta_i = \sum_{j=1}^{n} \alpha_{ij} \xi_j \qquad (i = 1, 2, \ldots, m). \tag{4.12}$$

The importance of this product may be seen from the fact that (4.12) is a system of m linear equations in the n unknowns $\xi_1, \xi_2, \ldots, \xi_n$. Thus the problem of solving (4.12) is equivalent to the problem of solving the matrix equation

$$Ax = y.$$

While this is merely a notational change, it does suggest that we attempt to apply matrix techniques to solve linear equations. This we shall do, theoretically in Section 6 and computationally in Chapter 3.

The following example should add some insight into the nature of the matrix-vector product Ax.

EXAMPLE 4.23. Let $A = (a_1, a_2, \ldots, a_n)$ be an $m \times n$ matrix-partitioned by columns. If $x \in \mathbb{R}^n$, then

$$Ax = (a_1, a_2, \ldots, a_n)\begin{pmatrix} \xi_1 \\ \xi_2 \\ \cdot \\ \cdot \\ \cdot \\ \xi_n \end{pmatrix} = \xi_1 a_1 + \xi_2 a_2 + \cdots + \xi_n a_n.$$

Thus Ax is a linear combination of the columns of A. Conversely if y is a linear combination of the columns of A, say,

$$y = \xi_1 a_1 + \xi_2 a_2 + \cdots + \xi_n a_n,$$

then

$$y = Ax,$$

where $x = (\xi_1, \xi_2, \ldots, \xi_n)^{\mathrm{T}}$. Thus the set of all products Ax is the same as the set of linear combinations of columns of A. This set is the subspace of $\mathbb{R}^n$ spanned by the columns of A and is called the *column space* of A.

If x is an m-vector, then x^{T} is a row vector with m components. Hence the product $x^{\mathrm{T}}A$ between x^{T} and any $m \times n$ matrix is well defined. If A is partitioned by rows, it is seen that the product $x^{\mathrm{T}}A$ is a linear combination of the rows of A whose coefficients are given by the components of x.

A fruitful way of looking at the operation of matrix multiplication is to regard one matrix as operating on the other. From this point of view, in forming the product of A and B we regard B as containing a prescription of operations to be performed on A to obtain the product.

EXAMPLE 4.24. Let $B = \text{diag}(\beta_1, \beta_2, \ldots, \beta_n)$. Then the product AB is the matrix obtained by multiplying the ith column of A by β_i, for each i. The product BA is the matrix obtained by multiplying the ith row of A by β_i, for each i.

From the example it is apparent that the prescription for forming the product AB differs from the one for forming BA. Specifically, if we are forming AB we shall say that B *postmultiplies* A, while if we are forming BA we shall say that B *premultiplies* A.

Postmultiplication may be thought of in terms of column operations. Let $B = (b_1, b_2, \ldots, b_n)$ be partitioned by columns. Then

$$AB = A(b_1, b_2, \ldots, b_n) = (Ab_1, Ab_2, \ldots, Ab_n).$$

Thus the ith column of AB is of the form Ab_i. From Example 4.23 we know that the vector Ab_i is a linear combination of the columns of A whose coefficients are the components of b_i. To sum up, *the* ith *column of the product* AB *may be formed by taking a linear combination of the columns of* A. *The coefficients of the linear combination are the elements of the* ith *column of* B.

Premultiplication can be similarly described in terms of row operations. Namely, *the* ith *row of the product* BA *is a linear combination of the rows of* A *whose coefficients are the elements of the* ith *row of* B.

EXAMPLE 4.25. Let B be upper triangular. Then the first column of AB is a multiple of the first column of A. The second column of AB is a linear combination of the first two columns of A. In general, the ith column of AB is a linear combination of the first i columns of A.

EXAMPLE 4.26. Many discussions of matrices center around the *elementary row operations*. These are simple operations that can be performed on the rows of a matrix A to yield a new matrix. They are

1. multiply the ith row of A by λ,
2. interchange rows i and j,
3. add λ times row i to row j.

Our point of view leads us to expect that the elementary row operations can be effected by premultiplying A by suitable matrices. This is indeed the case. The following list gives the matrices corresponding to the elementary row operations. The reader should verify that they produce the desired effect when they premultiply A.

1.

$$\begin{array}{c} \\ \\ \\ \\ i \\ \\ \\ \\ \end{array}\begin{array}{c} \begin{matrix} & & & i & & & \end{matrix} \\ \begin{bmatrix} 1 & & & \vdots & & & \\ & \ddots & & \vdots & & & \\ & & 1 & & & & \\ & \cdots & & \lambda & & & \\ & & & & 1 & & \\ & & & & & \ddots & \\ & & & & & & 1 \end{bmatrix} \end{array}.$$

2.

$$\begin{array}{c} \\ \\ \\ \\ i \\ \\ j \\ \\ \\ \\ \end{array}\begin{array}{c} \begin{matrix} & & & i & & j & & \end{matrix} \\ \begin{bmatrix} 1 & & & \vdots & & \vdots & & \\ & \ddots & & \vdots & & \vdots & & \\ & & 1 & & & & & \\ & \cdots & & 0 & \cdots & 1 & & \\ & & & \vdots & & \vdots & & \\ & \cdots & & 1 & \cdots & 0 & & \\ & & & & & & 1 & \\ & & & & & & & \ddots \\ & & & & & & & & 1 \end{bmatrix} \end{array}.$$

3.

$$\begin{array}{c} \\ \\ \\ i \\ \\ j \\ \\ \\ \end{array}\begin{array}{c} \begin{matrix} & & i & & j & \end{matrix} \\ \begin{bmatrix} 1 & & \vdots & & \vdots & & \\ & \ddots & \vdots & & \vdots & & \\ & \cdots & 1 & \cdots & 0 & & \\ & & \vdots & & \vdots & & \\ & \cdots & \lambda & \cdots & 1 & & \\ & & & & & \ddots & \\ & & & & & & 1 \end{bmatrix} \end{array}.$$

Note that the matrix corresponding to an elementary row operation is the matrix obtained by performing the row operation on the identity. Matrices of the form 2 are called *elementary permutations* and will be denoted by I_{ij}.

EXERCISES

1. Compute the following:

(a) $\begin{pmatrix} 1 & 6 & 4 \\ -4 & 2 & 8 \end{pmatrix} + \begin{pmatrix} -2 & 0 & 1 \\ 2 & -3 & 4 \end{pmatrix}$, (b) $\begin{pmatrix} 1 & 2 \\ 0 & 1 \end{pmatrix} - \begin{pmatrix} 2 & -2 \\ 0 & 3 \end{pmatrix}$,

(c) $(1, 2, 3) + (3, 2, 1) + 2(1, 0) - 3(0, 1)$,

(d) $2\begin{pmatrix} 1 & 0 \\ 0 & 0 \end{pmatrix} + 4\begin{pmatrix} 0 & 1 \\ 0 & 0 \end{pmatrix} + 8\begin{pmatrix} 0 & 0 \\ 1 & 0 \end{pmatrix} + 16\begin{pmatrix} 0 & 0 \\ 0 & 1 \end{pmatrix}$.

2. Compute the following:

(a) $\begin{pmatrix} 1 & 2 \\ 3 & 4 \end{pmatrix}\begin{pmatrix} 4 & 3 \\ 2 & 1 \end{pmatrix}$, (b) $\begin{pmatrix} 8 & 0 & -1 \\ 2 & 4 & 1 \\ -3 & -2 & 1 \end{pmatrix}\begin{pmatrix} 1 \\ -2 \\ 3 \end{pmatrix}$,

(c) $\begin{pmatrix} 1 \\ 2 \\ 3 \end{pmatrix}(3, 2, 1)$, (d) $(3, 2, 1)\begin{pmatrix} 1 \\ 2 \\ 3 \end{pmatrix}$,

(e) $\begin{pmatrix} 0 & 0 & 1 \\ 0 & 1 & 0 \\ 1 & 0 & 0 \end{pmatrix}\begin{pmatrix} 6 & 2 & -1 \\ 1 & 4 & 6 \\ 3 & -5 & 4 \end{pmatrix}$, (f) $\begin{pmatrix} 1 & 0 & 0 \\ 0 & 1 & 0 \\ -3 & 0 & 1 \end{pmatrix}\begin{pmatrix} 1 & 4 & -1 \\ 2 & 5 & -2 \\ 3 & 6 & 3 \end{pmatrix}$.

3. Prove Theorem 4.3.

4. Show that $\lambda A = \operatorname{diag}(\lambda, \lambda, \ldots, \lambda)A$.

5. Let $A \in \mathbb{R}^{l \times m}$, $B \in \mathbb{R}^{m \times n}$, and $\lambda, \mu \in \mathbb{R}$. Show that $(\lambda A)(\mu B) = \lambda\mu AB$.

6. Prove Theorem 4.7.

7. Let A be strictly upper triangular of order n. Show that $A^n = 0$.

8. Find a matrix $A \neq 0$ and a vector $x \neq 0$ such that $Ax = 0$.

9. Complete the proofs of Theorems 4.10, 4.13, 4.20, and 4.22.

10. Let $A \in \mathbb{R}^{m \times n}$. Show that for $i = 1, 2, \ldots, n$, Ae_i is the ith column of A. Give an analogous expression for the ith row of A.

11. For any $A \in \mathbb{R}^{m \times n}$ show that $\alpha_{ij} = e_i^{\mathrm{T}} A e_j$.

12. Let $A, B \in \mathbb{R}^{m \times n}$. Prove that $A = B$ if and only if $Ax = Bx$ for all $x \in \mathbb{R}^n$.

13. Give an example to show that the product of two symmetric matrices need not be symmetric.

14. Let $A, M \in \mathbb{R}^{m \times n}$ with A symmetric. Show that MAM^{T} is symmetric.

15. Prove that A is tridiagonal if and only if A^{T} is tridiagonal.

16. Prove that if A is upper trapezoidal (Hessenberg), then A^{T} is lower trapezoidal (Hessenberg).

17. Show that if U is upper triangular, then $(AU)^{[k]} = A^{[k]}U^{[k]}$ for all k for which $A^{[k]}$ is defined. [*Hint*: Consider the partitions

$$A = \begin{pmatrix} A^{[k]} & A_{12} \\ A_{21} & A_{22} \end{pmatrix}, \qquad U = \begin{pmatrix} U^{[k]} & U_{12} \\ 0 & U_{22} \end{pmatrix},$$

which conform with respect to multiplication.]

18. Show that if L is lower triangular, then $(LA)^{[k]} = L^{[k]}A^{[k]}$ for all k for which $A^{[k]}$ is defined.

19. Show that if A is upper Hessenberg and T is upper triangular, then AT is upper Hessenberg.

20. Let $H = I - 2xx^{\mathrm{T}}$, where $x^{\mathrm{T}}x = 1$. Show that

    1. H is symmetric,
    2. H is *involutory*: $H^2 = I$,
    3. H is *orthogonal*: $H^{\mathrm{T}}H = I$.

    Matrices of the form of H are called *elementary Hermitian matrices, elementary reflectors*, or *Householder transformations.*

21. Show that for all $x \in \mathbb{R}^n$, $x^{\mathrm{T}}x \geq 0$; moreover $x^{\mathrm{T}}x = 0$ if and only if $x = 0$.

22. Let $A \in \mathbb{R}^{m \times n}$. Show that $x^{\mathrm{T}}A^{\mathrm{T}}Ax = 0$ if and only if $Ax = 0$.

23. Assuming the partitions conform, calculate

    (a) $(X, Y)^{\mathrm{T}}(X, Y)$, (b) $(X, Y)(X, Y)^{\mathrm{T}}$,

    (c) $(X, Y)^{\mathrm{T}}A(X, Y)$, (d) $(\xi, x^{\mathrm{T}})\begin{pmatrix} \alpha & a^{\mathrm{T}} \\ a & A \end{pmatrix}\begin{pmatrix} \xi \\ x \end{pmatrix}$,

    (e) $\begin{pmatrix} \alpha & a_2{}^{\mathrm{T}} \\ a_1 & A \end{pmatrix}\begin{pmatrix} \beta & b_2{}^{\mathrm{T}} \\ b_1 & B \end{pmatrix}$.

24. Let $x \in \mathbb{R}^n$ be partitioned in the form $x^{\mathrm{T}} = (\xi_1, x_1{}^{\mathrm{T}})$, where $x_1 = (\xi_2, \xi_3, \ldots, \xi_n)^{\mathrm{T}}$. Show that if $\xi_1 \neq 0$, there is a unique vector

$b \in \mathbb{R}^{n-1}$ such that

$$\begin{pmatrix} 1 & 0 \\ b & I_{n-1} \end{pmatrix} \begin{pmatrix} \xi_1 \\ x_1 \end{pmatrix} = \begin{pmatrix} \xi_1 \\ 0 \end{pmatrix}.$$

25. Show that if P is an elementary permutation (Example 4.26), then $P^2 = I$.

5. LINEAR TRANSFORMATIONS AND MATRICES

In Sections 3 and 4 we used the idea of a linear transformation from $\mathbb{R}^2$ into $\mathbb{R}^2$ to motivate the definition of matrices and their operations. In this section we shall reverse the process by showing how a linear transformation may be associated with a matrix. To each matrix operation there corresponds an operation with linear transformations, and this correspondence preserves the theorems about matrix operations. This allows us to identify matrices with linear transformations, using the terminology of either as it suits our purposes.

DEFINITION 5.1. A *linear transformation from* $\mathbb{R}^n$ *into* $\mathbb{R}^m$ is a function $f : \mathbb{R}^n \to \mathbb{R}^m$ that satisfies

$$f(\alpha x + \beta y) = \alpha f(x) + \beta f(y),$$

for all $x, y \in \mathbb{R}^n$ and $\alpha, \beta \in \mathbb{R}$.

EXAMPLE 5.2. The mapping $f : \mathbb{R}^3 \to \mathbb{R}^2$ defined by

$$f[(\xi_1, \xi_2, \xi_3)^{\mathrm{T}}] = (\xi_3, \xi_1)^{\mathrm{T}}$$

is a linear transformation.

EXAMPLE 5.3. The range of a linear transformation from $\mathbb{R}^n$ into $\mathbb{R}^m$ need not be all of $\mathbb{R}^m$. Let $f : \mathbb{R}^3 \to \mathbb{R}^3$ be the mapping defined by

$$f[(\xi_1, \xi_2, \xi_3)^{\mathrm{T}}] = (\xi_1, 0, \xi_3)^{\mathrm{T}}.$$

Then f is a linear transformation, but the range of f is the plane spanned by the ξ_1-axis and the ξ_3-axis. Geometrically, f projects the point (ξ_1, ξ_2, ξ_3) onto the ξ_1, ξ_3-plane along the ξ_2-axis (Fig. 1).

Although a linear transformation $f : \mathbb{R}^n \to \mathbb{R}^m$ may not be onto $\mathbb{R}^m$, its range must be a subspace of $\mathbb{R}^m$. This is easy to prove directly from the

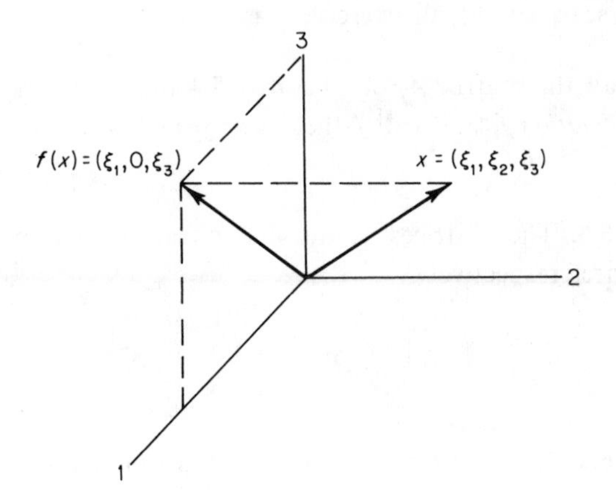

Fig. 1

definition of linear transformation. However, this and many other theorems follow from the correspondence between matrices and linear transformations, which we now proceed to establish.

THEOREM 5.4. Let $f : \mathbb{R}^n \to \mathbb{R}^m$ be a linear transformation. Then there is a unique $m \times n$ matrix A_f such that

$$f(x) = A_f x \tag{5.1}$$

for all $x \in \mathbb{R}^n$. Conversely if A_f is an $m \times n$ matrix, then the function f defined by (5.1) is a linear transformation from $\mathbb{R}^n$ into $\mathbb{R}^m$.

PROOF. First note that if B_f is any other matrix satisfying (5.1), then for all $x \in \mathbb{R}^n$

$$B_f x = f(x) = A_f x,$$

and hence $B_f = A_f$ (see Exercise 4.12). Thus if A_f exists, it is unique.

To construct A_f let $a_i = f(e_i)$ $(i = 1, 2, \ldots, n)$ and $A_f = (a_1, a_2, \ldots, a_n)$. Then

$$\begin{aligned} f(x) &= f(\xi_1 e_1 + \xi_2 e_2 + \cdots + \xi_n e_n) \\ &= \xi_1 f(e_1) + \xi_2 f(e_2) + \cdots + \xi_n f(e_n) \\ &= \xi_1 a_1 + \xi_2 a_2 + \cdots + \xi_n a_n = A_f x, \end{aligned}$$

so that A_f satisfies (5.1).

The converse is left as an exercise. ■

We shall call the matrix A_f of Theorem 5.4 the *matrix representing the linear transformation f*, and call f the *linear transformation induced by the matrix A_f*.

EXAMPLE 5.5. The matrices of the linear transformation of Examples 5.2 and 5.3 are, respectively,

$$\begin{pmatrix} 0 & 0 & 1 \\ 1 & 0 & 0 \end{pmatrix} \quad \text{and} \quad \begin{pmatrix} 1 & 0 & 0 \\ 0 & 0 & 0 \\ 0 & 0 & 1 \end{pmatrix}.$$

EXAMPLE 5.6. A *linear functional* on $\mathbb{R}^n$ is a function $\omega : \mathbb{R}^n \to \mathbb{R}$ that satisfies

$$\omega(\alpha x + \beta y) = \alpha\omega(x) + \beta\omega(y).$$

In view of our identification of $\mathbb{R}^1$ with $\mathbb{R}$, a linear functional is a linear transformation of $\mathbb{R}^n$ into $\mathbb{R}^1$ and hence is represented by an $1 \times n$ matrix or row vector. In other words, *if $\omega : \mathbb{R}^n \to \mathbb{R}$ is a linear functional, then there is a unique vector w such that*

$$w^{\mathrm{T}}x = \omega(x)$$

for all $x \in \mathbb{R}^n$. In more abstract settings this result is known as the Riesz–Fischer theorem. The product $w^{\mathrm{T}}x$ is called the *inner product* of w and x (cf. Example 4.18).

It is possible to define a sum, a scalar product, and a product among linear transformations in such a way that their matrix representations are related by the corresponding matrix operations. It should be noted that this correspondence determines the definition of the operation. For example, suppose we are given a sum $f + g$ between two linear transformations. Let A_f, A_g, and A_{f+g} be the matrices representing f, g, and $f + g$. Then the correspondence requires that

$$A_{f+g} = A_f + A_g. \tag{5.2}$$

Since the matrix sum is defined only among matrices of equal dimensions, say $m \times n$, it follows that f, g, and $f + g$ must all map $\mathbb{R}^n$ into $\mathbb{R}^m$. Moreover,

$$(f + g)(x) = A_{f+g}x = (A_f + A_g)x = A_fx + A_gx = f(x) + g(x). \tag{5.3}$$

Thus the sum $f + g$ of f and g must be the mapping of $\mathbb{R}^n$ into $\mathbb{R}^m$ that satisfies (5.3). Any other definition consistent with (5.2) must come down to this.

We now turn to the formal definition of the operations among linear transformations. For brevity, the definitions are embedded in the theorems that state their relations to the corresponding matrix operation. Throughout, the notation A_f will be used to denote the matrix representing the linear transformation f.

THEOREM 5.7. Let $f, g : \mathbb{R}^n \to \mathbb{R}^m$ be linear transformations. Then the function $f + g : \mathbb{R}^n \to \mathbb{R}^m$ defined by

$$(f + g)(x) = f(x) + g(x), \qquad x \in \mathbb{R}^n, \tag{5.4}$$

is a linear transformation. Moreover,

$$A_{f+g} = A_f + A_g. \tag{5.5}$$

PROOF. To show that $f + g$ is linear, let $x, y \in \mathbb{R}^n$ and $\alpha, \beta \in \mathbb{R}$. Then

$$\begin{aligned}(f + g)(\alpha x + \beta y) &= f(\alpha x + \beta y) + g(\alpha x + \beta y)\\ &= \alpha f(x) + \beta f(y) + \alpha g(x) + \beta g(y)\\ &= \alpha[f(x) + g(x)] + \beta[f(y) + g(y)]\\ &= \alpha(f + g)(x) + \beta(f + g)(y).\end{aligned}$$

To establish (5.5), let a_i, b_i, and c_i denote the ith columns of A_f, A_g, and A_{f+g}, respectively. Then from the proof of Theorem 5.4 it follows that $a_i = f(e_i)$, $b_i = g(e_i)$, and $c_i = (f + g)(e_i)$. Hence from (5.5)

$$c_i = (f + g)(e_i) = f(e_i) + g(e_i) = a_i + b_i.$$

Thus

$$(c_1, c_2, \ldots, c_n) = (a_1, a_2, \ldots, a_n) + (b_1, b_2, \ldots, b_n),$$

which is a partitioned form of (5.5). ■

THEOREM 5.8. Let $f : \mathbb{R}^n \to \mathbb{R}^m$ be a linear transformation and let $\lambda \in \mathbb{R}$. Then the function λf defined by

$$(\lambda f)(x) = \lambda f(x), \qquad x \in \mathbb{R}^n,$$

is a linear transformation. Moreover

$$A_{\lambda f} = \lambda A_f.$$

PROOF. The proof is left as an exercise. ■

THEOREM 5.9. Let $f : R^n \to R^m$ and $g : R^m \to R^l$ be linear transformations. Then the function $g \circ f$ defined by

$$(g \circ f)(x) = g[f(x)], \qquad x \in R^n,$$

is a linear transformation. Moreover,

$$A_{g \circ f} = A_g A_f. \tag{5.6}$$

PROOF. Let $x, y, \in R^n$ and $\alpha, \beta \in R$. Then

$$\begin{aligned}(g \circ f)(\alpha x + \beta y) = g[f(\alpha x + \beta y)] &= g[\alpha f(x) + \beta f(y)] \\ &= \alpha g[f(x)] + \beta g[f(y)] \\ &= \alpha (g \circ f)(x) + \beta (g \circ f)(y),\end{aligned}$$

which establishes the linearity of $g \circ f$. To establish (5.6), let a_i and c_i denote the ith columns of A_f and $A_{g \circ f}$, respectively. Then

$$c_i = (g \circ f)(e_i) = g[f(e_i)] = g(a_i) = A_g a_i.$$

Thus

$$\begin{aligned}A_{g \circ f} = (c_i, c_2, \ldots, c_n) &= (A_g a_1, A_g a_2, \ldots, A_g a_n) \\ &= A_g(a_1, a_2, \ldots, a_n) = A_g A_f. \quad ■\end{aligned}$$

To complete the correspondence between matrices and linear transformations, we list the transformations corresponding to the zero and identity matrices. The proofs of the following theorems are trivial and are left as exercises.

THEOREM 5.10. Let $z : R^n \to R^m$ be the function defined by

$$z(x) = 0, \qquad x \in R^n.$$

Then z is the linear transformation represented by the $m \times n$ zero matrix.

THEOREM 5.11. Let $i_n : \mathbb{R}^n \to \mathbb{R}^n$ be the function defined by

$$i_n(x) = x, \qquad x \in \mathbb{R}^n.$$

Then i_n is the linear transformation represented by I_n.

Theorems 5.4, 5.7–5.11 establish a complete correspondence or *isomorphism* between matrices and linear transformations with respect to all the operations except transposition. Thus for any fact about matrices that can be expressed in terms of these operations there is a corresponding fact about linear transformations. For example, the truth of the equation

$$A + 0 = A$$

implies that of

$$f + z = f$$

and vice versa. Because of this correspondence we shall often speak of matrices as if they were linear transformations that actually operate on vectors.

EXAMPLE 5.12. In the language of linear transformations, the matrix

$$A = \begin{pmatrix} 1 & 0 & 0 \\ 0 & 1 & 0 \\ 0 & 0 & -1 \end{pmatrix}$$

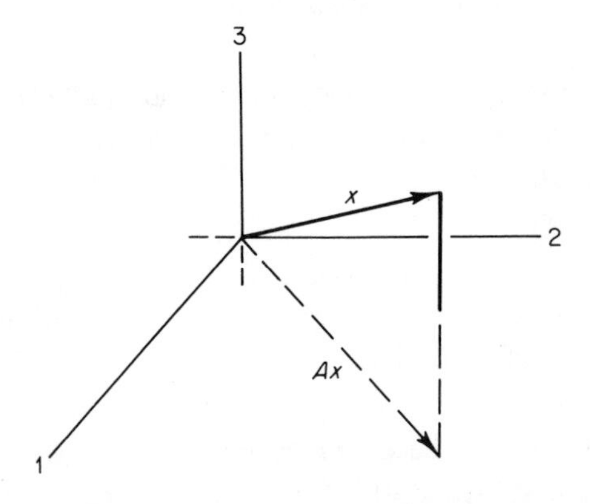

Fig. 2

reflects the vector x in the plane spanned by the ξ_1-axis and the ξ_2-axis (Fig. 2).

EXAMPLE 5.13. Let

$$A = \begin{pmatrix} \cos\theta & -\sin\theta \\ \sin\theta & \cos\theta \end{pmatrix}.$$

It is easily verified that the vector Ax may be obtained from the vector x by rotating x through an angle θ about the origin. For this reason the matrix A is called a *rotation* (Fig. 3).

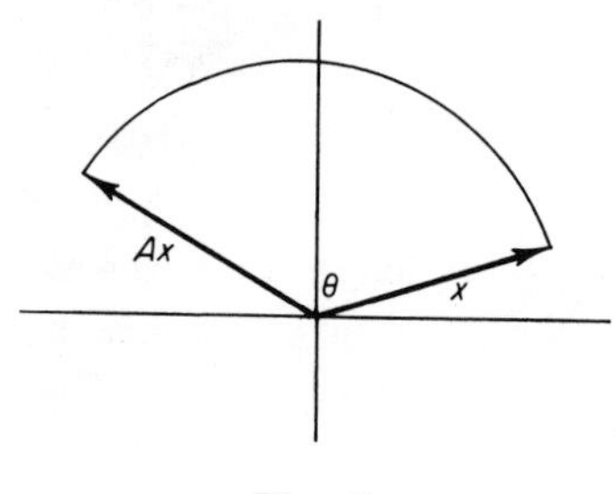

Fig. 3

The range of a linear transformation may be characterized in terms of the matrix representing it.

THEOREM 5.14. Let $f: \mathbb{R}^n \to \mathbb{R}^m$ be a linear transformation. Then the range of f is the subspace of $\mathbb{R}^m$ spanned by the columns of A_f.

PROOF. By Theorem 5.4, the range of f is the set of all vectors of the form $A_f x$, $x \in \mathbb{R}^n$. We have seen in Example 4.23 that this is the subspace spanned by the columns of A. ■

Thus the space spanned by the columns of a matrix have a natural interpretation as the range of the associated linear transformation. So important is this space that is has been given a name.

DEFINITION 5.15. Let A be an $m \times n$ matrix. The *column space* of A, written $\mathcal{R}(A)$, is the subspace of $\mathbb{R}^m$ spanned by the columns of A. The *row space* of A is $\mathcal{R}(A^{\mathrm{T}}) \subset \mathbb{R}^n$. The *rank of* A, written $\operatorname{rank}(A)$, is the dimension of $\mathcal{R}(A)$:

$$\operatorname{rank}(A) = \dim[\mathcal{R}(A)]. \tag{5.7}$$

It might seem more natural to call the number defined by (5.7) the "column rank" of A and define a row rank as $\dim[\mathcal{R}(A^{\mathrm{T}})]$, the dimension of the row space of A. However, it turns out that the row and column spaces have the same dimension, so that the "row rank" and the "column rank" are the same. The proof of this result will be deferred to Section 7.

Another important space may be associated with linear transformations, and hence matrices.

THEOREM 5.16. Let $f: \mathbb{R}^n \to \mathbb{R}^m$ be a linear transformation. Then the set of all $x \in \mathbb{R}^n$ such that $f(x) = 0$ is a subspace of $\mathbb{R}^n$.

PROOF. Let $f(x) = f(y) = 0$ and $\alpha, \beta \in \mathbb{R}$. Then

$$f(\alpha x + \beta y) = \alpha f(x) + \beta f(y) = \alpha \cdot 0 + \beta \cdot 0 = 0.$$

Thus the set of all $x \in \mathbb{R}^n$ such that $f(x) = 0$ is closed under linear combinations and is a subspace. ■

Since $f(x) = 0$ if and only if $A_f x = 0$, we may make the following definition.

DEFINITION 5.17. Let A be an $m \times n$ matrix. The *null space of A*, written $\mathcal{N}(A)$, is the space of all x such that $Ax = 0$. The *nullity of A*, written $\operatorname{null}(A)$, is $\dim[\mathcal{N}(A)]$.

EXAMPLE 5.18. Let

$$A = \begin{pmatrix} 1 & 0 & 1 \\ 0 & 1 & 1 \end{pmatrix}.$$

Then it is easy to verify that

$$\operatorname{rank}(A) = 2, \qquad \operatorname{null}(A) = 1,$$
$$\operatorname{rank}(A^{\mathrm{T}}) = 2, \qquad \operatorname{null}(A^{\mathrm{T}}) = 0.$$

In particular, note that $\operatorname{null}(A) \neq \operatorname{null}(A^{\mathrm{T}})$.

EXERCISES

1. Describe geometrically the effects of the transformations represented by the following matrices:

$$\text{(a)} \begin{pmatrix} \cos\theta & 0 & -\sin\theta \\ 0 & 1 & 0 \\ \sin\theta & 0 & \cos\theta \end{pmatrix}, \qquad \text{(b)} \begin{pmatrix} -1 & 0 & 0 \\ 0 & -1 & 0 \\ 0 & 0 & 1 \end{pmatrix},$$

$$\text{(c)} \begin{pmatrix} -1 & 0 & 0 \\ 0 & -1 & 0 \\ 0 & 0 & 0 \end{pmatrix}, \qquad \text{(d)} \begin{pmatrix} \lambda & 0 & 0 \\ 0 & \lambda & 0 \\ 0 & 0 & \lambda \end{pmatrix}.$$

2. Prove that $\text{null}(A) = 0$ if and only if A has linearly independent columns.

3. Let $A \in \mathbb{R}^{m\times n}$. Show that if $n > m$, then $\text{null}(A) \neq 0$.

4. Let $A \in \mathbb{R}^{m\times n}$. Show that $\text{rank}(A) = 1$ if and only if $A = xy^{\mathrm{T}}$ for some nonzero $x \in \mathbb{R}^m$ and $y \in \mathbb{R}^n$.

5. Show that

$$|\,\text{rank}(A) - \text{rank}(B)\,| \leq \text{rank}(A + B) \leq \text{rank}(A) + \text{rank}(B).$$

[*Hint*: The first inequality is an elementary consequence of the second.]

6. LINEAR EQUATIONS AND INVERSES

As we saw in Section 4, the system of m linear equations

$$\begin{aligned} \alpha_{11}\xi_1 + \alpha_{12}\xi_2 + \cdots + \alpha_{1n}\xi_n &= \beta_1 \\ \alpha_{21}\xi_1 + \alpha_{22}\xi_2 + \cdots + \alpha_{2n}\xi_n &= \beta_2 \\ &\vdots \\ \alpha_{m1}\xi_1 + \alpha_{m2}\xi_2 + \cdots + \alpha_{mn}\xi_n &= \beta_m \end{aligned} \tag{6.1}$$

in the unknowns $\xi_1, \xi_2, \ldots, \xi_n$ can be written in the matrix form

$$Ax = b. \tag{6.2}$$

In this section we shall derive conditions under which (6.2) has a solution and discuss the closely related notion of an inverse matrix. The results of this section are not constructive; they do not tell us how to compute a solution x of (6.2). Computational methods for solving linear systems will be treated in Chapter 3.

There is some terminology associated with the system (6.1) and the matrix equation (6.2). Let $A \in \mathbb{R}^{m \times n}$. If $m > n$, the system is said to be *overdetermined* (there are more equations than unknowns). If $m < n$, the system is said to be *underdetermined* (there are more unknowns than equations). If $b = 0$, the system is said to be *homogeneous.*

The first question to be treated is the existence of a solution of (6.2). A classic criterion is contained in the following theorem.

THEOREM 6.1. Equation (6.2) has a solution x if and only if

$$\operatorname{rank}[(A, b)] = \operatorname{rank}(A). \tag{6.3}$$

PROOF. Let there exist an x such that $Ax = b$. Then b is a linear combination of the columns of A and hence lies in $\mathcal{R}(A)$. It follows that $\mathcal{R}[(A, b)] = \mathcal{R}(A)$, and $\operatorname{rank}[(A, b)] = \operatorname{rank}(A)$.

Conversely, note that $\mathcal{R}(A) \subset \mathcal{R}[(A, b)]$. Hence if (6.3) is satisfied, $\mathcal{R}(A)$ and $\mathcal{R}[(A, b)]$ have the same dimension and are therefore equal (see Exercise 2.11). Hence $b \in \mathcal{R}(A)$; that is, b is a linear combination of the columns of A. By Example 4.23 this means that $b = Ax$ for some $x \in \mathbb{R}^n$. ■

COROLLARY 6.2. If $\operatorname{rank}(A) = m$, then (6.2) always has a solution.

PROOF. Since the columns of A and (A, b) are m-vectors $m = \operatorname{rank}(A) \le \operatorname{rank}[(A, b)] \le m$, which implies $\operatorname{rank}(A) = \operatorname{rank}[(A, b)] = m$. ■

The theorem makes it easy to construct systems that do not have solutions. For example, take

$$A = \begin{pmatrix} 1 & 1 \\ 1 & 1 \end{pmatrix}$$

and $b = (1, 0)^{\mathrm{T}}$. Then $\operatorname{rank}(A) = 1 \ne 2 = \operatorname{rank}[(A, b)]$.

The question of uniqueness is best discussed in terms of *affine subspaces.* Specifically, let $\mathcal{S}$ be a subspace of $\mathbb{R}^n$ and let $x \in \mathbb{R}^n$. Then the affine subspace $x + \mathcal{S}$ is the set $x + \mathcal{S} = \{x + s : s \in \mathcal{S}\}$. Geometrically, the set $x + \mathcal{S}$ is obtained from $\mathcal{S}$ by moving $\mathcal{S}$ away from the origin along x (Fig. 1).

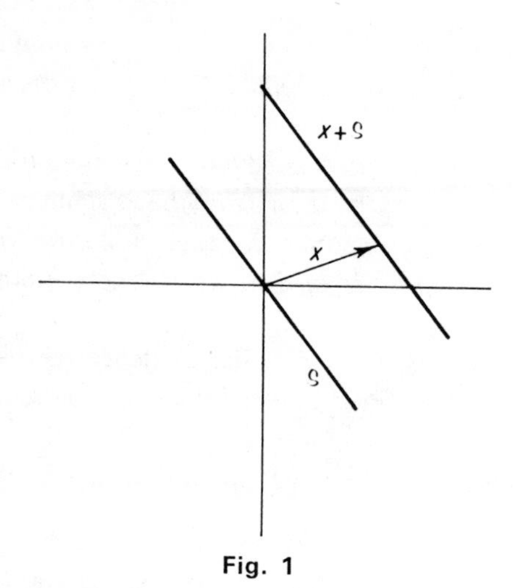

Fig. 1

THEOREM 6.3. Let x be a solution of (6.2). Then the set of all solutions of (6.2) is $x + \mathfrak{N}(A)$.

PROOF. If y satisfies (6.2), then $A(y - x) = 0$. Hence $(y - x) \in \mathfrak{N}(A)$ and $y = x + (y - x) \in x + \mathfrak{N}(A)$. Conversely if $y \in x + \mathfrak{N}(A)$, then $y = x + z$ for some z satisfying $Az = 0$. Hence $Ay = Ax + Az = Ax = b$. ∎

Thus, given a solution of (6.2), we may obtain all others by adding vectors in the null space of A. The only way for a solution to be unique is for the null space of A to consist of only the zero vector. This proves

COROLLARY 6.4. A solution of (6.2) is unique if and only if $\text{null}(A) = 0$.

The homogeneous equation $Ax = 0$ always has the trivial solution $x = 0$. By Theorem 6.3 the set of all solutions is the subspace $0 + \mathfrak{N}(A) = \mathfrak{N}(A)$. Thus, the homogeneous equation $Ax = 0$ has a nontrivial solution if and only if $\text{null}(A) \neq 0$. Now if $n > m$, then $\text{null}(A) \neq 0$ (Exercise 5.3). Using the terminology introduced above, we have the following theorem.

THEOREM 6.5. An underdetermined system of homogeneous equations always has a nontrivial solution.

If $\alpha \neq 0$, the scalar equation $\alpha\xi = \beta$ can be solved by dividing by α to give $\xi = \alpha^{-1}\beta$. This suggests that it might be possible to express the solution of (6.2) in the form

$$x = A^{-1}b \tag{6.4}$$

for some suitable matrix A^{-1}.

The requirement that (6.2) have a unique solution expressible in the form (6.4) restricts the dimensions of A. First, if the solution is unique, we must have $\text{null}(A) = 0$, from which it follows that $m \geq n$. Second, the representation (6.4) implies that (6.2) has a solution for all $b \in \mathbb{R}^m$. This means that the n columns of A must span $\mathbb{R}^m$ and hence that $n \geq m$. Thus $n = m$. Accordingly, we shall restrict our attention to square matrices.

EXAMPLE 6.6. The restriction that $n \geq m$ cannot be relaxed, since (6.4) implies that (6.2) must always have a solution. However, we can relax the restriction that the solution be unique and require only that (6.4) produce one of the solutions. For example, let

$$A = \begin{pmatrix} 1 & 0 & 0 \\ 0 & 1 & 1 \end{pmatrix}$$

and

$$C = \begin{pmatrix} 1 & 0 \\ 0 & 1 \\ 0 & 0 \end{pmatrix}.$$

Then $AC = I$. Hence if

$$x = Cb,$$

we have $Ax = ACb = Ib = b$. Thus the matrix C serves the role of A^{-1} in (6.4). However, C is not the only matrix that will do this. For example, the matrix

$$D = \begin{pmatrix} 1 & 0 \\ 1 & 2 \\ -1 & -1 \end{pmatrix}$$

also satisfies $AD = I$, and will serve equally well for A^{-1} in (6.4).

From the above example, we see that the critical property of A^{-1} is that $AA^{-1} = I$. Such a matrix is called a *right inverse* of A. In Theorem 6.8 below we shall show that if A is square and $AA^{-1} = I$, then $A^{-1}A = I$. In other words, a right inverse for a square matrix is also a left inverse, and conversely. Thus for square matrices we may drop the "left" and "right" and speak simply of inverses.

Even when A is square, it need not have an inverse. For example, if

$$A = \begin{pmatrix} 1 & 0 \\ 0 & 0 \end{pmatrix},$$

then, whatever the matrix C, AC must have its second row zero and cannot be equal to I.

Necessary and sufficient conditions for A to have an inverse are given in the following theorem.

THEOREM 6.7. Let A be square of order n. Then the following statements are equivalent:

1. $\text{rank}(A) = n$,
2. $\text{null}(A) = 0$,
3. A has linearly independent columns,
4. A has linearly independent rows,
5. there is a matrix A^{-1} of order n satisfying

$$A^{-1}A = AA^{-1} = I_n. \tag{6.5}$$

PROOF. Conditions 1, 2, and 3 are obviously equivalent. To show that part 3 implies part 5, let $f : \mathbb{R}^n \to \mathbb{R}^n$ be the linear transformation represented by the matrix A. Now by part 1 and Corollary 6.2, the equation $Ax = b$ always has a solution. By part 2 and Corollary 6.4, the solution is unique. This amounts to saying that f is 1–1 and onto. Hence f has an inverse g satisfying

$$g \circ f = f \circ g = i_n. \tag{6.6}$$

We claim that g is linear. To show this, let $x_1, x_2 \in \mathbb{R}^n$ and $y_1 = g(x_1)$, $y_2 = g(x_2)$ so that

$$x_i = f(y_i) \qquad (i = 1, 2).$$

Then

$$f(\alpha_1 y_1 + \alpha_2 y_2) = \alpha_1 f(y_1) + \alpha_2 f(y_2) = \alpha_1 x_1 + \alpha_2 x_2.$$

Hence

$$\begin{aligned} \alpha_1 g(x_1) + \alpha_2 g(x_2) = \alpha_1 y_1 + \alpha_2 y_2 &= g[f(\alpha_1 y_1 + \alpha_2 y_2)] \\ &= g(\alpha_1 x_1 + \alpha_2 x_2). \end{aligned}$$

Since g is a linear transformation of $\mathbb{R}^n$ into $\mathbb{R}^n$ it is represented by a matrix A^{-1} of order n. Then Equation (6.6) is equivalent to $A^{-1}A = AA^{-1} = I_n$, which is just (6.5).

To show that part 5 implies part 3, suppose that $Ax = 0$. Then $0 = A^{-1}0 = A^{-1}Ax = I_n x = x$. Thus if $Ax = 0$, then $x = 0$, which is equivalent to saying that the columns of A are linearly independent.

Identical arguments with rows instead of columns will establish the equivalence of parts 4 and 5. ■

Theorem 6.7 leaves two questions unanswered. First, is the matrix A^{-1} of 6.7.5 unique? Second, even if A^{-1} is unique, can A have a one-sided inverse; that is, can there be a matrix B, distinct from A^{-1}, such that $AB = I$ but $BA \neq I$? The following theorem answers these questions.

THEOREM 6.8. Let $A, B \in \mathbb{R}^{n \times n}$. If $AB = I$, then $BA = I$. Moreover, if $AC = I$, then $B = C$.

PROOF. We show first that if $AB = I$, then A has linearly independent rows. Suppose that $x^T A = 0$. Then $0 = 0B = x^T AB = x^T I = x^T$. Hence if $x^T A = 0$, $x = 0$, and A has linearly independent rows. Now from 6.7.5, A has an inverse A^{-1} satisfying (6.5). Hence

$$B = IB = A^{-1}AB = A^{-1}I = A^{-1}.$$

Thus $BA = A^{-1}A = I$. Moreover, if $AC = I$, then C is also equal to A^{-1}, which is equal to B. ■

Similarly, if $BA = I$, then $B = A^{-1}$. The answers to the questions raised above are thus: the matrix B cannot satisfy one of the equations $AB = I$ and $BA = I$ without satisfying the other, and there is at most one such matrix. This justifies the following definition.

DEFINITION 6.9. The square matrix A is *nonsingular* if it has an *inverse* A^{-1} satisfying (6.5). Otherwise, A is said to be *singular*.

If A is nonsingular and $Ax = b$, then $x = A^{-1}Ax = A^{-1}b$. Conversely, if $x = A^{-1}b$, then $Ax = AA^{-1}b = b$. In other words, *if the linear equation $Ax = b$ has a nonsingular matrix, then its unique solution can be written in the form $x = A^{-1}b$.*

An exhortation is in order here. In certain problems (perhaps one of the exercises) the reader will find himself presented with the equation $Ax = b$, when what he needs is the vector x. In such cases, the temptation to write $x = A^{-1}b$ and proceed is almost irresistible. However, it should be resisted until the reader has verified that A is nonsingular, for otherwise the results will not be validly derived (though they may be correct). In particular, A must be square to be nonsingular.

We turn now to the question of what matrix operations preserve nonsingularity. The following theorem is trivial, and its proof is left to the reader.

THEOREM 6.10. Let A be nonsingular. Then A^{-1} is nonsingular and

$$(A^{-1})^{-1} = A.$$

Likewise A^{T} is nonsingular and

$$(A^{\mathrm{T}})^{-1} = (A^{-1})^{\mathrm{T}}.$$

In the literature one sometimes finds the notation $A^{-\mathrm{T}}$ for $(A^{\mathrm{T}})^{-1}$, and we shall use it in this book.

The sum of nonsingular matrices is not necessarily nonsingular. For example, the matrix I is nonsingular, while $0 = I - I$ is singular. In general, whenever the reader feels urged to write something like $(A + B)^{-1}$, he should stop and check that $A + B$ is indeed nonsingular.

The product of two nonsingular matrices is nonsingular. In fact a little more is true.

THEOREM 6.11. Let A and B be of order n. Then AB is nonsingular if and only if A and B are nonsingular, and

$$(AB)^{-1} = B^{-1}A^{-1}.$$

PROOF. It is easy to verify that if A and B are nonsingular, then $B^{-1}A^{-1}$ is an inverse for AB. Conversely suppose AB is nonsingular. Then $B(AB)^{-1}$ is an inverse of A. Also $(AB)^{-1}A$ is an inverse of B. ■

Theorem 6.11 may be paraphrased as follows: if A is nonsingular and B is of rank n, then AB is of rank n. In other words, premultiplication by a nonsingular matrix preserves the rank of B. This is generally the case, as we shall show in Theorem 6.13. However, before proving this theorem, we shall consider the properties of subspaces that have been transformed by a matrix.

Let $\mathcal{S} \subset \mathbb{R}^n$ and A be an $m \times n$ matrix. We can define a subset $A\mathcal{S}$ of $\mathbb{R}^m$ by

$$A\mathcal{S} = \{Ax : x \in \mathcal{S}\}.$$

The following theorem gives some of the properties of $\mathcal{S}$ that are preserved under multiplication by A.

THEOREM 6.12. Let $A \in \mathbb{R}^{m\times n}$. If $\mathcal{S} \subset \mathbb{R}^n$ is a subspace, then $A\mathcal{S}$ is a subspace. If $\mathcal{S}$ is generated by C, then $A\mathcal{S}$ is generated by AC. If A is nonsingular and $\mathcal{B}$ is a basis for $\mathcal{S}$, then $A\mathcal{B}$ is a basis for $A\mathcal{S}$; hence $\dim(A\mathcal{S}) = \dim(\mathcal{S})$.

PROOF. To show A is a subspace, let $x', y' \in A\mathcal{S}$. Then $x' = Ax$ and $y' = Ay$ for some $x, y \in \mathcal{S}$, and

$$\alpha x' + \beta y' = \alpha Ax + \beta Ay = A(\alpha x + \beta y).$$

Since $\alpha x + \beta y \in \mathcal{S}$, $\alpha x' + \beta y' \in A\mathcal{S}$, and $A\mathcal{S}$ is a subspace.

Let $\mathcal{S}$ be generated by $\mathcal{C}$, and let $x' = Ax$ be a member of $A\mathcal{S}$. Now $x = \alpha_1 c_1 + \alpha_2 c_2 + \cdots + \alpha_k c_k$ for some $c_1, c_2, \ldots, c_k \in \mathcal{C}$. Hence

$$x' = Ax = \alpha_1 Ac_1 + \alpha_2 Ac_2 + \cdots + \alpha_k Ac_k. \tag{6.7}$$

Since $Ac_1, Ac_2, \ldots, Ac_k \in A\mathcal{C}$, Equation (6.7) expresses x' as a linear combination of elements of $A\mathcal{C}$. Hence $A\mathcal{C}$ generates $A\mathcal{S}$.

Let $\mathcal{B} = \{b_1, b_2, \ldots, b_k\}$ be a basis for $\mathcal{S}$ and A be nonsingular. Since $\mathcal{B}$ generates $\mathcal{S}$, $A\mathcal{S}$ generates $A\mathcal{S}$, and we need only show that the elements of $A\mathcal{B}$ are linearly independent. In other words we must show that if $(Ab_1, Ab_2, \ldots, Ab_k)x = 0$, then $x = 0$. However, if $B = (b_1, b_2, \ldots, b_k)$, then $(Ab_1, Ab_2, \ldots, Ab_k) = AB$. Now suppose that $ABx = 0$. Then $0 = A^{-1}ABx = Bx$. Since the columns of B are linearly independent, $x = 0$. ■

From Theorem 6.12 it follows that many important matrix properties are preserved under multiplication by nonsingular matrices.

THEOREM 6.13. Let A be an $m\times n$ matrix and let B and C be nonsingular matrices of orders m and n, respectively. Then

1. $B^{-1}\mathcal{R}(BA) = \mathcal{R}(A) = \mathcal{R}(AC)$,
2. $\text{rank}(BA) = \text{rank}(A) = \text{rank}(AC)$,
3. $\mathcal{N}(BA) = \mathcal{N}(A) = C\mathcal{N}(AC)$,
4. $\text{null}(BA) = \text{null}(A) = \text{null}(AC)$.

PROOF. Parts 2 and 4 obviously follow, respectively, from parts 1 and 3 via Theorem 6.12. We shall prove the equalities $B^{-1}\mathcal{R}(BA) = \mathcal{R}(A)$ and $\mathcal{N}(A) = C\mathcal{N}(AC)$, leaving the other two as an exercise. The first equality is equivalent to $\mathcal{R}(BA) = B\mathcal{R}(A)$. Let $\mathcal{A}$ consist of the columns of A. Then $\mathcal{R}(A)$ is generated by $\mathcal{A}$. By Theorem 6.12, $B\mathcal{R}(A)$ is generated by $B\mathcal{A}$, and this space is obviously $\mathcal{R}(BA)$. To show the second equality, let $Ax = 0$. Then $AC(C^{-1}x) = 0$ and $C^{-1}x \in \mathcal{N}(AC)$. Thus $x \in C\mathcal{N}(AC)$. Conversely let $x \in C\mathcal{N}(AC)$. Then $C^{-1}x \in \mathcal{N}(AC)$. That is $0 = AC(C^{-1}x) = Ax$, and $x \in \mathcal{N}(A)$. ■

EXERCISES

1. Show that the matrix $A \in \mathbb{R}^{n\times n}$ is nonsingular if and only if the homogeneous equation $Ax = 0$ has no nontrivial solution.

2. Let $A \in \mathbb{R}^{m\times n}$. Show that $A^{\mathrm{T}}A$ is nonsingular if and only if A has linearly independent columns (see Exercise 4.22).

3. Show that if A has linearly independent rows and linearly independent columns, then A is square and nonsingular.

4. Let $A = \text{diag}(\lambda_1, \lambda_2, \ldots, \lambda_n)$. Show that A is nonsingular if and only if $\lambda_i \neq 0$ $(i = 1, 2, \ldots, n)$. What is the inverse of A?

5. Show that if A is nonsingular and $\lambda \neq 0$, then $(\lambda A)^{-1} = \lambda^{-1}A^{-1}$.

6. Let $u, v \in \mathbb{R}^n$ and $\sigma \neq 0$. Suppose that $v^{\mathrm{T}}u - \sigma^{-1} \neq 0$. Show that $I - \sigma uv^{\mathrm{T}}$ is nonsingular and has an inverse given by

$$I - (\sigma + \tau - \sigma\tau v^{\mathrm{T}}u)uv^{\mathrm{T}},$$

where $\sigma^{-1} + \tau^{-1} = v^{\mathrm{T}}u$.

7. Let $B \in \mathcal{R}^{n\times n}$, $S \in \mathcal{R}^{k\times k}$, and $U, V \in \mathcal{R}^{n\times k}$. Assuming that B, S, and $V^{\mathrm{T}}B^{-1}U - S^{-1}$ are nonsingular, show that

$$(B - USV^{\mathrm{T}})^{-1} = B^{-1} - B^{-1}UTV^{\mathrm{T}}B^{-1},$$

where

$$S^{-1} + T^{-1} = V^{\mathrm{T}}B^{-1}U.$$

8. Let $A \in \mathcal{R}^{m\times n}$. The matrix $B \in \mathcal{R}^{n\times m}$ is a *left inverse* of A if $BA = I$. Show that A has a left inverse if and only if the columns of A are linearly independent and that it is unique if and only if the rows of A are linearly independent. [*Hint*: $(A^{\mathrm{T}}A)^{-1}A^{\mathrm{T}}$ is such an inverse.]

9. Define the notion of right inverse, and discuss its existence and uniqueness.

10. Let $\mathcal{G} = \{g_1, g_2, \ldots, g_n\}$ be a basis for $\mathcal{R}^n$. Given any vector $x \in \mathcal{R}^n$ there is a unique vector $w \in \mathcal{R}^n$ whose components $\omega_1, \omega_2, \ldots, \omega_n$ are the components of x with respect to $\mathcal{G}$ (that is, $x = \omega_1 g_1 + \omega_2 g_2 + \cdots + \omega_n g_n$). Show that the function $g : \mathcal{R}^n \to \mathcal{R}^n$ defined by $g(x) = w$ is a nonsingular linear transformation. What matrix represents g?

11. Just as a vector may have different components with respect to a basis, so may a linear transformation $f : \mathcal{R}^n \to \mathcal{R}^m$ have a different matrix representation with respect to bases $\mathcal{G}$ and $\mathcal{H}$ for $\mathcal{R}^n$ and $\mathcal{R}^m$, respectively. Let $x \in \mathcal{R}^n$. By Exercise 6.10 the mapping that carries x onto its vector w of components with respect to $\mathcal{G}$ is a linear transformation represented, say, by the matrix G. Likewise if $y \in \mathcal{R}^m$, there is a matrix H that transforms y into its vector z of components with respect to $\mathcal{H}$. Then if $y = f(x)$, the function mapping w onto z is a linear transformation. It is represented by the matrix HAG^{-1}, where A is the matrix representing f. The matrix HAG^{-1} is called the matrix representing f with respect to the bases $\mathcal{G}$ and $\mathcal{H}$.

7. A MATRIX REDUCTION AND SOME CONSEQUENCES

In this section we shall introduce a theme which, with variations, reappears throughout this book. Namely, given a problem involving a matrix A, we shall use matrix operations to reduce A to a simple form, from which the given problem may be readily solved. Such techniques will be used in Chapter 3 to solve linear systems, in Chapter 5 to solve the linear least squares problem, and in Chapters 6 and 7 to discuss and solve the eigenvalue problem. To convince himself of the utility of such techniques, the reader might glance ahead to the introductory material in Section 3.2.

The specific purpose of this section is to establish some important theorems about rank and nullity; for example, $\operatorname{rank}(A) = \operatorname{rank}(A^T)$. From Theorem 6.13, we know that if X and Y are nonsingular matrices, then Y^TAX has the same rank as A and $(Y^TAX)^T$ has the same rank as A^T. We shall construct nonsingular matrices X and Y so that it is obvious that Y^TAX and $(Y^TAX)^T$ have the same rank.

The reductions of this and subsequent sections simplify a matrix by introducing various systematic patterns of zeros. For example, in Chapter 3 we shall consider algorithms for reducing a matrix to upper trapezoidal form. The reduction of this section concentrates the nonzero elements into a non singular leading principal submatrix. In reading the proof of the following theorem, the reader will find it helpful to write down explicitly the dimensions of each matrix as it is introduced.

THEOREM 7.1. Let $A \in \mathbb{R}^{m\times n}$. Then there are nonsingular matrices X and Y, of order n and m, respectively, such that

$$Y^TAX = \begin{pmatrix} C & 0 \\ 0 & 0 \end{pmatrix}, \tag{7.1}$$

where C is nonsingular.

PROOF. Let $\operatorname{null}(A) = n - r$ (it will turn out that r is the rank of A), and let $x_{r+1}, x_{r+2}, \ldots, x_n$ form a basis for $\mathfrak{N}(A)$. Then if $X_2 = (x_{r+1}, \ldots, x_n)$, $AX_2 = 0$. By Theorem 2.17, there are vectors $x_1, x_2, \ldots, x_r$ such that $\{x_1, x_2, \ldots, x_r, x_{r+1}, \ldots, x_n\}$ is a basis for all $\mathbb{R}^n$. Let $X_1 = (x_1, x_2, \ldots, x_r)$ and $X = (X_1, X_2)$. Then X is nonsingular (its n columns span $\mathbb{R}^n$). Moreover, the columns of $B = AX_1$ are linearly independent. To see this, let $Bv = 0$. Then $A(X_1v) = 0$, and $X_1v \in \mathfrak{N}(A)$. By construction, no linear combination of $x_1, x_2, \ldots, x_r$ can lie in $\mathfrak{N}(A)$ unless it is the zero vector. Hence $X_1v = 0$, and, since the columns of X_1 are linearly independent, $v = 0$.

Now

$$AX = (AX_1, AX_2) = (B, 0).$$

This completes the first half of the reduction. The second half is accomplished by doing the same thing to B^T. In fact, let the $m - l$ columns of Y_2 span $\mathfrak{N}(B^T)$, and let the columns of $Y = (Y_1, Y_2)$ form a basis for $\mathbb{R}^m$. Then Y is nonsingular, $B^T Y_2 = 0$, and $B^T Y_1$ has linearly independent columns. Setting $C^T = B^T Y_1$, we have

$$Y^T A X = \begin{pmatrix} Y_1^T \\ Y_2^T \end{pmatrix}(B, 0) = \begin{pmatrix} C & 0 \\ 0 & 0 \end{pmatrix},$$

where C is an $l \times r$ matrix with linearly independent rows.

It remains to show that C is nonsingular. Consider the matrix

$$Y^T B = \begin{pmatrix} C \\ 0 \end{pmatrix}.$$

Since Y^T is nonsingular and B has linearly independent columns, it follows from Theorem 6.13 that

$$\begin{pmatrix} C \\ 0 \end{pmatrix},$$

and hence C, has linearly independent columns. However, C has linearly independent rows. Hence C is square and nonsingular (see Exercise 6.3). ∎

In the proof, it may happen that $r = m$ or $r = n$. In this case, some of the zero matrices appearing in (7.1) will be void. For example, if A is nonsingular, we may take $X = Y = I$, so that $C = A$ and all the zero matrices are void.

The proof of the theorem is a fine example of the use of partitioned matrices. At no stage have we had to manipulate individual elements of A. However, there is a price to be paid for this simplicity; the proof gives no hint of how to compute the matrices X, Y, and C. In Chapter 3, where we shall develop algorithms that in principle will allow us to compute the decomposition (7.1), the individual elements of A will come very much to the fore.

Theorem 7.1 has important consequences for the rank or nullity of a matrix.

COROLLARY 7.2. Let $A \in \mathbb{R}^{m \times n}$. Then

1. $\operatorname{rank}(A) = \operatorname{rank}(A^T)$,
2. $\operatorname{rank}(A) + \operatorname{null}(A) = n$.

PROOF. In Theorem 7.1, let C be of order r and let

$$Z = \begin{pmatrix} C^{-1} & 0 \\ 0 & I_{m-r} \end{pmatrix}.$$

Then

$$E = ZY^{\mathrm{T}}AX = \begin{pmatrix} I_r & 0 \\ 0 & 0 \end{pmatrix}.$$

Since ZY^{T} and X are nonsingular, $\operatorname{rank}(E) = \operatorname{rank}(A)$, $\operatorname{null}(E) = \operatorname{null}(A)$, and $\operatorname{rank}(E^{\mathrm{T}}) = \operatorname{rank}(A^{\mathrm{T}})$. Hence it suffices to establish the result for the matrix E.

Now it is easy to verify that the m-vectors $e_1, e_2, \ldots, e_r$ form a basis for $\mathcal{R}(E)$, the n-vectors $e_1, e_2, \ldots, e_r$ form a basis for $\mathcal{R}(E^{\mathrm{T}})$, and the n-vectors $e_{r+1}, e_{r+2}, \ldots, e_n$ form a basis for $\mathcal{N}(E)$. Hence $\operatorname{rank}(E) = \operatorname{rank}(E^{\mathrm{T}}) = r$, and $\operatorname{null}(E) = n - r$. ■

Corollary 7.2 may be used to establish an important inequality concerning the rank of a product of matrices.

THEOREM 7.3. Let $A \in \mathbb{R}^{l\times m}$ and $B \in \mathbb{R}^{m\times n}$. Then

$$\operatorname{rank}(AB) \leq \operatorname{rank}(A) \tag{7.2}$$

and

$$\operatorname{rank}(AB) \leq \operatorname{rank}(B). \tag{7.3}$$

PROOF. We establish (7.3) first. Since both AB and B have n columns, it follows from 7.2.2 that (7.3) is equivalent to

$$n - \operatorname{null}(AB) \leq n - \operatorname{null}(B),$$

or

$$\operatorname{null}(B) \leq \operatorname{null}(AB). \tag{7.4}$$

Now if $x \in \mathcal{N}(B)$, then $Bx = 0$ and $ABx = 0$. Hence $x \in \mathcal{N}(AB)$, and $\mathcal{N}(B) \subset \mathcal{N}(AB)$, which is equivalent to (7.4). To establish (7.2) note that (7.3) implies that $\operatorname{rank}(B^{\mathrm{T}}A^{\mathrm{T}}) \leq \operatorname{rank}(A^{\mathrm{T}})$. Then (7.2) follows from 7.2.1. ■

EXERCISES

1. Prove that the equation $Ax = b$ has a solution if and only if $y^{\mathrm{T}}A = 0$ implies $y^{\mathrm{T}}b = 0$.

2. A matrix is of *full rank* if its rows or its columns are linearly independent. Show that if $A \in \mathbb{R}^{m\times n}$ and $\text{rank}(A) = r$, then $A = UV^{\text{T}}$ where U and V are of full rank r. [*Hint*: In (7.1) make the following partitions

$$Y^{-\text{T}} = (U_1, U_2), \qquad X^{-\text{T}} = (W_1, W_2),$$

where $U_1 \in \mathbb{R}^{m\times r}$ and $W_1 \in \mathbb{R}^{n\times r}$. Then $U = U_1$ and $V^{\text{T}} = CW_1^{\text{T}}$.]

NOTES AND REFERENCES

The material in this chapter is for the most part the content of an elementary linear algebra course with special emphasis on matrix structure and operations. There are, however, two substantial differences from the usual linear algebra course: the development is tied to the real field, and the notion of abstract vector space is largely ignored. The first difference is more apparent than real. The definitions and theorems of this chapter remain valid for any field of scalars, a fact which we shall exploit in Chapter 6 where we must work with vectors and matrices having complex elements. The second difference was dictated by the desire that the book be accessible to the widest possible audience.

A similar, but less rapidly paced, treatment of linear algebra may be found in the book by Noble (1969). An excellent abstract treatment is given by Halmos (1958). On a more advanced level, much material basic to numerical linear algebra is contained in Wilkinson's "The Algebraic Eigenvalue Problem" (1965a)* and Householder's work (1964). Householder's book has in addition an extensive bibliography and a fine set of very difficult exercises. Householder is also the originator of the Latin–Greek notational convention used in this book.

* J. H. Wilkinson (1965). "The Algebraic Eigenvalue Problem." Oxford Univ. Press (Clarendon), London and New York will hereafter be referred to as (AEP).

PRACTICALITIES

CHAPTER 2

Many existence proofs in mathematics are constructive in the sense that they specify procedures by which the object whose existence is to be established may actually be constructed. In matrix theory, such a constructive proof may suggest an algorithm for computing the object (perhaps the inverse of a matrix or the solution of a system of linear equations). However, there is a wide gap between mathematical algorithm and a reliable computer program for solving a given problem. For example, rounding errors may cause an otherwise unobjectionable algorithm to produce inaccurate solutions, or an algorithm may require an impossibly large amount of computer memory. This chapter is devoted to a discussion of some of the practical considerations that may make the difference between an ineffective and an effective algorithm.

In Section 1 we discuss the effects of rounding errors on computations. One of the recent advances in numerical analysis has been to place the subject of rounding error in its proper prospective, thereby removing a good deal of the pessimism that once surrounded the subject. This advance has been achieved largely by the introduction of the notions of well- and ill-conditioned problems. These notions are also discussed in Section 1.

In Section 2 we introduce an informal language, called INFL, which we shall use throughout this book to describe matrix algorithms. Section 3 is devoted to some of the programming considerations that are important for matrix algorithms. These considerations are illustrated with simple algorithms for matrix addition and multiplication. It is suggested that the reader treat sections 2 and 3 as a unit, reading them together first and then returning to the description of INFL in Section 2.

1. ERRORS, ARITHMETIC, AND STABILITY

The algorithms to be described in the remaining chapters of this book will usually be executed on a high-speed digital computer where the individual operations of addition, multiplication, and division are performed with rounding error. On a desk calculator it is possible to monitor each operation as it is performed and perhaps detect a breakdown in the calculation. At present this is impossible on a computer, and from the user's point of view the algorithm must be regarded as a black box that takes a problem and, after a large number of inexact operations, returns what purports to be an answer. It is therefore necessary to give *a priori* assessments of the effect of rounding errors on the algorithms.

It is beyond the scope of this book to give detailed error analyses. However, we shall frequently discuss the numerical properties of our algorithms. Such discussions presuppose some acquaintance with computer arithmetic and with the distinction between ill-conditioned problems and unstable algorithms. Since the error analyses are essentially unaffected by the number system in which the computations are performed, all discussion and examples will be confined to the decimal system, even though many computers operate in the binary system.

Since we shall be dealing with approximate quantities, the first problem is to give a suitable definition of the error in an approximation. Let β be an approximation to α. A natural measure of the accuracy of β is the number $|\alpha - \beta|$. However, this number is meaningless unless the size of α is known. For example, consider the two cases

$$\begin{aligned} \alpha &= 1.234, & \alpha &= 0.002, \\ \beta &= 1.233, & \beta &= 0.001, \\ |\alpha - \beta| &= 10^{-3}, & |\alpha - \beta| &= 10^{-3}. \end{aligned}$$

In both cases $|\alpha - \beta|$ is 10^{-3}, but only in the first case should we say that

β is a good approximation to α. In the second case, α is so small that the small difference between α and β represents a significant change. Thus it is clear that what is important is the ratio of $|\alpha - \beta|$ to α.

This example motivates the following definitions of error. The terminology adopted here is mildly nonstandard.

DEFINITION 1.1. Let $\alpha, \beta \in \mathbb{R}$, with β regarded as an approximation to α. The *residual* of β is the number

$$\alpha - \beta.$$

The *absolute error* or *error* in β is the number

$$|\alpha - \beta|.$$

If $\alpha \neq 0$, the *relative residual* of β is the number

$$\frac{\alpha - \beta}{\alpha},$$

and the *relative error* is the number

$$\left|\frac{\alpha - \beta}{\alpha}\right|.$$

Thus if the residual of β is ε, then $\beta = \alpha - \varepsilon$. In other words, if β has small absolute error, than β may be obtained from α by subtracting a small quantity. Similarly if β has relative residual ϱ, then $\beta = \alpha(1 - \varrho)$. In other words, if β has small relative error, then β may be obtained from α by multiplying by a quantity very near unity.

EXAMPLE 1.2. Some insight into the nature of relative error may be had by examining the accompanying table of approximations to $\alpha = 3.526437$. An examination of the relative errors suggests the following

β	Relative error
3.526430	2.0×10^{-6}
3.526400	1.0×10^{-5}
3.526000	1.2×10^{-4}
3.520000	1.8×10^{-3}
3.500000	7.5×10^{-3}
3.000000	1.5×10^{-1}

rule: *If the relative error in* β *is approximately* 10^{-n}, *then* α *and* β *agree to about n significant figures, and conversely.*

Most numerical computations on computers are performed in *floating-point* arithmetic. The floating-point number is simply a computer representation of the familiar scientific notation in which a number is expressed as a fraction times a power of 10. The actual form of a floating-point number varies widely from computer to computer. To illustrate the common features, we make the following definition.

DEFINITION 1.3. A *t-digit decimal floating-point number* or *word* is an ordered pair (m, c) where m is a t-digit decimal fraction with $-1 < m < 1$ and c is a decimal integer. The number m is the *mantissa* of the floating-point number (m, c) and the number c is its *characteristic*. The value of (m, c) is $m \cdot 10^c$. If $.1 \leq |m| < 1$, then (m, c) is said to be *normalized.*

Thus $(-.0342, 5)$ is an unnormalized, four-digit, floating-point number whose value is -3420. A normalized floating-point number always has a nonzero digit in the leading position of its fraction; for example, $(.34753, -1)$ is a five-digit, normalized, floating-point number whose value is .034753.

If the floating-point number (m, c) has value α, it is said to *represent* α. The utility of floating-point numbers lies in the fact that they can represent a wide range of real numbers. However, not all real numbers can be represented by a floating-point number. For example, the number .1947 cannot be represented by a three-digit floating-point number, although it can be represented as a four-digit floating-point number in the form $(.1947, 0)$. The fraction $\frac{1}{3}$, whose decimal expansion is nonterminating, can be represented by no decimal floating-point number (however, it may be represented by a base 3 floating-point number).

The number of digits in a floating-point number is called the *precision* of the number. On most computers, the floating-point numbers have a fixed precision. Many computers also have the ability to manipulate floating-point numbers with about twice the usual precision. Such floating-point numbers are called *double precision* numbers and computations involving them are said to be done in double precision. It should be stressed that the number of digits in a double precision number may vary from computer to computer, and that one computer's double precision may be another's single precision. Performing a calculation in double precision usually reduces the effects of rounding error but increases the computer time and storage required for the calculation.

When working with floating-point numbers of fixed precision, it is customary to identify them with their values. Thus if α is a real number that can be represented by a floating-point number of the proper precision, we shall refer to the floating-point number α.

Computers with floating-point hardware are provided with a set of instructions for manipulating floating-point numbers. These instructions mimic the operations of addition, subtraction, multiplication, and division. However, these operations cannot be performed exactly. For example, if we are working to t digits, the exact product of two floating-point numbers will generally require $2t$ digits for its representation and hence must be represented inexactly as a t-digit floating-point number. The error introduced by the inexactness of the floating-point arithmetic operations is called *rounding error*. In an extensive calculation, there is a real possibility that rounding errors will accumulate and contaminate the answer. It is therefore desirable that any specific algorithm be analyzed to show that rounding error will not effect the results unduly. We shall see a little later that the key word here is unduly.

A rounding-error analysis requires that one have bounds for the errors in the floating-point operations. For most computers, these errors can be described (somewhat conservatively) as follows. Let $\mathrm{fl}(\alpha)$ denote the rounded value of the real number α. Similarly let $\mathrm{fl}(\alpha\beta)$, $\mathrm{fl}(\alpha/\beta)$, and $\mathrm{fl}(\alpha + \beta)$ denote the *computed* product, quotient, and sum of the floating-point numbers α and β. Then for t-digit computations on a given computer there is a number μ of order unity such that:

1. $\mathrm{fl}(\alpha) = \alpha(1 + \varrho)$, where $|\varrho| \leq \mu \cdot 10^{-t}$,
2. $\mathrm{fl}(\alpha\beta) = \alpha\beta(1 + \varrho)$, where $|\varrho| \leq \mu \cdot 10^{-t}$,
3. if $\beta \neq 0$,

$$\mathrm{fl}\left(\frac{\alpha}{\beta}\right) = \left(\frac{\alpha}{\beta}\right)(1 + \varrho),$$

   where $|\varrho| \leq \mu \cdot 10^{-t}$,

4. $\mathrm{fl}(\alpha + \beta) = \alpha(1 + \varrho) + \beta(1 + \sigma)$, where $|\varrho|, |\sigma| \leq \mu \cdot 10^{-t}$.

These bounds say that the operations of rounding, multiplication, and division introduce only small relative errors into the calculation. The results of an addition can be quite inaccurate; however, the computed sum of α and β is the exact sum of numbers that are very near α and β. For most purposes, this result is good enough; for in practice the numbers α and β will themselves be computed and already have errors that are larger than those introduced by the addition.

A frequently occurring calculation is the computation of the inner product

$$x^{\mathrm{T}}y = \xi_1\eta_1 + \xi_2\eta_2 + \cdots + \xi_n\eta_n. \tag{1.1}$$

In multiplying two single precision numbers some computers calculate the full $2t$-digit product and round only when the product is transferred to storage. Since the double precision product is available for free it seems natural to perform the additions in (1.1) in double precision. This procedure is known as the *double precision accumulation of inner products*. It gives a very accurate result that may cost little more than the single precision calculation, depending on the relative speeds of multiplication and double precision addition.

In our definition of a floating-point number we placed no limitations on the size of the characteristic. In practice, however, all computers restrict the range of their characteristic. If the characteristic must lie between, say, -100 and 100, then no number whose absolute value is greater than 10^{100} or less than 10^{-100} has a floating-point representation. On such a computer it is impossible to calculate directly the square of, say, 10^{60} or 10^{-60}. When an operation would produce a number with too large or too small a characteristic, it is said to suffer respectively from *overflow* or *underflow*. Obviously any well-constructed algorithm must specify what to do in the event of overflow or underflow.

We now give two examples of calculations in which rounding error produces inaccurate results. All calculations are done in four-digit arithmetic.

EXAMPLE 1.4. The smallest root ϱ_1 of the quadratic equation

$$\xi^2 - 6.433\xi + .009474 = 0 \tag{1.2}$$

is, to five figures, 1.4731×10^{-3}. A numerical expression for ϱ_1 is given by the quadratic formula

$$\varrho_1 = \frac{6.433 - \sqrt{(6.433)^2 - 4(.009474)}}{2}. \tag{1.3}$$

In four-digit arithmetic, an evaluation of (1.3) might go as follows:

$$\begin{aligned}
&1)\ \alpha = \mathrm{fl}[(6.433)^2] = 41.38\\
&2)\ \beta = \mathrm{fl}[4(.009474)] = 0.03800\\
&3)\ \gamma = \mathrm{fl}(\alpha - \beta) = 41.34\\
&4)\ \delta = \mathrm{fl}(\sqrt{\gamma}) = 6.430\\
&5)\ \varepsilon = \mathrm{fl}(6.433 - \delta) = 0.003\\
&6)\ \bar{\varrho}_1 = \mathrm{fl}(\varepsilon/2) = 0.0015
\end{aligned}$$

Thus the computed value $\bar{\varrho}_1$ fails to agree with ϱ_1 in its second significant figure.

EXAMPLE 1.5. The set of two linear equations

$$\begin{aligned}
3.000\xi_1 + 4.127\xi_2 &= 15.41\\
1.000\xi_1 + 1.374\xi_2 &= 5.147
\end{aligned} \tag{1.4}$$

has for its solution $\xi_1 = 13.6658$ and $\xi_2 = -6.2$. If the first equation is divided by 3.000 and subtracted from the second, the result is

$$\left(1.374 - \frac{4.127}{3}\right)\xi_2 = 5.147 - \frac{15.41}{3}.$$

Thus

$$\xi_2 = \frac{5.147 - \dfrac{15.41}{3}}{1.374 - \dfrac{4.127}{3}}. \tag{1.5}$$

A stepwise evaluation of (1.5) proceeds as follows.

$$\begin{aligned}
&1)\ \alpha = \mathrm{fl}(15.41/3) = 5.140\\
&2)\ \beta = \mathrm{fl}(5.147 - \alpha) = 0.007\\
&3)\ \gamma = \mathrm{fl}(4.127/3) = 1.376\\
&4)\ \delta = \mathrm{fl}(1.374 - \gamma) = -0.002\\
&5)\ \xi_2 = \mathrm{fl}(\beta/\delta) = -3.500
\end{aligned}$$

Again the computed value fails to agree with the true value.

Superficially both examples appear to be similar. In both, a short and seemingly reasonable calculation yields bad results. In both, the failure is signaled by a subtraction of nearly equal quantities that cancels most of the significant figures (step 5 in Example 1.4 and steps 2 and 4 in Example 1.5).

In spite of their similarities, however, the two computations fail for quite different reasons. The failure in Example 1.4 was caused by the unfortunate representation (1.3) of ϱ_1, whose computation requires the cancellation of significant figures. This cancellation is absent in the alternate expression

$$\varrho_1 = \frac{2(.009474)}{6.433 + \sqrt{(6.433)^2 - 4(.009474)}}, \tag{1.6}$$

whose four-digit realization gives the approximation $\bar{\varrho}_1 = 1.473$, a perfectly satisfactory answer. On the other hand, there is good reason to believe that no rearrangement of the computations will allow us to solve the problem of Example 1.5 accurately in four-digit arithmetic, for the system

$$3.000\xi_1 + 4.122\xi_2 = 15.41$$
$$1.000\xi_1 + 1.374\xi_2 = 5.147$$

which differs from (1.4) only in the fourth place of the coefficient 4.127, is inconsistent and has no solution. We would thus expect that rounding errors in the fourth place would cause the computed solution to vary wildly, no matter how it is arrived at. The difference between the examples is, then, that the first presents a bad way of solving a reasonable problem while the second presents an unreasonable problem.

If the reader is, for the moment, willing to use the word "near" without specifying exactly what it means, we can make this distinction more precise. Many numerical problems can be described as follows. We are given a mathematically defined function $f : \mathcal{S} \subset \mathbb{R}^m \to \mathbb{R}^n$. The m components of the argument x of f are the data that determine the problem and the n components of $f(x)$ are the answers. The numerical problem is then to compute an approximation to $f(x)$ given x. For example, the data in Example 1.4 is the vector $x = (\alpha, \beta)^{\mathrm{T}}$ of coefficients of the equation $\xi^2 + \alpha\xi + \beta = 0$, and $f(x)$ is the smallest root of the equation.

The nature of the function f limits the accuracy attainable by the computations, for in practice only an approximation x^* to the data is known, and even at best we can calculate only $f(x^*)$. Now for some classes of problems $f(x)$ and $f(x^*)$ may differ greatly. Such a problem is said to be *ill conditioned.* If we attempt to solve an ill-conditioned problem starting with inexact data, the solution will be inexact no matter how it is computed. The problem of Example 1.5 is ill conditioned.

The choice and implementation of an algorithm for solving the mathematical problem associated with f amounts to defining a new function f^* that, given data x, yields an approximate solution $f^*(x)$. We do not expect f^* to solve ill-conditioned problems more accurately than the data warrants; however, we should be very unhappy if f^* introduced larger inaccuracies of its own. This suggests that we require the following of f^*: For any $x \in \mathfrak{S}$ there is a nearby $x^* \in \mathfrak{S}$ such that $f(x^*)$ is near $f^*(x)$. In other words we require that the algorithm yield a solution that is near the exact solution of a slightly perturbed problem. Algorithms with this property are called *stable.* Equation (1.3) represents an unstable way of computing the root ϱ_1 of (1.2).

Some insight into the meaning of this definition of stability can be gained by considering what happens when a stable algorithm is used to solve a well-conditioned problem (Fig. 1). By the stability property the computed

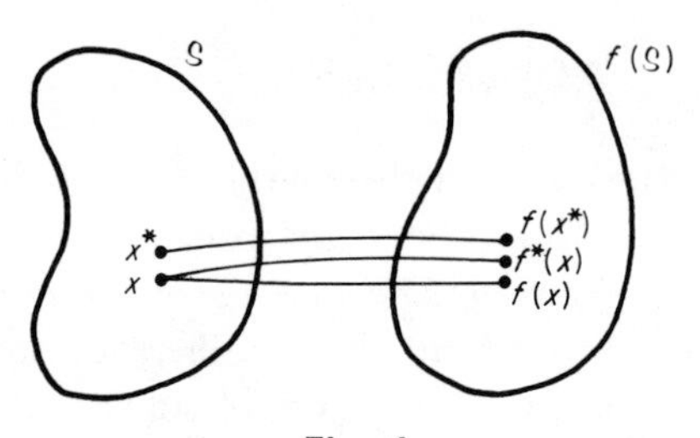

Fig. 1

solution $f^*(x)$ must be near some exact solution $f(x^*)$, where x^* is near x. Since the problem is well conditioned and x^* is near x, $f(x^*)$ is near $f(x)$. Thus $f^*(x)$ is near $f(x^*)$. In other words, if a stable algorithm is applied to a well-conditioned problem, the computed solution is near the exact solution.

When a stable algorithm is used to solve an ill-conditioned problem (Fig. 2), the computed solution and the exact solution need not agree at all, since there is no guarantee that $f(x^*)$ and $f(x)$ are near one another. However, since $f(x^*)$ and $f^*(x)$ are near, the difference between $f(x)$ and $f^*(x)$ must be about equal to the difference between $f(x)$ and $f(x^*)$. In other words, if an algorithm is stable, no matter how big the difference between the computed solution and the exact solution, almost the same

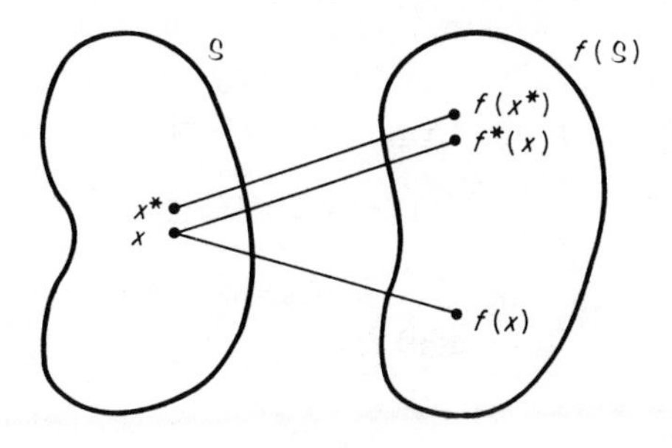

Fig. 2

error could be achieved by perturbing the initial data slightly. If this hypothetical perturbation is smaller than the inaccuracies already in the data, we can say that the errors introduced by the algorithm are "no more than are warranted by the data." Thus in Example 1.5 the rounding of the data to four places has already changed the solution in its first place, and no algorithm, however stable, can be expected to restore this lost accuracy.

EXAMPLE 1.6. The function $f : \mathbb{R}^2 \to \mathbb{R}$ defined by

$$f(x) = \xi_1 + \xi_2$$

is associated with the problem of computing the sum of two numbers. This problem may or may not be ill conditioned, depending on the data. For example

$$1.947 - 1.943 = 0.004$$

while

$$1.946 - 1.944 = 0.002.$$

Thus small relative errors in the components of $x = (1.947, -1.943)^{\mathrm{T}}$ produce large relative errors in $f(x)$. On the other hand, the sum of 2.321 and -1.023 is quite insensitive to small relative errors in the data, and the problem of computing the sum is well conditioned for $x = (2.321, -1.023)^{\mathrm{T}}$.

Suppose now that the sum is computed in floating-point arithmetic. The function f^* associated with this algorithm is

$$f^*(x) = \mathrm{fl}(\xi_1 + \xi_2).$$

From the floating-point error bounds we know that

$$f^*(x) = \xi_1(1 + \varrho_1) + \xi_2(1 + \varrho_2),$$

where ϱ_1 and ϱ_2 are small. Thus, given any vector x, there is a nearby vector $x^* = (\xi_1(1 + \varrho_1), \xi_2(1 + \varrho_2))^T$ such that

$$f^*(x) = f(x^*).$$

Hence floating-point arithmetic, as described here, is a stable way of computing the sum of two numbers.

The above example is a simple illustration of the technique of *backward rounding-error analysis.* In this technique one uses the error bounds for floating-point arithmetic to show that the computed solution of a given problem is the exact solution of a slightly perturbed problem. This is sufficient to ensure that the algorithm that did the computations is stable. A less trivial example is given in Appendix 3, where it is shown that the computed solution $\bar{x}$ of the triangular system

$$Tx = b$$

satisfies

$$(T + E)\bar{x} = b,$$

for some small matrix E.

Proving that an algorithm is stable does not guarantee that it will produce accurate answers. For this one must show that the problem being solved is well conditioned, which is usually done by producing a bound on the size of $f(x) - f(x^*)$ as a function of $x - x^*$. When the function f is differentiable, an estimate of its condition may be obtained by evaluating the partial derivatives $\partial\phi_i/\partial\xi_j$, for if ξ_j is perturbed by a small quantity $\delta\xi_j$, then ϕ_i will be perturbed by

$$\delta\phi_i \cong \frac{\partial\phi_i}{\partial\xi_i}\,\delta\xi_j.$$

Thus if $\partial\phi_i/\partial\xi_j$ is large, the problem may be ill conditioned. If one is interested in relative errors, the relation

$$\frac{\delta\phi_i}{\phi_i} \cong \left(\frac{\xi_j}{\phi_i}\,\frac{\partial\phi_i}{\partial\xi_j}\right)\frac{\delta\xi_j}{\xi_j}$$

shows that it is the numbers

$$\frac{\xi_j}{\phi_i}\,\frac{\partial\phi_i}{\partial\xi_j}$$

that determine the condition of the problem.

EXAMPLE 1.7. Let the quadratic equation

$$\xi^2 + \alpha\xi + \beta = 0$$

have the roots ϱ_1 and ϱ_2. We wish to determine the sensitivity of the root ϱ_1 to perturbations in β. Differentiating

$$\varrho_1^2 + \alpha\varrho_1 + \beta = 0$$

implicitly with respect to β gives

$$2\varrho_1 \frac{\partial\varrho_1}{\partial\beta} + \alpha \frac{\partial\varrho_1}{\partial\beta} + 1 = 0$$

or

$$\frac{\partial\varrho_1}{\partial\beta} = -\frac{1}{2\varrho_1 + \alpha}.$$

Since $\alpha = -(\varrho_1 + \varrho_2)$, we have

$$\frac{\partial\varrho_1}{\partial\beta} = \frac{1}{\varrho_2 - \varrho_1}.$$

This quantity is large when ϱ_1 is near ϱ_2, which warns us to beware of quadratic equations with poorly separated roots. Using the relation $\beta = \varrho_1\varrho_2$, we may obtain

$$\frac{\beta}{\varrho_1}\frac{\partial\varrho_1}{\partial\beta} = \frac{1}{1 - \dfrac{\varrho_1}{\varrho_2}},$$

which says that nearly equal roots are also a problem as far as relative errors are concerned.

The above example should serve to dispel the notion that an ill-conditioned problem is necessarily a discontinuous problem. The root ϱ_1 is always a continuous function of α and β, but if ϱ_1 is near ϱ_2 it will be ill conditioned.

One word of caution. In our discussion of ill conditioning and stability we have used words like "near" and "slightly perturbed" without specifying precisely what was meant. The idea of a norm, by means of which our use of these words can be made precise, will be introduced in Chapter 4. In the meantime the reader should remember that the notions of ill conditioning and stability depend critically on how one defines "near"; a change in the definition may change a well-conditioned problem into an ill-conditioned one or a stable algorithm into an unstable one.

EXERCISES

1. Let $A, B \in \mathbb{R}^{m \times n}$. Show that $\mathrm{fl}(A + B) = A' + B'$, where

$$\alpha'_{ij} = \alpha_{ij}(1 + \varrho_{ij}), \qquad |\varrho_{ij}| \leq \mu \cdot 10^{-t},$$

and

$$\beta'_{ij} = \beta_{ij}(1 + \sigma_{ij}), \qquad |\sigma_{ij}| \leq \mu \cdot 10^{-t}.$$

2. Show that if $|\varrho_1|, |\varrho_2|, \ldots, |\varrho_n| \leq \mu \cdot 10^{-t}$, then

$$(1 + \varrho_1)(1 + \varrho_2) \cdots (1 + \varrho_n) = (1 + \varrho)^n,$$

where $|\varrho| \leq \mu \cdot 10^{-t}$.

3. Let $\mathrm{fl}(\alpha_1 + \alpha_2 + \cdots + \alpha_n)$ denote the sum $\alpha_1 + \alpha_2 + \cdots + \alpha_n$ evaluated in floating-point arithmetic from left to right; that is

$$\mathrm{fl}(\alpha_1 + \alpha_2 + \cdots + \alpha_n) = \mathrm{fl}(\cdots \mathrm{fl}(\mathrm{fl}(\alpha_1 + \alpha_2) + \alpha_3) \cdots + \alpha_n).$$

Show that

$$\mathrm{fl}(\alpha_1 + \alpha_2 + \cdots + \alpha_n) = \alpha_1(1 + \varrho_1)^{n-1} + \alpha_2(1 + \varrho_2)^{n-1} + \alpha_3(1 + \varrho_3)^{n-2} + \cdots + \alpha_n(1 + \varrho_n),$$

where

$$|\varrho_i| \leq \mu \cdot 10^{-t}.$$

4. Let $|\varrho| \leq \mu \cdot 10^{-t}$. Show that if $n\mu \cdot 10^{-t} < .1$, then $(1 + \varrho)^n = 1 + n\varrho'$, where $|\varrho'| \leq 1.06\mu \cdot 10^{-t}$. Thus if

$$\mu' = 1.06\mu$$

and $|\varrho_1|, |\varrho_2|, \ldots, |\varrho_n| \leq \mu \cdot 10^{-t}$, it follows that

$$(1 + \varrho_1)(1 + \varrho_2) \cdots (1 + \varrho_n) = 1 + n\varrho,$$

where $|\varrho| \leq \mu' \cdot 10^{-t}$. The use of this relation in rounding-error analyses, rather than the slightly stronger relation of Exercise 1.2, usually gives simpler results.

5. Show that

$$\mathrm{fl}(\alpha_1 + \alpha_2 + \cdots + \alpha_n) = \alpha_1(1 + \varrho_1) + \alpha_2(1 + \varrho_2) + \cdots + \alpha_n(1 + \varrho_n),$$

where

$$|\varrho_1| \leq (n-1)\mu' \cdot 10^{-t}$$

and

$$|\varrho_i| \leq (n-i+1)\mu' \cdot 10^{-t} \qquad (i = 2, 3, \ldots, n).$$

6. Show that if $\mathrm{fl}(\alpha_1\beta_1 + \alpha_2\beta_2 + \cdots + \alpha_n\beta_n)$ is evaluated from left to right, then

$$\mathrm{fl}\left(\sum_{i=1}^{n} \alpha_i\beta_i\right) = \sum_{i=1}^{n} \alpha_i\beta_i(1 + \varrho_i),$$

where

$$|\varrho_1| \leq n\mu' \cdot 10^{-t},$$
$$|\varrho_i| \leq (n-i+2)\mu' \cdot 10^{-t} \qquad (i = 2, 3, \ldots, n).$$

7. Show that if $|\alpha + \beta| = \tau(|\alpha| + |\beta|)$, then

$$\frac{|\mathrm{fl}(\alpha+\beta) - (\alpha+\beta)|}{|\alpha+\beta|} \leq \frac{\mu}{\tau} 10^{-t}.$$

When is a sum computed with low relative error?

8. Let $A \in \mathbb{R}^{l\times m}$ and $B \in \mathbb{R}^{m\times n}$. Let $\alpha = \max\{|\alpha_{ij}| : i = 1, 2, \ldots, l;\ j = 1, 2, \ldots, m\}$ and $\beta = \max\{|\beta_{ij}| : i = 1, 2, \ldots, m;\ j = 1, 2, \ldots, n\}$. Show that

$$\mathrm{fl}(AB) = AB + E$$

where $|\varepsilon_{ij}| \leq m^2\alpha\beta\mu' 10^{-t}$.

9. Assume that double precision arithmetic satisfies the relations on page 72 with bound $\mu \cdot 10^{-2t}$. Let $\mathrm{fl}_2(\sum_{i=1}^{n} \alpha_i\beta_i)$ denote the result of computing the inner product $\sum_{i=1}^{n} \alpha_i\beta_i$ in double precision and rounding it to single precision. Show that

$$\mathrm{fl}_2\left(\sum_{i=1}^{n} \alpha_i\beta_i\right) = (1+\sigma)[\alpha_1\beta_1(1+\varrho_1)^n + \alpha_2\beta_2(1+\varrho_2)^n + \alpha_3\beta_3(1+\varrho_3)^{n-1} + \cdots + \alpha_n\beta_n(1+\varrho_n)^2],$$

where $|\sigma| \leq \mu \cdot 10^{-t}$ and $|\varrho_i| \leq \mu \cdot 10^{-2t}$. Thus if $1.06n \cdot 10^{-t} < 1$, then

$$\mathrm{fl}_2\left(\sum_{i=1}^{n} \alpha_i\beta_i\right) = \sum_{i=1}^{n} \alpha_i\beta_i(1 + \tau_i),$$

where $|\tau_i| \le 2\mu' \cdot 10^{-t}$. In other words, *if* $a, b \in \mathbb{R}^n$ *and the inner product* $a^{\mathrm{T}}b$ *is accumulated in double precision and rounded to single precision, then the error in the result is essentially independent of* n.

10. In order to avoid overflows or underflows it is sometimes necessary to scale the vectors a and b before calculating their inner product. Let $x = \mathrm{fl}(\gamma a)$, $y = \mathrm{fl}(\delta b)$, and $\pi = \mathrm{fl}_2(x^{\mathrm{T}}y)$. Show, as in Exercise 1.9, that the error in μ is independent of the dimension of a and b. (However, the τ_i, which are almost equal in Exercise 1.9, may now be quite different.)

11. The polynomial $\phi(\xi) = \alpha_n\xi^n + \alpha_{n-1}\xi^{n-1} + \cdots + \alpha_0$ may be evaluated by synthetic division as follows:

$$\phi_n = \alpha_n,$$

$$\phi_{i-1} = \mathrm{fl}(\xi\phi_i + \alpha_{i-1}) \qquad (i = n, n-1, \ldots, 1).$$

The computed approximation to $\phi(\xi)$ is then ϕ_0. Show that

$$\phi_0 = \alpha_n(1 + \varrho_n)\xi^n + \alpha_{n-1}(1 + \varrho_{n-1})\xi^{n-1} + \cdots + \alpha_0(1 + \varrho_0),$$

where

$$|\varrho_n| \le 2n\mu' \cdot 10^{-t},$$

$$|\varrho_i| \le (2i + 1)\mu' \cdot 10^{-t} \qquad (i = 0, 1, \ldots, n-1).$$

Is this algorithm for evaluating $\phi(\xi)$ stable?

12. Let $A \in \mathbb{R}^{2\times 2}$ and $Ax = b$. Show that the approximation to ξ_1 obtained from the procedure

$$\bar{\xi}_1 = \mathrm{fl}\left[\frac{\beta_1\alpha_{22} - \beta_2\alpha_{12}}{\alpha_{11}\alpha_{22} - \alpha_{12}\alpha_{21}}\right]$$

is the first component of the solution of a system $A'x = b'$, where the elements of A' and b' differ from those of A and b by small relative errors.

NOTES AND REFERENCES

A very readable historical survey of the concepts discussed in this chapter has been given by Wilkinson (1971). The first modern rounding-error analysis of Gaussian elimination was given by von Neumann and Goldstein

(1947) and served to dispel the pessimism that had earlier surrounded the subject. Although this paper implicitly contains the notions of ill-conditioning and stability, Givens (1954) was the first person to demonstrate explicitly the stability of an algorithm by showing that the effects of rounding errors could be accounted for by a perturbation in the original problem. Wilkinson extended this technique of backward error analysis, applying it to algorithms for finding zeros of polynomials (1959), for solving linear equations (1961), and for the algebraic eigenvalue problem (AEP). Much of this material is contained in his book "Rounding Errors in Algebraic Processes" (1963), which remains the best introduction to the subject.

Only in the last half of the 1950's did computers with floating-point hardware come into general use, although floating-point arithmetic performed by subroutines had been used earlier. The elimination of the scaling problems associated with fixed-point arithmetic and the simplicity of the error bounds for floating-point arithmetic [which are due to Wilkinson (1960, 1963)] cleared away much of the detail that had previously cluttered rounding-error analyses.

The definition of stability given here differs slightly from the usual one, which requires that one be able to find an x^* near x such that $f(x^*) = f(x)$. With this more stringent definition the algorithm for inverting a matrix that returns the rounded inverse is not stable. For example, the rounded inverse of

$$A = \begin{pmatrix} 7.000 & 7.000 \\ 8.000 & 8.001 \end{pmatrix}$$

is

$$10^{+3}\begin{pmatrix} 1.143 & -1.000 \\ -1.143 & 1.000 \end{pmatrix},$$

which is singular and hence not the inverse of any matrix, much less one near A. W. Kahan has used the definition given here in some lectures on rounding error.

2. AN INFORMAL LANGUAGE

One of the chief purposes of this book is to describe algorithms for performing various matrix computations such as solving systems of equations or calculating inverses. The descriptions will usually be accompanied by a derivation of the method under consideration and will be couched in standard mathematical parlance. While it is possible to code a working program starting from such a description, the difficulties involved in extracting the necessary information from several pages of theorems and discussion present an almost insurmountable barrier to the novice. Accordingly, we will conclude our discussions with a concise summary of the proposed algorithm. The purpose of this section is to give a brief description of the informal language that we shall use for these summaries.

There are several programming languages, such as FORTRAN, ALGOL, or PL/I, that are reasonably well suited for coding matrix computations. However, for a number of reasons the resulting code may not very closely resemble the original description of the algorithm. In the first place, the full complement of mathematical notation is not available to these languages, so that simple mathematical expressions must often be coded in expanded form. For example, the mathematical statement

$$\gamma_{ij} = \sum_{k=1}^{n} \alpha_{ik}\beta_{kj} \qquad (i, j = 1, 2, \ldots, n), \tag{2.1}$$

which tells how to form the matrix product $C = AB$, would appear in FORTRAN as

```
    DØ 10 I = 1,N
      DØ 10 J = 1,N
        C(I,J) = 0.
        DØ 10 K = 1,N                                      (2.2)
          C(I,J) = C(I,J) + A(I,K) * B(K,J)
 10 CØNTINUE
```

It is easy enough for a person conversant with FORTRAN to see that (2.1) and (2.2) amount to the same thing, but he should not have to do it while at the same time trying to understand the algorithm of which (2.1) is a part.

In the second place, the languages mentioned above are not rich enough in different types of variables. For example, the FORTRAN syntax does not permit lower case Latin or Greek letters. Thus any description using our conventions regarding scalars, vectors, and matrices must lose something in its translation to code [e.g., compare the use of A, B, and C in (2.2) for α, β, and γ in (2.1)]. Moreover, these languages have at best primitive provisions for manipulating vectors and matrices. For example, both (2.1) and (2.2) are expanded ways of specifying the computation of $C = AB$. If it can be assumed that the reader already knows how to compute a matrix product, there is no reason to use these longer expressions in a description of an algorithm.

A third difficulty is introduced by the need for handling the computer's storage efficiently. Since computations with matrices may involve large arrays, it is often desirable to place new results in locations occupied by numbers that are no longer needed. For example, the inverse of a matrix might overwrite the matrix itself. The coding of an algorithm with considerable overwriting usually results in the code's diverging from the description of the algorithm. For example, the following simple description is typical of others that will come later.

> We compute $\gamma = \alpha + \beta$ and $\delta = \beta + \gamma$. Since α and β are not used in subsequent calculations, we may overwrite α with γ and β with δ. (2.3)

A FORTRAN realization of this description might go as follows:

```
A = A + B
B = B + A
```
(2.4)

Already in the first statement of (2.4), the connection with the original description has been blurred, and in the second statement it has been lost. In subsequent statements the reader must remember that references to A and B in (2.4) really refer to γ and δ in (2.3). It is true that in this case the difficulty can be overcome by coding

```
EQUIVALENCE (A,C),(B,D)
C = A + B
D = B + C
```

However, this option will not be possible in more complicated situations.

For these reasons, we shall not use one of the standard programming languages to summarize our algorithms. Rather we shall use an INFormal Language, called INFL, which we now describe. INFL is a loose collection of conventions designed to communicate algorithms to an informed reader.

The conventions can be altered at will, provided that the alterations will not confuse the prospective reader.

As we shall use it, an INFL description will consist of a prolog and a sequence of numbered statements. The main function of the prolog is to specify the input and the output of the algorithm being described. The prolog may also be used to explain the meaning of the variables used in the description or to make any other explanations that might clarify the statements to follow.

Any statement is acceptable to INFL, provided that it can be expected to make sense to the reader. For example, the statement

Go to the first traffic light on Kingston Highway

is acceptable, provided the reader knows where to start from. On the other hand, the statement

Go to the last traffic light on Kingston Highway

is unacceptable, unless it is expected that the reader will go all the way to Kingston to find out which is the last traffic light.

Statements are numbered sequentially: 1), 2), 3), ... A sequence of statements explaining a given statement may be indented below the given statement; for example,

1) Go to Jack's house
    1) Turn left on 16th Street
    2) Go to the third house on the left
    3) Stop
2) Go to Jill's house
    1) Go to the end of 16th Street
    2) Turn right
    3) Go to the end of the block
    4) Stop

(2.5)

Subordinate statements of this kind will be referred to by concatenating the sequence of statement numbers leading to it. For example, the statement "Go to the end of the block" in (2.5) is statement 2.3.

Considerable flexibility is allowed in statement numbering. For example, the following sequence of statements is meaningful, and hence allowable.

$$
\begin{array}{l}
1)\ \ \delta = \sqrt{\beta^2 - 4\gamma} \\
2)\ \ \text{If } \beta > 0, \text{ then} \\
\quad 1t)\ \ \varrho_1 = (-\beta - \delta)/2 \\
\quad 2t)\ \ \varrho_2 = \gamma/\varrho_1 \\
\quad \text{otherwise} \\
\quad 1f)\ \ \varrho_1 = (-\beta + \delta)/2 \\
\quad 2f)\ \ \varrho_2 = \gamma/\varrho_1
\end{array}
\tag{2.6}
$$

Here the additional letters "t" and "f" obviously stand for true and false.

For the purposes of this book we shall use three principal kinds of variables: scalars, vectors, and matrices, denoted as usual by lower case Greek letters, lower case Latin letters, and upper case Latin and Greek letters. In addition, we will use lower case Latin and Greek letters as integer variables for subscripting and indexing.

Each scalar variable is associated with a location or address in a hypothetical computer. This location may contain a number called the value of the scalar variable. Likewise with each vector or matrix variable there is associated an array of locations that contain the values of the components of the variables.

It should be noted that we make no assumptions concerning how the elements of vectors and matrices are stored in the computer. For example, the elements of a matrix may be stored on one or more auxiliary storage devices, such as a disk or tape. In this case, the location of the element will point to the device rather than to the computer's high-speed memory. Nor is the vector–matrix notation meant to suggest a necessary connection with how the algorithm is implemented in a particular programming language. It is indeed true that in FORTRAN the matrix A will usually be contained in a doubly subscripted FORTRAN array called A with the element α_{ij} associated with `A(I,J)`. However, A could as well be contained in a linear array T according to the correspondence

$$
\begin{array}{lll}
\alpha_{11} - \mathrm{T}(1) & \alpha_{12} - \mathrm{T}(2) & \cdots\ \alpha_{1n} - \mathrm{T}(\mathrm{N}) \\
\alpha_{21} - \mathrm{T}(\mathrm{N}+1) & \alpha_{22} - \mathrm{T}(\mathrm{N}+2) & \cdots\ \alpha_{2n} - \mathrm{T}(2 * \mathrm{N}). \\
\vdots & \vdots & \vdots \\
\alpha_{m1} - \mathrm{T}(\mathrm{M} * \mathrm{N} - \mathrm{N}+1) & \alpha_{m2} - \mathrm{T}(\mathrm{M} * \mathrm{N} - \mathrm{N}+2) & \cdots\ \alpha_{mn} - \mathrm{T}(\mathrm{M} * \mathrm{N})
\end{array}
$$

We shall make a loose distinction between a matrix or vector variable as a collection of numbers forming a matrix or vector and such a variable as an array of locations into which we may store various numbers. In particular, it will often prove convenient to make the array A bigger than the matrix A in order to store additional information. For example, the matrix A and the vector b may be stored in the array A according to the scheme

$$\begin{matrix} \alpha_{11} & \alpha_{12} & \cdots & \alpha_{1n} & \beta_1 \\ \alpha_{21} & \alpha_{22} & \cdots & \alpha_{2n} & \beta_2 \\ \vdots & \vdots & & \vdots & \vdots \\ \alpha_{m1} & \alpha_{m2} & \cdots & \alpha_{mn} & \beta_m . \end{matrix} \tag{2.7}$$

The ith component of b may then be obtained by referencing β_i or $\alpha_{i,n+1}$, whichever is more convenient.

Any well-formed mathematical expression taking on a definite value is permitted in INFL. So is any reasonable verbal expression, such as "the first odd prime number." In evaluating expressions, we shall identify a variable with its value, so that $a + b$ is an expression whose value is the sum of the vectors a and b.

We shall use three replacement operators to modify the values and locations of variables. Let v be a variable and e be an expression with a value. Then the statement

$$v = e$$

means that the value of e is to be placed in the location (or locations) of v. This replacement operator is equivalent to the FORTRAN operator $=$ or the ALGOL operator $:=$. Its use is illustrated in (2.6).

The second operator serves as a dynamic equivalence statement. Let u and v be variables. Then the statement

$$u \leftarrow v$$

means that the variable v is to be associated with the location of u and the value of v is to be placed in that location. The statement $u \leftarrow v$ may be read loosely as "v overwrites u"; but there is more to it than this, for u and v may subsequently be used interchangeably.

The operator $\leftarrow$ will be used most frequently in conjunction with the operator $=$ in statements of the form

$$u \leftarrow v = e,$$

where e is an expression and u and v are variables. The meaning of such a statement is that the value of e is stored in the location of u and that this value may be obtained or modified by referencing either u or v. For example, the description (2.3) might be coded in INFL as

$$1)\quad \alpha \leftarrow \gamma = \alpha + \beta$$
$$2)\quad \beta \leftarrow \delta = \beta + \gamma$$

Note that this INFL code tells us that the numbers γ and δ are to overwrite α and β while at the same time it allows us to use the variables γ and δ in subsequent calculations.

As a second example, note that the arrangement of the array A in (2.7) could be specified by the INFL statement

$$1)\quad \alpha_{i,n+1} \leftarrow \beta_i \qquad (i = 1, 2, \ldots, m)$$

We shall also permit statements of the form $v \leftarrow e$ where v is a variable and e is an expression. Such a statement is defined to be equivalent to the statement $v = e$; however, the statement $v \leftarrow e$ will usually carry the implication that some old information currently contained in v is being destroyed. For example, the sum $\sigma = \sum_{i=1}^{n} \xi_i$ may be accumulated as follows.

$$1)\quad \sigma = 0$$
$$2)\quad \sigma \leftarrow \sigma + \xi_i \qquad (i = 1, 2, \ldots, n)$$

The third replacement operator allows us to interchange the values of two variables. Specifically, if u and v are variables, the statement

$$u \leftrightarrow v$$

means interchange the values in the locations of u and v. The operator $\leftrightarrow$ will be most frequently used to interchange rows and columns of a matrix. For example, the statement

$$1)\quad \alpha_{ik} \leftrightarrow \alpha_{jk} \qquad (k = 1, 2, \ldots, n)$$

interchanges rows i and j of A. If $A = (a_1, a_2, \ldots, a_n)$ is partitioned by columns, the first and second columns of A may be interchanged by the statement

$$1)\quad a_1 \leftrightarrow a_2$$

Looping may be accomplished in INFL with the For statement. Its general form is

```
| For   (indexing information)
| 1)    (Statement)
| 2)    (Statement)
| .
| .
| .
| n)    (Statement)
|____
```

The L-shaped line delimits the *scope* of the For statement. The indexing information specifies a sequence of values that a set of indexing variables is to take. The effect of the For statement is to execute the statements within its scope for each set of values of the indexing variables in the order specified by the indexing information. For example, the following sequence of statements describes the computation of the inner product $\gamma = x^{\mathrm{T}}y$.

1) $\gamma = 0$
2) | For $i = 1, 2, \ldots, n$
   | 1) $\gamma = \gamma + \xi_i \eta_i$

Ordinarily there will be only one indexing variable in a For statement. When there is more than one, it will usually happen that it does not matter in which order the indexing variables take on their values. For example, the following statements compute the matrix product $C = AB$.

1) | For $i, j = 1, 2, \ldots, n$
   | 1) $\gamma_{ij} = \sum_{k=1}^{n} \alpha_{ik}\beta_{kj}$

However, if we wish to specify that the product C be computed by rows, then we must use two nested For statements:

1) | For $i = 1, 2, \ldots, n$
   | 1) | For $j = 1, 2, \ldots, n$
   | | 1) $\gamma_{ij} = \sum_{k=1}^{n} \alpha_{ik}\beta_{kj}$

When the set of values specified in the indexing information of a For loop is null, the loop is not executed (as in ALGOL, but not FORTRAN). Such a situation is usually signaled by an inconsistency of notation. Thus the For statement

$$\text{For} \quad i = k, k+1, \ldots, n-1$$

will not be executed if $k = n$. In this connection we adopt the useful convention that an inconsistent sum is equal to zero and an inconsistent product is equal to unity.

Occasionally we may wish to restart a For loop with the next value of the indexing information. This is accomplished by a statement of the form

$$\text{Step (statement number)} \tag{2.8}$$

The statement number in (2.8) refers to a For statement containing (2.8) in its scope. The effect of the statement is to restart the For statement with the next value of its indices. If a For statement contains only a single index a step statement may refer to it by specifying the index rather than the statement number. The following INFL code replaces the nonzero components of x by their reciprocals and leaves the zero components undisturbed.

1) | For $k = 1, 2, \ldots, n$
   1) If $\xi_k = 0$, step k
   2) $\xi_k \leftarrow \xi_k^{-1}$

This completes the description of the INFL conventions that we shall use in this book. In using this informal language, there are two points that the reader should keep in mind. First, INFL is a language designed to be executed by a human being, not a computer. The output may be a program for a computer or it may be the human being's increased comprehension of an algorithm. It follows that the amount of detail in an INFL description will depend on the audience to which the description is addressed. What might be adequate for one specialist speaking to another might be inexcusably sketchy as the answer to a homework assignment.

The second point is that an INFL description can be made easy or difficult to program. By expanding matrix operations into their component arithmetical operations and avoiding the replacement operator $\leftarrow$, we can make an INFL description correspond as closely as we like to a FORTRAN or ALGOL program. However, such a description will lack the readability for which INFL was introduced. The descriptions in this book will vary in detail and programmability according to the purposes of the moment.

EXERCISES

1. Let $\varphi(\xi) = \alpha_n \xi^n + \alpha_{n-1}\xi^{n-1} + \cdots + \alpha_1\xi + \alpha_0$, and let σ be given. The coefficients of the polynomial $\varphi(\xi + \sigma) \equiv \beta_n \xi^n + \beta_{n-1}\xi^{n-1} + \cdots + \beta_0$ may be computed as follows. Let $\beta_i^{(-1)} = \alpha_i$ $(i = 0, 1, \ldots, n)$, and $\beta_n^{(j)} = \alpha_n$ $(j = 0, 1, 2, \ldots, n)$. Let

$$\beta_{i-1}^{(j)} = \sigma\beta_i^{(j)} + \beta_{i-1}^{(j-1)} \qquad (i = n, n-1, \ldots, j+1; \; j = 0, 1, 2, \ldots, n-1).$$

   Then $\beta_i = \beta_i^{(i)}$ $(i = 0, 1, \ldots, n)$. Write an INFL program to compute the β_i from the α_i in such a way that no more than $n + 1$ storage locations are used.

2. Write an INFL program to rearrange the components of the vector x in ascending algebraic order; in ascending order of magnitude.

3. Let $x \in \mathbb{R}^n$. Consider the following INFL program.

```
1)  τ = ξ_1
2) | For  i = 1, 2, ..., n − 1
   | 1)   ξ_i ← η_i = ξ_{i+1}
   |____
3)  ξ_n ← η_n = τ
```

   What is y? Where is it stored?

4. Let A be of order n. What is the effect of the following INFL code?

```
1) | For  i = 1, 2, ..., n
   | 1) | For  j = i + 1, i + 2, ..., n
   |    | 1)   α_ij ↔ α_ji
   |    |____
   |_________
```

   For what values of i is 1.1 executed?

5. Write an INFL program to compute the roots $\varrho_1 = \xi_i + i\eta_i$ and $\varrho_2 = \xi_2 + i\eta_2$ of the quadratic equation $\alpha\varrho^2 + \beta\varrho + \gamma$, where $\alpha \neq 0$. [*Hint*: The naive use of the quadratic formula can be disastrous. See Example 2.1.4 and the discussion following.]

6. Find the error in the following FORTRAN realization of the INFL statement $\xi_1 \leftarrow \sum_{i=1}^{n} \xi_i$.

```
   X(1) = 0.
   DØ 10 I = 1,N
10 X(1) = X(1) + X(I)
```

   This mistake is easy to make.

7. Write an INFL program to find all the prime numbers less than n and store them in the array p.

NOTES AND REFERENCES

In the lecture notes that preceded this book, the INFL conventions were introduced as they were needed. The addition of Section 3 on coding matrix operations made it necessary to collect these conventions in a single section, which unfortunately gives INFL the appearance of a formal language.

The symbol $\leftarrow$ is conventionally used to denote simple overwriting. The utility of the extended meaning will become apparent in the sequel. The convention contains one minor ambiguity. The INFL statement

$$\alpha \leftarrow \beta$$

has two meanings depending on whether β is regarded as variable, in which case α and β are identified in subsequent calculations, or β is regarded as an arithmetic expression whose value is the value of the variable β, in which case the statement is equivalent to $\alpha = \beta$. This ambiguity is easily resolved by context.

The For statement is adapted from a notation used by Wendroff (1969).

3. CODING MATRIX OPERATIONS

In programming from a mathematical description of a problem, one is faced with the problem of turning mathematical equations into hard code. When the quantities involved are scalars, this usually amounts to a straightforward translation (even here a little care may eliminate extra computations). However, when matrices and vectors are involved, considerably more finesse is required. For example, computations may be saved by treating specially the zero elements of a triangular matrix. The purpose of this section is to illustrate some of these considerations with coding examples from matrix addition and multiplication. We shall be especially concerned with three points: operation counts, economization of storage, and the treatment of zero elements.

We begin considering the problem of computing the matrix sum

$$C = A + B,$$

where $A, B, C \in \mathbb{R}^{m \times n}$. The definition of matrix sum yields the following simple algorithm.

ALGORITHM 3.1. Given the $m \times n$ matrices A and B this algorithm produces their sum $C = A + B$.

1) For $i = 1, 2, \ldots, m$
   1) For $j = 1, 2, \ldots, n$
      1) $\gamma_{ij} = \alpha_{ij} + \beta_{ij}$

It is important to know how much work is required for the execution of an algorithm. One way of estimating the work is to count the number of times the algorithm executes a particular operation. For example, in Algorithm 3.1 the statement 1.1.1 contains exactly one addition. It is executed n times for every time statement 1.1 is executed. However, statement 1.1 is executed m times. Hence the algorithm requires mn additions.

In view of this *operation count* for Algorithm 3.1, it might be felt that an estimate of the running time for the algorithm could be obtained by multiplying the time required for an addition by mn. Unfortunately this process does not work in practice; for in statement 1.1.1 the computer must retrieve the numbers α_{ij} and β_{ij} from storage and store back the number γ_{ij}. This involves an auxiliary computation of the locations of α_{ij}, β_{ij}, and γ_{ij} from the subscripts i and j. These auxiliary computations may require a good deal more time than the addition in statement 1.1.1.

None the less, a count of the arithmetic operations in an algorithm provides useful information. The auxiliary computations in Algorithm 3.1 must be performed once each time an addition is performed, that is mn times. Thus, while the running time for the algorithm is not the product of mn and the addition time, it is proportional to mn. This tells us, for example, that the running time will increase linearly with n. Moreover, operation counts serve as a rough basis of comparison for different algorithms. We could not accept without further justification an algorithm for matrix addition that required mn^2 additions, since, other things being equal, it would require far more time than Algorithm 3.1 when n is large.

When all the matrices involved in an algorithm are contained in the high-speed storage of a computer, it is usually the case that the time required for the execution of the algorithm is proportional to the number of arithmetic operations generated by the algorithm. For this reason we shall give operation counts for our algorithms; however, the reader is warned that such counts are not an absolute measure of the worth of an algorithm. Considerations of storage or numerical stability may make an algorithm with a high operation count the preferable one.

Since matrices with even moderate dimensions have a large number of elements, matrix algorithms can easily exceed a computer's storage capacity. For example, if A, B, and C are of order 100, Algorithm 3.1 will require 30,000 locations. When A or B is not required for future use, relief may be sought by overwriting the unneeded array with $A + B$. The following algorithm, which overwrites B, requires no further comment.

ALGORITHM 3.2. Given $A, B \in \mathbb{R}^{m \times n}$ this algorithm overwrites B with $A + B$.

1) For $i = 1, 2, \ldots, m$
    1) For $j = 1, 2, \ldots, n$
        1) $\beta_{ij} \leftarrow \alpha_{ij} + \beta_{ij}$

We turn now to the computation of the matrix product

$$C = AB.$$

As for matrix addition, the definition of the product yields a straightforward algorithm.

ALGORITHM 3.3. Given $A \in \mathbb{R}^{l \times m}$ and $B \in \mathbb{R}^{m \times n}$, this algorithm produces the product $C = AB$.

1) For $i = 1, 2, \ldots, l$
    1) For $j = 1, 2, \ldots, n$
        1) $\gamma_{ij} = \sum_{k=1}^{m} \alpha_{ik}\beta_{kj}$

An operation count for Algorithm 3.3 is easy to derive. The statement 1.1.1 requires m multiplications and $m - 1$ additions for its execution. It must be executed n times for every one time 1.1 is executed, and 1.1 must be executed l times. Hence Algorithm 3.3 requires lmn multiplications and $(m - 1)ln$ additions. If $l = m = n$, this means that we require n^3 multiplications and $n^3 - n^2$ additions. Since, for large n, n^3 is very much greater than n^2, we drop the n^2 term and say that the algorithm requires about n^3 additions. It should be noted that the number of additions and multiplications are roughly the same. This is often the case with matrix algorithms, and in the sequel, unless there is special reason, we will report only the number of multiplications required by an algorithm.

The problem of conserving storage by overwriting, say, B with AB is not as simple as it was for addition. In the first place, when A and B are not square, the product AB may require many more locations to store than A and B together, but even when A and B are square, say of order n, we cannot simply replace the symbol γ_{ij} in statement 1.1.1 of Algorithm 3.3 by β_{ij}, for β_{ij} may be needed later to compute the other elements of the product. For example, suppose we attempt to compute the product AB of the 2×2 matrices A and B by the statements

$$
\begin{aligned}
&1)\quad \beta_{11} \leftarrow \alpha_{11}\beta_{11} + \alpha_{12}\beta_{21}\\
&2)\quad \beta_{21} \leftarrow \alpha_{21}\beta_{11} + \alpha_{22}\beta_{21}\\
&3)\quad \beta_{12} \leftarrow \alpha_{11}\beta_{12} + \alpha_{12}\beta_{22}\\
&4)\quad \beta_{22} \leftarrow \alpha_{21}\beta_{12} + \alpha_{22}\beta_{22}
\end{aligned}
$$

Then the (2, 1)-element of the product will be in error, since β_{11} has been overwritten by the (1, 1)-element of the product. Likewise the (2, 2)-element of the product will be in error.

However, we can overwrite B with AB provided we are willing to use a little additional storage. Let $B = (b_1, b_2, \ldots, b_n)$ be partitioned by columns. Then

$$AB = (Ab_1, Ab_2, \ldots, Ab_n).$$

Now once Ab_1 has been computed, the column b_1 is no longer needed to form the product. Hence the components of Ab_1 may overwrite those of b_1. This suggests that we form AB as follows.

$$
\begin{array}{ll}
1) & \text{For } j = 1, 2, \ldots, n \\
 & \quad 1)\quad b_j \leftarrow Ab_j
\end{array}
\tag{3.1}
$$

Even this scheme requires some care. We cannot overwrite the components of b_j by those of Ab_j as they are formed, since the first component of b_j is required to compute all the components of Ab_j. Thus we must compute Ab_j in a scratch array before storing it in b_j. This is done as follows.

$$
\begin{array}{l|l}
1) & \text{For } j = 1, 2, \ldots, n \\
 & 1) \quad s = Ab_j \\
 & 2) \quad b_j \leftarrow s
\end{array}
\tag{3.2}
$$

[Incidently, there is nothing wrong with the INFL code (3.1); it is merely less explicit than (3.2).] A more detailed description of this technique is given in the following algorithm.

ALGORITHM 3.4. Given $A, B \in \mathbb{R}^{n\times n}$, this algorithm overwrites B with the product AB. A one-dimensional scratch array s is required.

$$
\begin{array}{l|l}
1) & \text{For } j = 1, 2, \ldots, n \\
 & 1) \quad \sigma_i = \sum_{k=1}^{n} \alpha_{ik}\beta_{kj} \qquad (i = 1, 2, \ldots, n) \\
 & 2) \quad \beta_{ij} \leftarrow \sigma_i \qquad (i = 1, 2, \ldots, n)
\end{array}
$$

In practice algorithms such as Algorithm 3.4 may be implemented as a subroutine. In languages such as ALGOL or PL/I the scratch array s offers no difficulties. Storage for s is allocated when the subroutine is entered, and the locations are freed on return. In FORTRAN, however, the size of the array s must be specified at compile time, and the storage allocated for s is unavailable to other routines. For example, if one expected matrices mostly of order 10 but possibly of order 100, then one would have to set aside 100 locations for s, of which only 10 would be frequently used. To avoid this inefficient use of storage, it is customary to require the user to furnish a scratch array as input to the subroutine. However, if several scratch arrays are required, this will clutter the argument list of the subroutine. Simplicity can sometimes be restored by using extra rows and columns of the input matrices as scratch arrays. In Algorithm 3.4, for example, we could accumulate the σ_i in an $(n + 1)$th column of B. The modified algorithm would proceed as follows.

ALGORITHM 3.5. Given $A, B \in \mathbb{R}^{n\times n}$, this algorithm overwrites B with the product AB. The array B is assumed to have $n+1$ columns.

1) For $j = 1, 2, \ldots, n$
   1) $\beta_{i,n+1} = \sum_{k=1}^{n} \alpha_{ik}\beta_{kj} \qquad (i = 1, 2, \ldots, n)$
   2) $\beta_{ij} \leftarrow \beta_{i,n+1} \qquad (i = 1, 2, \ldots, n)$

We turn now to the treatment of matrices with specially distributed zero elements. The main point to be observed is that arithmetic operations involving the zero elements may often be omitted, with a resulting saving of operations. For example, consider the problem of computing

$$C = AT$$

where A and T are of order n and T is upper triangular. The (i, j)-element of C is given by

$$\gamma_{ij} = \sum_{k=1}^{n} \alpha_{ik}\tau_{kj}. \tag{3.3}$$

However, for $k = j+1, j+2, \ldots, n$, the element τ_{kj} is zero. Hence

$$\gamma_{ij} = \sum_{k=1}^{j} \alpha_{ik}\tau_{kj}. \tag{3.4}$$

If (3.4) rather than (3.3) is used to define the product AT, the result is the following algorithm.

ALGORITHM 3.6. Given $A, T \in \mathbb{R}^{n\times n}$ with T upper triangular, this algorithm calculates the product $C = AT$.

1) For $i = 1, 2, \ldots, n$
   1) For $j = 1, 2, \ldots, n$
      1) $\gamma_{ij} = \sum_{k=1}^{j} \alpha_{ik}\tau_{kj}$

There are two reasons for preferring Algorithm 3.6 to Algorithm 3.3 when T is upper triangular. The most obvious reason is that Algorithm 3.6 requires about half the number of arithmetic operations. In fact, statement 1.1.1 requires j multiplications, where j is the index in statement 1.1.

Since 1.1 is executed for $j = 1, 2, \ldots, n$, it requires

$$1 + 2 + 3 + \cdots + n = \frac{n(n+1)}{2}$$

multiplications. Finally, statement 1.1 is executed once for each of the n values of i in statement 1. Hence, the total number of multiplications required by the algorithm is

$$\frac{n^2(n+1)}{2} \sim \frac{n^3}{2},$$

half the number required by Algorithm 3.3.

A less obvious advantage of Algorithm 3.6 is that it in no way uses or alters the elements in the lower half of the array T. Thus we are free to use this storage in any convenient way. For example, it could be used to hold a strictly lower triangular matrix L according to the scheme

$$\begin{matrix} \tau_{11} & \tau_{12} & \tau_{13} & \cdots & \tau_{1n} \\ \lambda_{21} & \tau_{22} & \tau_{23} & \cdots & \tau_{2n} \\ \lambda_{31} & \lambda_{32} & \tau_{33} & \cdots & \tau_{3n} \\ \vdots & \vdots & \vdots & & \vdots \\ \lambda_{n1} & \lambda_{n2} & \lambda_{n3} & \cdots & \tau_{nn} \end{matrix} .$$

The presence of the numbers λ in the array T will not affect the performance of Algorithm 3.6. In later chapters we shall make good use of the extra storage afforded by triangular matrices.

The presence of triangular matrices sometimes makes overwriting easy. The following algorithm forms the product of two upper triangular matrices S and T, overwriting T with the result. The reader should convince himself that it actually works. Note that no additional arrays are required. For generalizations, see Exercise 3.5.

ALGORITHM 3.7. Given the upper triangular matrices S and T of order n, this algorithm overwrites T with the product ST.

1) For $i = 1, 2, \ldots, n$
    1) For $j = i, i+1, \ldots, n$
        1) $\tau_{ij} \leftarrow \sum_{k=i}^{j} \sigma_{ik}\tau_{kj}$

An operation count for Algorithm 3.7 may be calculated as follows. Statement 1.1.1 requires

$$j - i + 1$$

multiplications. Since, according to 1.1, it is executed for $j = i, i + 1, \ldots, n$, statement 1.1 requires

$$\sum_{j=i}^{n} (j - i + 1)$$

multiplications. Finally, since 1.1 is executed for $i = 1, 2, \ldots, n$, the entire algorithm requires

$$\sum_{i=1}^{n} \sum_{j=i}^{n} (j - i + 1) \tag{3.5}$$

multiplications.

It is possible to evaluate the sum (3.5) using standard formulas to obtain the operation count. However, we are only interested in the term involving the highest power of n in the final result. To obtain this, we may approximate the sum (3.5) by an iterated integral; namely, we replace the sums in (3.5) by integrals with the summation limits as limits of integration:

$$\int_1^n \int_i^n (j - i + 1)\, dj\, di = \frac{n^3}{6} - \frac{n}{2} + \frac{1}{3} \sim \frac{n^3}{6}.$$

Thus for large n, Algorithm 3.7 requires about $n^3/6$ multiplications, one sixth the number required for Algorithm 3.3.

We conclude this section with an example of a matrix product in which one of the matrices is presented in a factored form. Let $A, B \in \mathbb{R}^{n \times n}$, and suppose we can write A in the form

$$A = xy^{\mathrm{T}},$$

where, $x, y \in \mathbb{R}^n$. We wish to compute the product $C = AB$.

A naive scheme for computing this product is to form the matrix A from the vectors x and y and use Algorithm 3.3 to form the product. Aside from the n^2 multiplications required to form A, this scheme has the drawback that it requires n^2 extra storage locations to contain the full matrix A. Moreover, the n^3 multiplications required by Algorithm 3.3 is, in this case, unnecessarily large.

A more efficient algorithm may be derived as follows. Let $B = (b_1, b_2, \ldots, b_n)$ and $C = (c_1, c_2, \ldots, c_n)$ be partitioned by columns. Then

$$c_j = Ab_j = xy^{\mathrm{T}}b_j = (y^{\mathrm{T}}b_j)x \qquad (j = 1, 2, \ldots, n).$$

Thus to compute the jth column of c we need only compute the scalar $y^{\mathrm{T}}b_j$ and multiply it into x. The following algorithm accomplishes this.

ALGORITHM 3.8. Given $B \in \mathbb{R}^{n \times n}$ and $x, y \in \mathbb{R}^n$, this algorithm computes the product $C = (xy^{\mathrm{T}})B$.

```
1) | For  j = 1, 2, ..., n
   | 1)   σ = Σ_{i=1}^{n} η_i β_ij
   | 2)   γ_ij = σ ξ_i      (i = 1, 2, ..., n)
   |____
```

Note that no additional storage is required. The algorithm requires $2n^2$ multiplications, a considerable improvement over the n^3 multiplications required by the naive scheme proposed above.

EXERCISES

In all the following exercises, give operations counts.

1. Let $A, B \in \mathbb{R}^{n \times n}$. Describe an algorithm that overwrites A with AB.

2. Let $x, y \in \mathbb{R}^m$ and $A \in \mathbb{R}^{m \times n}$. Write an INFL program to overwrite A with $(I + xy^{\mathrm{T}})A$. [*Hint*: $(I + xy^{\mathrm{T}})A = A + x(y^{\mathrm{T}}A)$.]

3. Let $D = \mathrm{diag}(\delta_1, \delta_2, \ldots, \delta_n)$ and $A \in \mathbb{R}^{m \times n}$. Write an INFL program to compute AD and overwrite A with the results.

4. Let $D = \mathrm{diag}(\delta_1, \delta_2, \ldots, \delta_m)$ and $A \in \mathbb{R}^{m \times n}$. Write an INFL program to compute DA and overwrite A with the results.

5. Let $A, T \in \mathbb{R}^{n \times n}$, and let T be upper triangular. Write an INFL program to overwrite A with AT. Use no scratch arrays. [*Hint*: Compute the product $C = AT$ in the following order; $\gamma_{nn}, \gamma_{n-1,n}, \ldots, \gamma_{1n}, \gamma_{n,n-1}, \gamma_{n-1,n-1}, \ldots, \gamma_{1,n-1}, \ldots$.] Do the same for TA. For AL and LA, where $L \in \mathbb{R}^{n \times n}$ is lower triangular.

6. Let $L, H \in \mathbb{R}^{n \times n}$ with L lower triangular and H lower Hessenberg. Write an INFL program to compute the lower Hessenberg matrix $P = LH$. Do not assume that the upper parts of the array L and H contain zeros. Write a second INFL program in which P overwrites H.

7. If T is strictly upper triangular or order n, T^{n-1} has no more than one nonzero element. Write an INFL program to compute it.

8. Let $L \in \mathbb{R}^{n \times n}$ have the form

$$L = \begin{pmatrix} 1 & 0 & \cdots & 0 & 0 & \cdots & 0 \\ 0 & 1 & \cdots & 0 & 0 & \cdots & 0 \\ \vdots & \vdots & & \vdots & \vdots & & \vdots \\ 0 & 0 & \cdots & 1 & 0 & \cdots & 0 \\ 0 & 0 & \cdots & \mu_{k+1} & 1 & \cdots & 0 \\ \vdots & \vdots & & \vdots & \vdots & & \vdots \\ 0 & 0 & \cdots & \mu_n & 0 & & 1 \end{pmatrix} k.$$

(the column containing the μ's is column k)

Write an INFL program to compute LA, where $A \in \mathbb{R}^{n \times m}$, and one to compute BL, where $B \in \mathbb{R}^{m \times n}$.

9. Let $R \in \mathbb{R}^{n \times n}$ have the form

$$R = \begin{pmatrix} 1 & 0 & \cdots & 0 & 0 & \cdots & 0 & \cdots & 0 \\ 0 & 1 & \cdots & 0 & 0 & \cdots & 0 & \cdots & 0 \\ \vdots & \vdots & & \vdots & \vdots & & \vdots & & \vdots \\ 0 & 0 & \cdots & 1 & 0 & \cdots & 0 & \cdots & 0 \\ 0 & 0 & \cdots & 0 & \varrho_1 & \cdots & \varrho_2 & \cdots & 0 \\ \vdots & \vdots & & \vdots & \vdots & & \vdots & & \vdots \\ 0 & 0 & \cdots & 0 & \varrho_3 & \cdots & \varrho_4 & \cdots & 0 \\ \vdots & \vdots & & \vdots & \vdots & & \vdots & & \vdots \\ 0 & 0 & \cdots & 0 & 0 & \cdots & 0 & \cdots & 1 \end{pmatrix}.$$

Write an INFL program to compute RA, where $A \in \mathbb{R}^{n \times m}$, and to compute BR, where $B \in \mathbb{R}^{m \times n}$.

10. Let $A, U \in \mathbb{R}^{n \times n}$ with A symmetric. Then A can be written in the form $L + L^{\mathrm{T}}$ where L is lower triangular, and the product $U^{\mathrm{T}}AU$ can be written in the form $U^{\mathrm{T}}AU = U^{\mathrm{T}}LU + (U^{\mathrm{T}}LU)^{\mathrm{T}}$. This suggests the following algorithm for computing $U^{\mathrm{T}}AU$.

    1) $A \leftarrow L$
    2) $A \leftarrow LU$
    3) $A \leftarrow U^{\mathrm{T}}A$
    4) $A \leftarrow A + A^{\mathrm{T}}$

    Write an INFL program implementing this algorithm in detail. [See Bartels and Stewart (1970).]

11. Let $A, L \in \mathbb{R}^{n \times n}$ with A symmetric and L lower triangular. Show that, because LAL^{T} is symmetric, the elements of $LAL^{\mathrm{T}} = L(AL^{\mathrm{T}})$ can be calculated if only the upper part of AL^{T} is known. Use this fact to write an INFL program that overwrites the upper half of A with the upper half of LAL^{T} and leaves the lower half undisturbed. Use no additional storage.

12. Let U and V be upper triangular of order n. Suppose that the diagonal and first $l - 1$ superdiagonals of U are zero and the diagonal and first $k - 1$ superdiagonals of V are zero. Write an INFL program to compute UV taking advantage of the zero elements. [*Hint*: Partition U and V in the forms:

$$U = \overset{\begin{matrix} \underline{l} & \underline{n-(k+l)} & \underline{k} \end{matrix}}{\begin{pmatrix} 0 & U_{12} & U_{13} \\ 0 & 0 & U_{23} \\ 0 & 0 & 0 \end{pmatrix}} \begin{matrix} n-(k+l) \\ k \\ l \end{matrix},$$

$$V = \overset{\begin{matrix} \underline{k} & \underline{l} & \underline{n-(k+l)} \end{matrix}}{\begin{pmatrix} 0 & V_{12} & V_{13} \\ 0 & 0 & V_{23} \\ 0 & 0 & 0 \end{pmatrix}} \begin{matrix} l \\ n-(k+l). \\ k \end{matrix}]$$

NOTES AND REFERENCES

Although the efficient handling of matrices in the high-speed storage of a computer is not a deep subject (it requires little more than a thorough-going parsimony on the part of the algorithmist or programmer), very little attention is paid to it in elementary texts. The object of this section has been to make the reader aware of the considerations of efficiency that must enter any practical implementation of a matrix algorithm. Perhaps the best way to increase one's skill in this art is to examine in detail well–coded programs. In this connection the algorithms in the "Handbook for Automatic Computation," Volume II edited by Wilkinson and Reinsch (1971) are particularly recommended.*

The idea of replacing a sum such as (3.5) with a corresponding integral is part of the folklore on the subject of operation counts. The dominant term of the integral will be the same as the dominant term of the sum, so that for large values of the parameters the two approaches yield essentially equivalent results.

* J. M. Wilkinson and C. Reinsch, eds. (1971). "Handbook for Automatic Computation, Vol. II, Linear Algebra." Springer-Verlag, Berlin and New York will be referred to as HACLA hereafter.

THE DIRECT SOLUTION OF LINEAR SYSTEMS CHAPTER 3

In Theorem 1.6.1 we stated a necessary and sufficient condition for the equation

$$Ax = b \tag{1}$$

to have a solution. As we noted then, this theorem and its proof do not suggest how a solution may be computed. In this chapter we shall be concerned with deriving algorithms for solving (1). The algorithms have in common the idea of factoring A in the form

$$A = R_1 R_2 \cdots R_k,$$

where each matrix R_i is so simple that the equation $Ry = c$ can be easily solved. The equation $Ax = b$ can then be solved by taking $y_0 = b$ and for $i = 1, 2, \ldots, k$ computing y_i as the solution of the equation

$$R_i y_i = y_{i-1}.$$

Obviously y_k is the desired solution x.

The matrices R_i arising in the factorization of A are often triangular. Hence Section 1 will be devoted to algorithms for solving triangular systems. Sections 2 and 3 will be devoted to algorithms for factoring A, and in Section 4 we will apply these factorizations to the solution of linear systems. The last section of this chapter will be devoted to a discussion of the numerical properties of the algorithms.

When A is nonsingular, the methods of this chapter can be adapted to give algorithms for computing A^{-1}. However, we shall be at pains to stress that in most applications the computation of A^{-1} is both unnecessary and inordinately time consuming. For example, if it is desired to compute a solution of (1), the methods of this chapter will be cheaper than first computing A^{-1} and then forming $A^{-1}b$.

1. TRIANGULAR MATRICES AND SYSTEMS

In this section we shall give algorithms for inverting triangular matrices and solving triangular systems of equations. There are several reasons for considering this special case first. In the first place, our subsequent algorithms for solving general linear systems will presuppose the ability to solve triangular systems. Second, the algorithms for triangular systems are comparatively simple. Third, the algorithms furnish good examples in a simple setting of some of the practical considerations concerning storage and operation counts that were discussed in Section 2.3. We shall restrict our attention to upper triangular matrices, the modifications for lower triangular matrices being obvious.

The first problem is to determine when T is nonsingular. We begin our investigations with the following formula for inverting a *block triangular matrix*.

THEOREM 1.1. Let A and C be of order l and m, respectively, and B be an $l\times m$ matrix. Then the matrix

$$T=\begin{pmatrix} A & B \\ 0 & C \end{pmatrix}$$

is nonsingular if and only if A and C are nonsingular. In this case

$$T^{-1}=\begin{pmatrix} A & B \\ 0 & C \end{pmatrix}^{-1}=\begin{pmatrix} A^{-1} & -A^{-1}BC^{-1} \\ 0 & C^{-1} \end{pmatrix}. \tag{1.1}$$

PROOF. If A and C are nonsingular, the matrix on the right-hand side of (1.1) is well defined, and it is easily verified that it is an inverse for T. Conversely suppose that, say, C is singular. Then there is a nonzero vector x such that $x^{\mathrm{T}}C = 0$. If 0 denotes the zero row vector with l components, then $(0, x^{\mathrm{T}}) \neq 0$ and

$$\begin{aligned}(0, x^{\mathrm{T}})T &= (0, x^{\mathrm{T}})\begin{pmatrix} A & B \\ 0 & C \end{pmatrix} \\ &= (0A + x^{\mathrm{T}}0, 0B + x^{\mathrm{T}}C) = 0.\end{aligned}$$

Hence T is singular. The case where A is singular is treated similarly. ■

Theorem 1.1 has a rather useful theoretical consequence. Note that the matrix A is the leading principal submatrix of order l of T; that is $A = T^{[l]}$ (see the discussion following Definition 1.3.11). From (1.1), the matrix A^{-1} is the leading principal submatrix of order l of T^{-1}. In other words, *if T is upper triangular, then* $(T^{[l]})^{-1} = (T^{-1})^{[l]}$. A similar result is true of the trailing submatrices.

Theorem 1.1 allows us to describe the inverse of a triangular matrix.

THEOREM 1.2. Let T be upper triangular. Then T is nonsingular if and only if its diagonal elements are nonzero. In this case T^{-1} is upper triangular.

PROOF. The proof is by induction on the order of T. The assertion is obviously true for matrices of order 1. Suppose the theorem is true of all triangular matrices of order less than $n > 1$ and T is of order n. Then T can be partitioned in the form

$$T = \begin{pmatrix} T_{n-1} & t_n \\ 0 & \tau_{nn} \end{pmatrix},$$

where T_{n-1} is the leading principal submatrix of order $n - 1$ of T and $t_n = (\tau_{1n}, \tau_{2n}, \ldots, \tau_{n-1,n})^{\mathrm{T}}$. Now T_{n-1} and τ_{nn} are upper triangular matrices of order less than n. They have no nonzero diagonal entries if and only if T has no nonzero diagonal elements. By the induction hypothesis they are nonsingular if and only if they have no nonzero diagonal elements. Finally by Theorem 1.1, T is nonsingular if and only if T_{n-1} and τ_{nn} are nonsingular.

To see that T^{-1} is upper triangular, note that by Theorem 1.1

$$T^{-1} = \begin{pmatrix} T_{n-1}^{-1} & -\tau_{nn}^{-1}\ T_{n-1}^{-1}\ t_n \\ 0 & \tau_{nn}^{-1} \end{pmatrix}. \tag{1.2}$$

By the induction hypothesis T_{n-1}^{-1} is upper triangular. Hence T^{-1} is upper triangular. ■

We now turn to the problem of solving the linear system

$$Tx = b, \tag{1.3}$$

where T is a nonsingular upper triangular matrix of order n. This may be done as follows. The last equation of the system (1.3) is

$$\tau_{nn}\xi_n = \beta_n,$$

from which

$$\xi_n = \tau_{nn}^{-1}\beta_n. \tag{1.4}$$

In general, if we know $\xi_n, \xi_{n-1}, \ldots, \xi_{i+1}$, we can solve the ith equation of (1.3), which is

$$\tau_{ii}\xi_i + \tau_{i,i+1}\xi_{i+1} + \cdots + \tau_{in}\xi_n = \beta_i,$$

to obtain

$$\xi_i = \tau_{ii}^{-1}\Big(\beta_i - \sum_{j=i+1}^{n} \tau_{ij}\xi_j\Big). \tag{1.5}$$

The formulas (1.4) and (1.5) define an algorithm for the computation of the solution of (1.3).

ALGORITHM 1.3. Let $T \in \mathbb{R}^{n\times n}$ be upper triangular and nonsingular, and let $b \in \mathbb{R}^n$. This algorithm computes the solution of the equation $Tx = b$.

1) For $i = n, n-1, \ldots, 1$

   1) $\xi_i = \tau_{ii}^{-1}\Big(\beta_i - \sum_{j=i+1}^{n} \tau_{ij}\xi_j\Big)$

Algorithm 1.3 for solving an upper triangular system is sometimes called the method of *back substitution*. Mathematically, the algorithm can fail only when some $\tau_{ii} = 0$, in which case τ_{ii}^{-1} in statement 1.1 is not defined. By Theorem 1.2 this can happen only when T is singular. In other words, *if T is nonsingular, Algorithm* 1.3 *can be carried to completion.*

In practice, however, the calculations must be performed on a computer, usually in floating-point arithmetic. We shall postpone discussing the effects of rounding errors until Section 5; however, we note that in statement 1.1 of the algorithm an inner product must be computed and some accuracy may be gained by accumulating it in double precision. (Incidentally, note that in accordance with our INFL conventions no inner product is computed in the first step, when $i = n$.)

Another important practical consideration is the possibility of the algorithm failing because of overflow, even though T is nonsingular. For example, if

$$T = \begin{pmatrix} 10^{-60} & 1 \\ 0 & 10^{-60} \end{pmatrix}$$

and $b = (0, 1)^{\mathrm{T}}$, then $x = (-10^{120}, 10^{60})^{\mathrm{T}}$. On a computer whose floating-point word has a characteristic bounded by 100, the first component of x cannot be represented as a floating-point number, and any attempt to execute Algorithm 1.3 for this data will result in a overflow. At the very least, such an overflow should cause the computations to stop and a diagnostic message to be printed. On many computers, this is done by the system, which regards any floating-point overflow as a fatal error.

Like Algorithm 3.6 of the preceding chapter, Algorithm 1.3 makes no reference to the elements in the lower part of the array T. Thus this storage is free for other purposes, say to store the elements of a strictly lower triangular matrix.

Finally, note that Algorithm 1.3 requires about $n^2/2$ multiplications.

Algorithm 1.3 can be adapted to compute the inverse S of an upper triangular matrix T. Specifically, if $S = (s_1, s_2, \ldots, s_n)$, then the equation $TS = I$ is equivalent to the set of equations

$$Ts_k = e_k \qquad (k = 1, 2, \ldots, n).$$

Each of these equations may be solved by Algorithm 1.3 to yield the inverse S.

However, this is an inefficient procedure, for half of the elements of S are known to be zero and it is senseless to compute them. To circumvent this difficulty, let $s_k' = (\sigma_{1k}, \sigma_{2k}, \ldots, \sigma_{kk})^{\mathrm{T}}$. Then it is easily verified that

$$T^{[k]}s_k' = e_k \qquad (k = 1, 2, \ldots, n), \tag{1.6}$$

where e_k now is a k-vector. Equation (1.6) is an upper triangular system of order k for the nonzero elements of s_k which may be solved by Algorithm 1.3. If this is done for each k, there results an algorithm for inverting an upper triangular matrix. For reasons that will become apparent later, we solve the equations in the order $k = n, n - 1, \ldots, 1$ (cf. Algorithm 1.5).

ALGORITHM 1.4. Let $T \in \mathbb{R}^{n \times n}$ be a nonsingular upper triangular matrix. This algorithm computes T^{-1} and returns it in the upper half of the array S.

1) For $k = n, n-1, \ldots, 1$
   1) $\sigma_{kk} = \tau_{kk}^{-1}$
   2) $\sigma_{ik} = -\tau_{ii}^{-1} \sum_{j=i+1}^{k} \tau_{ij}\sigma_{jk} \qquad (i = k-1, k-2, \ldots, 1)$

In deriving this algorithm we have taken advantage of the fact that all components but the last of e_k are zero. Like Algorithm 1.3, this algorithm is mathematically well defined when T is nonsingular. Practically, overflow may prevent its complete execution. Some accuracy may be gained by accumulating inner products in double precision.

When it is not required to save the elements of the matrix T, they may be overwritten by the elements of $S = T^{-1}$. In fact, in Algorithm 1.4, once τ_{ik} has been used to compute σ_{ik} it is no longer needed, and hence we may overwrite τ_{ik} by σ_{ik}. This results in the following algorithm for inverting T in its own array.

ALGORITHM 1.5. Let $T \in \mathbb{R}^{n \times n}$ be a nonsingular upper triangular matrix. This algorithm overwrites T with its inverse S.

1) For $k = n, n-1, \ldots, 1$
   1) $\tau_{kk} \leftarrow \sigma_{kk} = \tau_{kk}^{-1}$
   2) $\tau_{ik} \leftarrow \sigma_{ik} = -\tau_{ii}^{-1} \sum_{j=i+1}^{k} \tau_{ij}\sigma_{jk} \qquad (i = k-1, k-2, \ldots, 1)$

Both Algorithms 1.4 and 1.5 require about $n^3/6$ multiplications which differs from the multiplication count for Algorithm 1.3 by a factor of $n/3$. This fact has important consequences for the solution of triangular systems. A naive scheme for solving, say, Equation (1.3) might go as follows.

1) $S = T^{-1}$
2) $x = Sb$

where the first step is accomplished by Algorithm 1.5. However, statement 1 requires $n^3/6$ multiplications, beside which the $n^2/2$ multiplications of statement 2 are negligible. Obviously, when n is large, this procedure is far more expensive than Algorithm 1.3. In other words, *if one requires only the solution of* (1.3), *one should not compute* T^{-1}.

It should be stressed that the inverse of a matrix is seldom required in matrix computations. For whenever we are asked to compute

$$x = T^{-1}b,$$

we can alternatively calculate the solution of

$$Tx = b.$$

Such recasting of a problem can save a good deal of computations, as the following examples show.

EXAMPLE 1.6. We wish to calculate the expression

$$\alpha = x^{\mathrm{T}}T^{-1}y,$$

where T is an upper triangular matrix of order n. Rather than inverting T, we may proceed as follows.

1) Solve the equation $Tz = y$
2) Calculate $\alpha = x^{\mathrm{T}}z$

The naive way of computing α via T^{-1} requires $n^3/6$ multiplications to invert T, beside which the remaining calculations are insignificant. The alternative procedure, on the other hand, requires only $n^2/2$ multiplications.

EXAMPLE 1.7. We wish to compute

$$C = T^{-1}B.$$

where T is of order n and $B \in \mathbb{R}^{n\times m}$. If we partition B and C by columns, then

$$(c_1, c_2, \ldots, c_m) = (T^{-1}b_1, T^{-1}b_2, \ldots, T^{-1}b_m),$$

whence $c_i = T^{-1}b_i$. Thus the columns of C can be found by solving the systems

$$Tc_i = b_i \qquad (i = 1, 2, \ldots, m).$$

If T is upper triangular, the naive computation requires $n^3/6$ multiplications for the inversion of T and another $mn^2/2$ multiplications to compute $C = T^{-1}B$, giving a total of $(\frac{1}{6}n + \frac{1}{2}m)n^2$ multiplications. The alternative procedure requires $mn^2/2$ multiplications. Comparing these operation counts, we see that the procedure involving T^{-1} always requires more work than the alternative procedure, although the difference becomes negligible for $m \gg n$.

EXERCISES

1. Show that if $T = D - U$, where D is a nonsingular diagonal matrix of order n and U is a strictly upper triangular matrix, then
$$T^{-1} = D^{-1}[I + UD^{-1} + (UD^{-1})^2 + \cdots + (UD^{-1})^{n-1}].$$
2. Describe an equivalent of Algorithm 1.1 for solving lower triangular systems.
3. Describe an equivalent of Algorithm 1.4 for inverting lower triangular matrices.
4. Write an INFL program to generate the inverse of an upper triangular matrix T row-wise. [*Hint*: Solve the systems $s_k^{\mathrm{T}}T = e_k^{\mathrm{T}}$ $(k = 1, 2, \ldots, n)$.]
5. A lower triangular matrix is a Stieltjes matrix if its nonzero elements lie on its diagonal and first subdiagonal. Describe an efficient algorithm for inverting a lower Stieltjes matrix L; for solving the system $Lx = b$.
6. Call a matrix of order n "upper Stieltjes of width k" if its nonzero elements lie on its diagonal and first $k - 1$ superdiagonals. Describe an efficient algorithm to invert an upper Stieltjes matrix U of width k; to solve the equation $Ux = b$.
7. Equation (1.2) shows that $(T^{[k]})^{-1}$ may be easily calculated from $(T^{[k-1]})^{-1}$. Describe an algorithm that computes T^{-1} by computing successively $(T^{[1]})^{-1}, (T^{[2]})^{-1}, \ldots, (T^{[n]})^{-1} = T^{-1}$.
8. A matrix T is block upper triangular if it can be partitioned in the form
$$T = \begin{pmatrix} T_{11} & T_{12} & \cdots & T_{1m} \\ 0 & T_{22} & \cdots & T_{2m} \\ \vdots & \vdots & & \vdots \\ 0 & 0 & \cdots & T_{mm} \end{pmatrix},$$
where each diagonal block T_{ii} is square. Prove that T is nonsingular

if and only if its diagonal blocks are nonsingular, and that its inverse is block triangular with the same partitioning. Assuming that the matrices T_{ii}^{-1} are known, generalize Algorithm 1.5 to find the inverse of a block upper triangular matrix. Give two INFL descriptions: one in terms of blocks and the other in terms of elements.

2. GAUSSIAN ELIMINATION

In this section we shall give algorithms for reducing a matrix to upper trapezoidal form. The basic algorithm, called Gaussian elimination, mimics the process of eliminating unknowns from a system of linear equations. As such it provides a direct method for solving linear equations. Moreover, the algorithms may be described in terms of matrix operations, and these descriptions in turn lead to the factorizations to be described in Section 3.

To illustrate these ideas we consider the following system of linear equations

$$\begin{aligned} 2\xi_1 + 4\xi_2 - 2\xi_3 &= 6 \\ \xi_1 - \xi_2 + 5\xi_3 &= 0 \, . \\ 4\xi_1 + \xi_2 - 2\xi_3 &= 2 \end{aligned} \tag{2.1}$$

If $\frac{1}{2}$ times the first equation is subtracted from the second and 2 times the first subtracted from the third, the result is the system

$$\begin{aligned} 2\xi_1 + 4\xi_2 - 2\xi_3 &= 6 \\ - 3\xi_2 + 6\xi_3 &= -3 \, , \\ - 7\xi_2 + 2\xi_3 &= -10 \end{aligned} \tag{2.2}$$

in which the variable ξ_1 does not appear in the second and third equations. Likewise the variable ξ_2 can be eliminated from the third equation by subtracting 7/3 times the second equation from the third to obtain

$$\begin{aligned} 2\xi_1 + 4\xi_2 - 2\xi_3 &= 6 \\ - 3\xi_2 + 6\xi_3 &= -3 \, . \\ - 12\xi_3 &= -3 \end{aligned} \tag{2.3}$$

The system (4.3) is upper triangular and can be solved by the techniques of Section 1. Obviously this idea of elimination can be extended to give a general method for solving systems of any order. However, there is more to the process than this.

Let $A = A_1$, A_2, and A_3 be the matrices of the systems (2.1), (2.2), and (2.3), respectively; e.g.,

$$A_2 = \begin{pmatrix} 2 & 4 & -2 \\ 0 & -3 & 6 \\ 0 & -7 & -2 \end{pmatrix}.$$

Then A_2 is obtained from A_1 by subtracting $\frac{1}{2}$ the first row of A_1 from the second row and 2 times the first row from the third. From the discussion of matrix multiplication in Section 1.4, we know that A_2 can be obtained by premultiplying A_1 by a suitable matrix, and in fact it is easy to verify that

$$A_2 = M_1 A_1, \tag{2.4}$$

where

$$M_1 = \begin{pmatrix} 1 & 0 & 0 \\ -\frac{1}{2} & 1 & 0 \\ -2 & 0 & 1 \end{pmatrix}.$$

Similarly

$$A_3 = M_2 A_2, \tag{2.5}$$

where

$$M_2 = \begin{pmatrix} 1 & 0 & 0 \\ 0 & 1 & 0 \\ 0 & -\frac{7}{3} & 1 \end{pmatrix}.$$

Combining (2.4) and (2.5), we see that

$$A_3 = M_2 M_1 A. \tag{2.6}$$

Thus to solve the system $Ax = b$ we need only calculate the vector $c = M_1 M_2 b$ and solve the upper triangular system

$$A_3 x = M_2 M_1 A x = M_2 M_1 b = c.$$

Moreover, the expression (2.6) can be rearranged to give a factorization of A into a lower and upper triangular matrix. Since the M_i are nonsingular, we have from (2.6)

$$A = M_1^{-1} M_2^{-1} A_3. \tag{2.7}$$

Since M_1 and M_2 are unit lower triangular, so is the product of their inverses. Hence, if we set $L = M_1^{-1}M_2^{-1}$ and $U = A_3$, Equation (2.7) becomes

$$A = LU,$$

where L is unit lower triangular and U is upper triangular. Since

$$M_1^{-1} = \begin{pmatrix} 1 & 0 & 0 \\ \frac{1}{2} & 1 & 0 \\ 2 & 0 & 1 \end{pmatrix}, \qquad M_2^{-1} = \begin{pmatrix} 1 & 0 & 0 \\ 0 & 1 & 0 \\ 0 & \frac{7}{3} & 1 \end{pmatrix},$$

L is itself the product of very simple triangular matrices.

Obviously the above discussion depends critically on the matrices M_i, which are called elementary lower triangular matrices. We begin our formal exposition of Gaussian elimination with a discussion of their properties.

DEFINITION 2.1. An *elementary lower triangular* matrix of order n and *index* k is a matrix of the form

$$M = I_n - me_k^{\mathrm{T}},$$

where

$$e_i^{\mathrm{T}}m = 0 \qquad (i = 1, 2, \ldots, k). \tag{2.8}$$

The conditions (2.8) say that the first k components of m are zero; that is, m has the form $m = (0, 0, \ldots, 0, \mu_{k+1}, \mu_{k+2}, \ldots, \mu_n)^{\mathrm{T}}$. In general an elementary lower triangular matrix has the form

$$M = \begin{pmatrix} 1 & 0 & \cdots & 0 & \cdots & 0 \\ 0 & 1 & \cdots & 0 & \cdots & 0 \\ \vdots & \vdots & & \vdots & & \vdots \\ 0 & 0 & \cdots & 1 & \cdots & 0 \\ 0 & 0 & \cdots & -\mu_{k+1} & \cdots & 0 \\ \vdots & \vdots & & \vdots & & \vdots \\ 0 & 0 & \cdots & -\mu_n & \cdots & 1 \end{pmatrix}.$$

Thus an elementary lower triangular matrix is an identity matrix with some additional nonzero elements in the kth column below the diagonal.

Elementary lower triangular matrices are easily inverted.

THEOREM 2.2. Let $M = I - me_k^{\mathrm{T}}$ be an elementary lower triangular matrix. Then

$$M^{-1} = I + me_k^{\mathrm{T}}.$$

PROOF. Let $X = I + me_k^{\mathrm{T}}$. Then

$$\begin{aligned} MX &= (I - me_k^{\mathrm{T}})(I + me_k^{\mathrm{T}}) \\ &= I - me_k^{\mathrm{T}} + me_k^{\mathrm{T}} - me_k^{\mathrm{T}}me_k^{\mathrm{T}} \\ &= I - m(e_k^{\mathrm{T}}m)e_k^{\mathrm{T}}. \end{aligned}$$

Since M is an elementary lower triangular matrix, $e_k^{\mathrm{T}}m = 0$. Hence $MX = I$ and X is the inverse of M. ∎

The computational significance of elementary lower triangular matrices is that they can be used to introduce zero components into a vector.

THEOREM 2.3. Let $e_k^{\mathrm{T}}x = \xi_k \neq 0$. Then there is a unique elementary lower triangular matrix M of index k such that

$$Mx = (\xi_1, \xi_2, \ldots, \xi_k, 0, 0, \ldots, 0)^{\mathrm{T}}. \tag{2.9}$$

PROOF. We seek M in the form $M = I - me_k^{\mathrm{T}}$. Since M is to be of index k, we must have

$$\mu_i = 0 \qquad (i = 1, 2, \ldots, k). \tag{2.10}$$

Now

$$Mx = (I - me_k^{\mathrm{T}})x = x - (e_k^{\mathrm{T}}x)m.$$

Since the last $n - k$ components of Mx are to be zero, we must have

$$\xi_i - \xi_k\mu_i = 0 \qquad (i = k+1, k+2, \ldots, n),$$

and if $\xi_k \neq 0$,

$$\mu_i = \frac{\xi_i}{\xi_k} \qquad (i = k+1, k+2, \ldots, n). \tag{2.11}$$

Thus if M exists, it is uniquely determined by (2.10) and (2.11). On the other hand, if $\xi_k \neq 0$, then the vector m can be determined from (2.10) and (2.11), and it is easy to verify that the associated matrix $M = I - me_k^{\mathrm{T}}$ satisfies (2.9). ■

Thus when $\xi_k \neq 0$, we can multiply x by an elementary lower triangular matrix chosen so that the last $n - k$ components of x are replaced by zeros and the other components are left unaltered. When $\xi_k = 0$, such a matrix does not exist, unless $\xi_{k+1}, \ldots, \xi_n$ are also zero, in which case any elementary lower triangular matrix of index k will do the job. It should be noted that Equations (2.10) and (2.11) define an algorithm for computing m and hence M.

In the method of Gaussian elimination for reducing a matrix to upper trapezoidal form, the matrix is premultiplied by a sequence of elementary lower triangular matrices, each chosen to introduce a column with zeros below the diagonal. The method revolves around Theorem 2.3, which allows us to calculate the necessary elementary lower triangular matrices. It should be stressed that any class of matrices satisfying something like Theorem 2.3 can be used for the reduction: the elementary lower triangular matrices are used here because they are easy to compute and multiply. An important alternative will be discussed in Chapter 5.

Let $A_1 = A$ be an $m \times n$ matrix. If $\alpha_{11}^{(1)} = \alpha_{11}$ is nonzero, then by Theorem 2.3 there is a unique elementary lower triangular matrix M_1 of index 1 that annihilates the last $n - 1$ elements of the first column of A_1. If A_1 is premultiplied by M_1, there results a matrix A_2 of the form

$$A_2 = M_1 A_1 = \begin{pmatrix} \alpha_{11}^{(1)} & \alpha_{12}^{(1)} & \alpha_{13}^{(1)} & & \alpha_{1n}^{(1)} \\ 0 & \alpha_{22}^{(2)} & \alpha_{23}^{(2)} & & \alpha_{2n}^{(2)} \\ 0 & \alpha_{32}^{(2)} & \alpha_{33}^{(2)} & \cdots & \alpha_{3n}^{(2)} \\ \vdots & \vdots & \vdots & & \vdots \\ 0 & \alpha_{m2}^{(2)} & \alpha_{m3}^{(2)} & \cdots & \alpha_{mn}^{(2)} \end{pmatrix}.$$

Note that the first row of A_2 is the same as the first row of A_1.

In general, suppose $A_k = M_{k-1} \cdots M_1 A_1$ has the form

$$A_k = \begin{pmatrix} A_{11}^{(k)} & A_{12}^{(k)} \\ 0 & A_{22}^{(k)} \end{pmatrix},$$

where $A_{11}^{(k)}$ is an upper triangular matrix of order $k - 1$ and for $i = 1, 2, \ldots, k - 1$ the ith row of A_k is the same as the ith row of A_i. For example, with $m = n = 5$ we have

$$A_3 = \begin{pmatrix} \alpha_{11}^{(1)} & \alpha_{12}^{(1)} & \alpha_{13}^{(1)} & \alpha_{14}^{(1)} & \alpha_{15}^{(1)} \\ 0 & \alpha_{22}^{(2)} & \alpha_{23}^{(2)} & \alpha_{24}^{(2)} & \alpha_{25}^{(2)} \\ 0 & 0 & \alpha_{33}^{(3)} & \alpha_{34}^{(3)} & \alpha_{35}^{(3)} \\ 0 & 0 & \alpha_{43}^{(3)} & \alpha_{44}^{(3)} & \alpha_{45}^{(3)} \\ 0 & 0 & \alpha_{53}^{(3)} & \alpha_{54}^{(3)} & \alpha_{55}^{(3)} \end{pmatrix}.$$

If $\alpha_{kk}^{(k)} \neq 0$, then there is an elementary lower triangular matrix M_k of index k that annihilates the last $m - k$ elements of the kth column of A_k. Such a matrix can be written in the form

$$M_k = \begin{pmatrix} I_{k-1} & 0 \\ 0 & M_k' \end{pmatrix},$$

where M_k' is an elementary lower triangular matrix of index 1.

Now

$$A_{k+1} = M_k A_k = \begin{pmatrix} I & 0 \\ 0 & M_k' \end{pmatrix} \begin{pmatrix} A_{11}^{(k)} & A_{12}^{(k)} \\ 0 & A_{22}^{(k)} \end{pmatrix} = \begin{pmatrix} A_{11}^{(k)} & A_{12}^{(k)} \\ 0 & M_k' A_{22}^{(k)} \end{pmatrix}.$$

Thus first $k - 1$ rows of A_{k+1} are the same as those of A_k. Since the first row of $A_{22}^{(k)}$ and $M_k' A_{22}^{(k)}$ are the same, the first k rows of A_k and A_{k+1} are the same. Also by the construction of M_k, $M_k' A_{22}^{(k)}$ has the form

$$\begin{pmatrix} x & x & \cdots & x \\ 0 & x & \cdots & x \\ \vdots & \vdots & & \vdots \\ 0 & x & \cdots & x \end{pmatrix}.$$

Hence the leading principal submatrix of order k of A_{k+1} is upper triangular. In other words, A_{k+1} is one step further along in the reduction.

The reduction will terminate when we run out of rows or columns. When $m > n$, this happens after the nth step. When $m \leq n$, the matrix is in lower trapezoidal form after the $(m - 1)$th step. For example, when $m = 1$, no reduction is required, since a row vector is trivially in upper trapezoidal form. Thus the reduction ends with the matrix A_{r+1}, where

$$r = \min\{m - 1, n\}.$$

The reduction of A is constructive in the sense that we can specify an algorithm for calculating the A_k and the M_k. Specifically, the kth column of A_k is the vector

$$a_k^{(k)} = (\alpha_{1k}^{(1)}, \alpha_{2k}^{(2)}, \ldots, \alpha_{kk}^{(k)}, \ldots, \alpha_{mk}^{(k)})^{\mathrm{T}},$$

and M_k is the elementary lower triangular matrix of index k such that

$$M_k a_k^{(k)} = (\alpha_{1k}^{(1)}, \alpha_{2k}^{(2)}, \ldots, \alpha_{kk}^{(k)}, 0, \ldots, 0)^{\mathrm{T}}.$$

By Theorem 2.3, $M_k = I - m_k e_k^{\mathrm{T}}$, where

$$m_k = (0, 0, \ldots, 0, \mu_{k+1,k}, \ldots, \mu_{mk})^{\mathrm{T}}$$

and

$$\mu_{ik} = \frac{\alpha_{ik}^{(k)}}{\alpha_{kk}^{(k)}} \qquad (i = k+1, k+2, \ldots, m). \tag{2.12}$$

Thus M_k is uniquely determined, provided $\alpha_{kk}^{(k)} \neq 0$.

In principle A_{k+1} may be calculated by forming M_k and computing the product $M_k A_k$. However, considerable savings in operations and storage may be obtained by taking advantage of the special form of M_k. In fact,

$$A_{k+1} = M_k A_k = (I - m_k e_k^{\mathrm{T}})A_k = A_k - m_k e_k^{\mathrm{T}} A_k.$$

Now e_k^{T} is the kth row of A_k. Hence the ith row of A_{k+1} may be formed by subtracting μ_{ik} times the kth row of A_k from the ith row of A_k. Since $\mu_{1k} = \mu_{2k} = \cdots = \mu_{kk} = 0$, only rows $k+1, k+2, \ldots, m$ are altered. In terms of elements this becomes

$$\alpha_{ik}^{(k+1)} = 0 \qquad (i = k+1, k+2, \ldots, m) \tag{2.13}$$

and

$$\alpha_{ij}^{(k+1)} = \alpha_{ij}^{(k)} - \mu_{ik}\alpha_{kj}^{(k)}$$
$$(i = k+1, k+2, \ldots, m; \quad j = k+1, k+2, \ldots, n). \tag{2.14}$$

Equations (2.12)–(2.14) completely specify the computations involved in the reduction. As far as storage is concerned, we can allow the elements of A_{k+1} to overwrite the corresponding elements of A_k. In applications, such as solving linear systems, it is important to know the matrices M_k. This can be done by saving the numbers μ_{ij}. After $\alpha_{ik}^{(k)}$ is used to compute

μ_{ik}, it is not used again, and in fact it is reduced to zero in the transformation from A_k to A_{k+1}. Hence, if we agree to store only the nonzero entries of A_{k+1}, the number μ_{ik} may be placed in the location originally occupied by α_{ik}.

These considerations may be summed up in the following algorithm.

ALGORITHM 2.4. (*Gaussian elimination*). Let the $m \times n$ matrix A be given with $m > 1$ and let $r = \min\{m - 1, n\}$. This algorithm overwrites A with the upper trapezoidal matrix A_{r+1}. The multipliers μ_{ik} overwrite α_{ik}.

1) For $k = 1, 2, \ldots, r$

   1) $\alpha_{ik} \leftarrow \mu_{ik} = \dfrac{\alpha_{ik}}{\alpha_{kk}}$ $\quad (i = k + 1, k + 2, \ldots, m)$

   2) $\alpha_{ij} \leftarrow \alpha_{ij} - \mu_{ik}\alpha_{kj}$ $\quad (i = k + 1, \ldots, m; \; j = k + 1, \ldots, n)$

For r large, the algorithm requires approximately

$$\frac{r^3}{3} - (m + n)\frac{r^2}{2} + mnr$$

multiplications. For $m = n$, this amounts to approximately $n^3/3$ multiplications.

The numbers $\alpha_{kk}^{(k)}$ $(k = 1, 2, \ldots, r)$ are called the *pivot elements* of the reduction. As it is given, Algorithm 2.4 can be carried to completion only if all the pivot elements are nonzero. For purposes of investigating the algorithm, it is convenient to have a condition on the original matrix under which the pivots are guaranteed to be nonzero. Such a condition is given in the following theorem.

THEOREM 2.5. The pivot elements $\alpha_{ii}^{(i)}$ $(i = 1, 2, \ldots, k)$, are nonzero if and only if the leading principal submatrices $A^{[i]}$ $(i = 1, 2, \ldots, k)$, are nonsingular.

PROOF. The proof is by induction on k. For $k = 1$, the theorem is trivially true since $A^{[1]} = \alpha_{11}^{(1)}$. For the induction step it is sufficient to assume that $A^{[1]}, \ldots, A^{[k-1]}$ are nonsingular and show that $A^{[k]}$ is nonsingular if and only if $\alpha_{kk}^{(k)}$ is nonzero.

By the induction hypothesis, $\alpha_{11}^{(1)}, \alpha_{22}^{(2)}, \ldots, \alpha_{k-1,k-1}^{(k-1)} \neq 0$. Hence Algorithm 2.4 may be carried out through its $(k - 1)$th step to give elementary lower triangular matrices M_i of index i $(i = 1, 2, \ldots, k-1)$ such that

$$A_k = M_{k-1}M_{k-2} \cdots M_1 A = \begin{pmatrix} A_{11}^{(k)} & A_{12}^{(k)} \\ 0 & A_{22}^{(k)} \end{pmatrix},$$

where $A_{11}^{(k)}$ is upper triangular with diagonal elements $\alpha_{11}^{(1)}, \ldots, \alpha_{k-1,k-1}^{(k-1)}$. It follows that $A_k^{[k]}$ is upper triangular with diagonal elements $\alpha_{11}^{(1)}, \ldots, \alpha_{k-1,k-1}^{(k-1)}, \alpha_{kk}^{(k)}$. Since $\alpha_{11}^{(1)}, \ldots, \alpha_{k-1,k-1}^{(k-1)} \neq 0$, $A_k^{[k]}$ is nonsingular if and only if $\alpha_{kk}^{(k)}$ is nonzero. Now since $M_1, \ldots, M_{k-1}$ are lower triangular, $A_k^{[k]} = M_{k-1}^{[k]} \cdots M_1^{[k]} A^{[k]}$ (Exercise 1.4.18). Since M_i is unit lower triangular, so is $M_i^{[k]}$. Thus $A^{[k]}$ is nonsingular if and only if $A_k^{[k]}$ is nonsingular, which, we have seen, is true if and only if $\alpha_{kk}^{(k)} \neq 0$. ■

Carried through its kth step, the method of Gaussian elimination produces elementary lower triangular matrices M_i of index i $(i = 1, 2, \ldots, k)$ such that $A_{k+1} = M_k M_{k-1} \cdots M_1 A$ is zero below its first k diagonal elements. Once started, the method proceeds deterministically, provided the pivots are nonzero. This suggests that the matrices $M_1, M_2, \ldots, M_k$ are uniquely determined by the requirement that $M_k M_{k-1} \ldots M_1 A$ be zero below its first k diagonal elements. The following theorem shows that this is indeed the case.

THEOREM 2.6. Let $A \in \mathbb{R}^{m\times n}$ and suppose that $A^{[1]}, A^{[2]}, \ldots, A^{[k]}$ are nonsingular. For $i = 1, 2, \ldots, k$ let M_i and N_i be elementary lower triangular matrices of index i. If $M_k M_{k-1} \cdots M_1 A$ and $N_k N_{k-1} \cdots N_1 A$ are both zero below their first k diagonal elements, then $M_i = N_i$ $(i = 1, 2, \ldots, k)$.

PROOF. The proof is by induction on k. For $k = 1$, M_1 and N_1 are elementary lower triangular matrices of index 1 that introduce zeros into the last $m - 1$ elements of the first column of A. Since $\alpha_{11} \neq 0$, this requirement uniquely determines M_1 and N_1 (Theorem 2.3) which are thereby equal.

For the induction step, assume that $A^{[1]}, \ldots, A^{[k]}$ are nonsingular and that $M_k \cdots M_1 A$ and $N_k \cdots N_1 A$ are both zero below their first k diagonal elements. Since M_k is an elementary lower triangular matrix of index k, so is M_k^{-1}, and it is easily verified that

$$M_{k-1} \cdots M_1 A = M_k^{-1} N_k \cdots N_1 A$$

is zero below its first $k - 1$ diagonal elements. Likewise $N_{k-1} \cdots N_1 A$ is zero below its first $k-1$ diagonal elements. By the induction hypothesis $M_i = N_i$ $(i = 1, 2, \ldots, k-1)$. Hence $M_{k-1} \cdots M_1 A = N_{k-1} \cdots N_1 A = A_k$, where A_k is the matrix resulting from $k-1$ steps of Gaussian elimination. By Theorem 2.5, $\alpha_{kk}^{(k)} \neq 0$. Hence M_k and N_k are the unique elementary lower triangular matrices of index k that introduce zeros below the kth diagonal of A_k. It follows that $M_k = N_k$. ∎

Algorithm 2.4 breaks down at step 1.1 when the pivot $\alpha_{kk} = 0$. If some $\alpha_{ik} \neq 0$ $(i = k+1, \ldots, m)$, then the reduction cannot be continued. However, if $\alpha_{ik} = 0$ $(i = k+1, \ldots, m)$, then A_k is already zero below its kth diagonal element, and any elementary lower triangular matrix of index k can be used for M_k, thus if a zero pivot emerges at step k, it may be possible to continue the reduction, but $M_k, M_{k+1}, \ldots, M_r$ are no longer unique.

EXAMPLE 2.7. We illustrate the various situations that can arise in the reduction with the following 2×2 examples. Note that $r = 1$ and hence α_{11} is the deciding factor in the reduction.

1. $\alpha_{11} \neq 0$.

$$A_1 = \begin{pmatrix} 1 & 2 \\ 1 & 3 \end{pmatrix}, \qquad M_1 = \begin{pmatrix} 1 & 0 \\ -1 & 1 \end{pmatrix}, \qquad A_2 = \begin{pmatrix} 1 & 2 \\ 0 & 1 \end{pmatrix}.$$

2. $\alpha_{11} \neq 0$.

$$A_1 = \begin{pmatrix} 1 & 2 \\ 1 & 2 \end{pmatrix}, \qquad M_1 = \begin{pmatrix} 1 & 0 \\ -1 & 1 \end{pmatrix}, \qquad A_2 = \begin{pmatrix} 1 & 2 \\ 0 & 0 \end{pmatrix}.$$

Here A_1 is singular. Nonetheless M_1 is uniquely determined.

3. $\alpha_{11} = 0$.

$$A_1 = \begin{pmatrix} 0 & 1 \\ 1 & 0 \end{pmatrix}.$$

The reduction cannot be carried out.

4. $\alpha_{11} = 0$.

$$A_1 = \begin{pmatrix} 0 & 1 \\ 0 & 2 \end{pmatrix}.$$

Here the reduction can be carried out, but it is not unique. In fact for any β we may set

$$M_1 = \begin{pmatrix} 1 & 0 \\ \beta & 1 \end{pmatrix},$$

giving

$$A_2 = \begin{pmatrix} 0 & 1 \\ 0 & 2+\beta \end{pmatrix}.$$

When a mathematical process fails to be defined for a particular value of a parameter, there is a good chance that the corresponding numerical process will be unstable when the parameter is near the offending value. We have seen that the above reduction of a matrix to trapezoidal form fails when the pivot element $\alpha_{kk}^{(k)}$ is zero. The following example illustrates the consequences of computing with a small pivot.

EXAMPLE 2.8. Let

$$A_1 = \begin{pmatrix} 0.001 & 2.000 & 3.000 \\ -1.000 & 3.712 & 4.623 \\ -2.000 & 1.072 & 5.643 \end{pmatrix}.$$

Then a four-digit realization of the reduction of A to upper triangular form will produce the following matrices.

$$M_1 = \begin{pmatrix} 1.000 & 0.000 & 0.000 \\ 1000. & 1.000 & 0.000 \\ 2000. & 0.000 & 1.000 \end{pmatrix},$$

$$A_2 = \begin{pmatrix} 0.001 & 2.000 & 3.000 \\ 0.000 & 2004. & 3005. \\ 0.000 & 4001. & 6006. \end{pmatrix},$$

$$M_2 = \begin{pmatrix} 1.000 & 0.000 & 0.000 \\ 0.000 & 1.000 & 0.000 \\ 0.000 & -1.997 & 1.000 \end{pmatrix},$$

$$A_3 = \begin{pmatrix} 0.001 & 2.000 & 3.000 \\ 0.000 & 2004. & 2005. \\ 0.000 & 0.000 & 5.000 \end{pmatrix}.$$

There is reason to suspect this calculation, for severe cancellation occurred in the calculation of the (3, 3)-element of A_3 by the formula

$$\mathrm{fl}[6006. - (1.997)(3005.)].$$

In fact the correct value of the (3, 3)-element rounded to four figures is 5.922, which disagrees with the computed value in the second figure. If this computed value is used in subsequent calculations, the results will in general be accurate to only one or two figures.

The above example suggests that the emergence of a small pivot in the course of Algorithm 2.4 may be a harbinger of disaster. If the problem at hand requires the matrices $M_1, M_2, \ldots, M_r, A_{r+1}$, there is very little that can be done other than redoing the calculations in higher precision. However, in many applications the reduction is an intermediate step in the solution of a larger problem and it may be possible to rearrange the problem so that the reduction is performed on a different matrix, one for which no small pivots emerge. For example, suppose the matrix in Example 2.8 had arisen in connection with solving the linear system

$$\begin{aligned} 0.001\xi_1 + 2.000\xi_2 + 3.000\xi_3 &= 1.000 \\ -1.000\xi_1 + 3.712\xi_2 + 4.623\xi_3 &= 2.000\,. \\ -2.000\xi_1 + 1.072\xi_2 + 5.643\xi_3 &= 3.000 \end{aligned}$$

If the first and second equations are interchanged, there results a linear system whose matrix

$$\begin{pmatrix} -1.000 & 3.712 & 4.623 \\ 0.001 & 2.000 & 3.000 \\ -2.000 & 1.072 & 5.643 \end{pmatrix}$$

can be reduced without difficulty.

The above considerations suggest that we modify Algorithm 2.4 to avoid small pivotal elements by interchanging the rows and columns of the matrix A. Specifically, if at the kth stage of the reduction, the element $\alpha_{kk}^{(k)}$ is unsatisfactorily small, we may choose some other element, say $\alpha_{\varrho_k,\gamma_k}^{(k)} \neq 0$, as the pivot. If we interchange rows k and ϱ_k and then columns k and γ_k, the result is to move $\alpha_{\varrho_k,\gamma_k}^{(k)}$ into the (k, k)-position, and the reduction may be continued with the new pivot. Obviously we must have $\varrho_k \geq k$ and $\gamma_k \geq k$, otherwise the interchanges will disturb the zeros previously introduced by the reduction.

In matrix terms the modified algorithm may be described as follows. At the kth stage, A_k is premultiplied and postmultiplied, respectively, by the elementary permutations I_{k,ϱ_k} and I_{k,γ_k} (cf. Example 1.4.26), and then premultiplied by an elementary lower triangular matrix M_k of index k chosen to introduce a new column of zeros. Thus

$$A_{k+1} = M_k I_{k,\varrho_k} A_k I_{k,\gamma_k}.$$

For simplicity of notation, we shall denote the elementary permutation I_{k,ϱ_k} and I_{k,γ_k} by P_k and Q_k, respectively, so that

$$A_{k+1} = M_k P_k A_k Q_k. \tag{2.15}$$

The incorporation of interchanges into Algorithm 2.4 complicates the expression for A_k. It might be thought that it would be considerably more difficult to analyze the modified algorithm. Fortunately, as the following theorem shows, the modified algorithm is equivalent to making all the interchanges first and then applying Algorithm 2.4. Of course in practice we must make the interchanges as we go along, since there is no way to know if a pivot is zero until it has been computed; but for theoretical purposes (say a rounding-error analysis) we may assume that all the interchanges have already been performed.

THEOREM 2.9. Let the modified algorithm be applied to A producing matrices Q_i, P_i, M_i, and A_{i+1} $(i = 1, 2, \ldots, k)$. Let

$$A' = P_k P_{k-1} \cdots P_1 A Q_1 Q_2 \cdots Q_k.$$

Then Algorithm 2.4 can be applied to A' through its kth step. If M_i' and A_{i+1}' $(i = 1, 2, \ldots, k)$, are the matrices produced by Algorithm 2.4, then

$$A_{k+1}' = A_{k+1}$$

and

$$M_i' = P_k P_{k-1} \cdots P_{i+1} M_i P_{i+1} \cdots P_{k-1} P_k. \tag{2.16}$$

PROOF. From Equation (2.15) it follows that

$$A_{k+1} = M_k P_k M_{k-1} P_{k-1} \cdots M_1 P_1 A Q_1 Q_2 \cdots Q_k.$$

Because $P_i^2 = I$, this equation is equivalent to

$$A_{k+1} = M_k' M_{k-1}' \cdots M_1' A', \tag{2.17}$$

where M_i' is defined by (2.16). Now it is easily verified that if M_k is an elementary lower triangular matrix of index k and $i, j > k$, then $I_{ij} M_k I_{ij}$ is also an elementary lower triangular matrix of index k. It follows that M_i' is an elementary lower triangular matrix of index k. The first k diagonal elements of A_{k+1} are nonzero and hence the first k leading principal submatrices of A' are nonsingular. Thus by Theorem 2.6, M_1', M_2', $\ldots$, M_k', and A_{k+1} are the result of applying k steps of Algorithm 2.4 to A'. ∎

There remains the problem of specifying a criterion for choosing the pivot element. Example 2.8 suggests that it may be desirable to choose the pivot as large as possible. In other words we wish to choose ϱ_k and γ_k, both not less than k, so that

$$| \alpha^{(k)}_{\gamma_k, \varrho_k} | \geq | \alpha^{(k)}_{ij} | \qquad (i = k, k+1, \ldots, m; \quad j = k, k+1, \ldots, n).$$

There is the possibility that this maximum element is zero; however, this can only happen when the last $m - k$ rows of A_k are zero, and in this case A_k is already in upper trapezoidal form. Incidentally, note that the above criterion does not completely specify ϱ_k and γ_k, for the maximum may be attained for more than one element. Thus, in contrast with Algorithm 2.4 the modification does not determine a unique sequence of calculations for reducing A to upper trapezoidal form.

The modified algorithm with the above choice of pivots is called Gaussian elimination with complete pivoting.

ALGORITHM 2.10. (*Gaussian elimination with complete pivoting*). Let $A \in \mathbb{R}^{m \times n}$ with $m > 1$. Let $r = \min\{m-1, n\}$. This algorithm overwrites A with the Gaussian decomposition of A with its rows and columns permuted. The multiplier μ_{ij} overwrites α_{ij}. The row and column interchange indices are ϱ_k and γ_k, respectively.

1) For $k = 1, 2, \ldots, r$
   1) Find $\varrho_k, \gamma_k \geq k$ such that
   $$| \alpha_{\varrho_k, \gamma_k} | = \max\{| \alpha_{ij} | : i, j \geq k\}$$
   2) If $\alpha_{\varrho_k, \gamma_k} = 0$, set $r = k - 1$ and end the calculations
   3) $\alpha_{kj} \leftrightarrow \alpha_{\varrho_k, j} \qquad (j = k, k+1, \ldots, n)$
   4) $\alpha_{ik} \leftrightarrow \alpha_{i, \gamma_k} \qquad (i = 1, 2, \ldots, m)$
   5) $\alpha_{ik} \leftarrow \mu_{ik} = \dfrac{\alpha_{ik}}{\alpha_{kk}} \qquad (i = k+1, k+2, \ldots, m)$
   6) $\alpha_{ij} \leftarrow \alpha_{ij} - \mu_{ik}\alpha_{kj} \qquad (i = k+1, \ldots, m; \quad j = k+1, \ldots, n)$

The algorithm can always be carried to completion. The final matrix is in upper trapezoidal form. For some k, depending on where the algorithm stops, its last $m - k$ rows are zero and its leading principal submatrix of order k is nonsingular. By Theorem 2.9, the same matrix may be obtained by applying k steps of Gaussian elimination to the A with its rows and columns suitably permuted. Thus Gaussian elimination with complete pivoting is a procedure for calculating the decomposition described in the following theorem.

THEOREM 2.11. Let A be an $m \times n$ matrix. Then if $A \neq 0$, there is a number $k \leq \min\{m, n\}$, elementary permutation matrices P_i, Q_i $(i = 1, 2, \ldots, r)$, and elementary lower triangular matrices M_i of index i such that

$$A_{r+1} = M_r M_{r-1} \cdots M_1 P_r P_{r-1} \cdots P_1 A Q_1 Q_1 \cdots Q_r$$

$$= \begin{pmatrix} A_{11}^{(r+1)} & A_{12}^{(r+1)} \\ 0 & 0 \end{pmatrix},$$

where $A_{11}^{(r+1)}$ is an $r \times r$ nonsingular upper triangular matrix.

In practice, step 1.1 of Algorithm 2.10 may consume a good deal of time, since it involves searching among $(m - k + 1)(n - k + 1)$ elements of A for the largest. A frequently used alternative is to search only the kth column for the largest element and perform a row interchange to bring it into the pivotal position. If, at the kth step, the pivot is zero, the required zeros are already in the kth column, and we may take $M_k = I$. This pivoting strategy is called Gaussian elimination with partial pivoting. It yields an upper trapezoidal matrix, which, however, has no nice distribution of zero and nonzero diagonal elements.

ALGORITHM 2.12. (*Gaussian elimination with partial pivoting*). Let A be an $m \times n$ matrix and $r = \min\{m - 1, n\}$. This algorithm overwrites A with the Gaussian decomposition of A with its rows permuted. The multiplier μ_{ij} overwrites α_{ij}. The row interchange indices are ϱ_k.

1) For $k = 1, 2, \ldots, r$
   1) Find $\varrho_k \geq k$ such that $|\alpha_{\varrho_k, k}| = \max\{|\alpha_{ik}| : i \geq k\}$
   2) If $\alpha_{\varrho_k, k} = 0$, step k
   3) $\alpha_{kj} \leftrightarrow \alpha_{\varrho_k, j}$ $\quad (j = k, k + 1, \ldots, n)$
   4) $\alpha_{ik} \leftarrow \mu_{ik} = \dfrac{\alpha_{ik}}{\alpha_{kk}}$ $\quad (i = k + 1, k + 2, \ldots, m)$
   5) $\alpha_{ij} \leftarrow \alpha_{ij} - \mu_{ik}\alpha_{kj}$ $\quad (i = k + 1, \ldots, m; \;\; j = k + 1, \ldots, n)$

THEOREM 2.13. Let A be an $m \times n$ matrix and let $r = \min\{m - 1, n\}$. Then there are elementary permutations P_i $(i = 1, 2, \ldots, r)$, and elementary lower triangular matrices M_i of index i $(i = 1, 2, \ldots, r)$, such that

$$A_{r+1} = M_r M_{r-1} \cdots M_1 P_r P_{r-1} \cdots P_1 A$$

is upper trapezoidal.

EXERCISES

1. Let $L \in \mathbb{R}^{n \times n}$ be unit lower triangular, and let M_k be the elementary lower triangular matrix of index k whose kth column is the same as the kth column of L. Show that $L = M_1 M_2 \cdots M_{n-1}$. Apply Theorem 2.2 to derive algorithms for inverting L and for solving the system $Lx = b$. How are these algorithms related to those of Exercises 1.2 and 1.3?

2. Let M be an elementary lower triangular matrix of index k and let $i, j > k$. Show that $M' = I_{ij} M I_{ij}$ is also an elementary lower triangular matrix of index k. How is M' obtained from M?

3. Let $A, M \in \mathbb{R}^{n \times n}$ with M an elementary lower triangular matrix of index 1. Give efficient INFL algorithms for overwriting A with AM^{-1}, AM^{T}, MAM^{-1}, and MAM^{T}.

4. Define the notion of an elementary upper triangular matrix of order n in a suitable way. Describe the important properties of such matrices.

5. Describe how a matrix can be reduced to lower trapezoidal form by postmultiplication by elementary upper triangular matrices. Give an INFL algorithm implementing this variation of Gaussian elimination. Discuss the incorporation of interchanges into the algorithm.

6. Give an efficient INFL program for reducing an upper Hessenberg matrix to upper triangular form by Gaussian elimination with partial pivoting.

7. Give an efficient INFL program for reducing a tridiagonal matrix to upper triangular form by Gaussian elimination (note that the result is Stieltjes matrix). Do the same for Gaussian elimination with partial pivoting. (The result is a Stieltjes matrix of width 3.) Also give algorithms in which the elements are suitably stored in linear arrays.

8. Give an efficient INFL program for reducing a band matrix of width $2k+1$ (cf. Exercise 1.3.6) to upper triangular form by Gaussian elimination with partial pivoting.

9. Let A be symmetric with $\alpha_{11} \neq 0$. After one step of Gaussian elimination A has the form

$$\begin{pmatrix} \alpha_{11} & a_1^{\mathrm{T}} \\ 0 & A_2 \end{pmatrix}.$$

Show that A_2 is symmetric.

10. Give an efficient INFL algorithm for reducing a symmetric matrix A to upper triangular form by Gaussian elimination. Assume that only the upper half of the matrix A is present in the array A. Do the same for a symmetric band matrix of width $2k+1$.

11. The matrix $A \in \mathbb{R}^{n\times n}$ is *diagonally dominant* if $|\alpha_{ii}| > \sum_{j\neq i} |\alpha_{ij}|$ $(i = 1, 2, \ldots, n)$. Prove that

    1. if A is diagonally dominant, then any principal submatrix of A is diagonally dominant,
    2. if A is diagonally dominant, then A is nonsingular.

    Conclude that Algorithm 2.4 will not fail when applied to a diagonally dominant matrix.

12. Let A be diagonally dominant and after one step of Gaussian elimination let A have the form

$$\begin{pmatrix} \alpha_{11} & a_1^{\mathrm{T}} \\ 0 & A_2 \end{pmatrix}.$$

Prove that A_2 is diagonally dominant. Conclude that for symmetric, diagonally dominant matrices Gaussian elimination and Gaussian elimination with partial pivoting amount to the same thing.

13. Describe how Gaussian elimination may be used to compute the decomposition of Theorem 1.7.1. (Note that as a practical procedure this approach has the drawback that one must be able to recognize when a number that has been contaminated by rounding error is zero.)

14. An elementary R-matrix of index k is a matrix of the form $R = I - re_k^{\mathrm{T}}$, where $e_k^{\mathrm{T}}r = 0$. Prove that

    1. $R^{-1} = I + re_k^{\mathrm{T}}$,
    2. $e_k^{\mathrm{T}}x = 0 \Rightarrow Rx = x$.

    (There is no standard terminology for such matrices.)

15. Prove that if $e_k^{\mathrm{T}} x \neq 0$, then there is an elementary R-matrix of index k such that

$$Rx = \xi_k e_k = (0, \ldots, 0, \xi_k, 0, \ldots, 0)^{\mathrm{T}}.$$

16. Gauss–Jordan reduction. Let $A \in \mathbb{R}^{n \times n}$. Determine elementary R-matrices R_i of index i such that $A_{k+1} = R_k R_{k-1} \cdots R_1 A$ has the form

$$A_{k+1} = \begin{pmatrix} D_{k+1} & A_{12}^{(k+1)} \\ 0 & A_{22}^{(k+1)} \end{pmatrix},$$

where D_{k+1} is diagonal of order k. Thus A_{n+1} is diagonal. Give an INFL program implementing this Gauss–Jordan reduction of A to diagonal form.

17. Show that the matrices $A_{22}^{(k+1)}$ in the Gauss–Jordan reduction are the same as the corresponding matrices resulting from Gaussian elimination. Hence derive conditions under which the Gauss–Jordan reduction can be carried to completion.

18. What kind of interchange strategies may be incorporated into the Gauss–Jordan reduction?

19. Strictly speaking, Algorithm 2.12 does not produce the Gaussian decomposition of A with its rows permuted, since the multipliers μ_{ij} do not appear in the correct order. Modify the algorithm so that the μ_{ij} do appear in the correct order. [*Hint*: only one minor change in one statement is required.]

20. The necessity of actually interchanging rows in Algorithm 2.12 can be circumvented by the following trick. Initialize an index vector l with the values $\lambda_1 = 1, \lambda_2 = 2, \ldots, \lambda_n = n$. Whenever a row interchange is required, interchange instead the corresponding components of l. Replace all references to α_{ij} by references to $\alpha_{\lambda_i, j}$. Give INFL code for this variant of Algorithm 2.12. Discuss the situation of the multipliers (cf. Exercise 2.19). Is there any significant difference in work between this variant and Algorithm 2.12?

21. Devise a variant of Algorithm 2.10 that avoids interchanges.

NOTES AND REFERENCES

A variant of Gaussian elimination for systems of order three appears in the Chinese work "Chiu-Chang Suan-Shu," composed about 250 B.C.

[see Boyer (1968, page 219)]. The method is readily motivated in terms of systems of linear equations, and in one form or another it often appears in linear algebra texts. Unfortunately many of these texts fail to describe the pivoting strategies that are so necessary to the numerical stability of the methods.

The point of view taken here in which Gaussian elimination is regarded as being accomplished by premultiplications by elementary lower triangular matrices originated with Turing (1948) and has been extensively exploited by Wilkinson (AEP, for example). This approach has the advantage of generalizing readily; any other computationally convenient class of matrices may be used to introduce zeros.

Elementary lower triangular matrices are a special case of a class of matrices that differ from an identity matrix by a matrix of rank unity. Householder (1964) calls all such matrices elementary matrices and shows how many of the transformations used in numerical linear algebra may be reduced to multiplications by elementary matrices.

Gaussian elimination lends itself naturally to the reduction of matrices with special distributions of zeros such as tridiagonal matrices, for it is easy to see what the pivoting and elimination operations will do to the zero elements. Martin and Wilkinson (1967, HACLA/I/6) give algorithms for efficiently reducing band matrices.

A natural extension of Gaussian elimination is to eliminate all the off-diagonal elements in a column at each step, which will finally yield a diagonal matrix. This method is known as the Gauss–Jordan method [the association of Jordan with the method is apparently a mistake; see Householder (1964, page 141)]. In most applications this method is unnecessarily expensive; however, Bauer and Reinsch (HACLA/I/3) have used it to invert positive definite matrices.

An excellent introduction to Gaussian elimination, as well as other algorithms in this chapter, has been given by Forsythe and Moler (1967).

3. TRIANGULAR DECOMPOSITION

In this section we shall consider algorithms for factoring a square matrix A into the product LU of a lower and an upper triangular matrix. Such a factorization is often called an LU decomposition of the matrix A. Since an LU decomposition expresses a matrix as the product of simpler matrices, it can be used to simplify calculations involving the matrix. For example, in the unlikely event that the inverse of A is required, it can be computed in the form $A^{-1} = U^{-1}L^{-1}$ by the techniques of Section 3.1. Throughout this section we shall assume that A is a square matrix of order n; however, some of the results may easily be generalized to rectangular matrices.

Actually we have already seen a method for computing an LU factorization of A, for if Gaussian elimination is applied to A, the result is a sequence $M_1, M_2, \ldots, M_{n-1}$ of unit lower triangular matrices such that

$$A_n = M_{n-1}M_{n-2} \cdots M_1A$$

is upper triangular. If we set

$$L = M_1^{-1}M_2^{-1} \cdots M_{n-1}^{-1},$$

then L is unit lower triangular and $A = LA_n$. Moreover, it is easy to verify that

$$L = \begin{pmatrix} 1 & 0 & \cdots & 0 \\ \mu_{21} & 1 & \cdots & 0 \\ \vdots & \vdots & & \vdots \\ \mu_{n1} & \mu_{n2} & \cdots & 1 \end{pmatrix}, \tag{3.1}$$

where the μ_{ij} are multipliers produced by Algorithm 2.4. Thus Algorithm 2.4 can be regarded as an algorithm for overwriting A with an LU decomposition, L being stored in the lower part of the array A and $U = A_n$ in the upper part.

The close connection of Gaussian elimination with an LU decomposition of A raises the question of the existence and uniqueness of such decompositions; for we have seen that the Gaussian decomposition may fail to exists or be unique. There is, in general, no unique LU factorization of a matrix. If $A = LU$ is an LU factorization of A and D is a nonsingular diagonal matrix, then $L' = LD$ is lower triangular and $U' = D^{-1}U$ is upper triangular. Hence

$$A = LU = LDD^{-1}U = L'U',$$

and $L'U'$ is also an LU decomposition. This example suggests the possibility of normalizing the LU decompositions of a matrix by inserting a diagonal matrix. We shall say that

$$A = LDU$$

is an LDU decomposition of A provided that L is unit lower triangular, D is diagonal, and U is unit upper triangular. The question of the existence and uniqueness of LDU decompositions may be answered in terms of the leading principal submatrices of A.

THEOREM 3.1. Let $A \in \mathbb{R}^{n \times n}$. Then A has a unique LDU decomposition if and only if $A^{[1]}, A^{[2]}, \ldots, A^{[n-1]}$ are nonsingular.

PROOF. We prove the theorem under the hypothesis that A is nonsingular, leaving the case of singular A as a (rather difficult) exercise.

We first show that if A has an LDU decomposition, then it is unique. Let $A = L_1 D_1 U_1$ and $A = L_2 D_2 U_2$ be LDU decompositions of A. Since A is nonsingular, so are D_1 and D_2. From the equation $L_1 D_1 U_1 = L_2 D_2 U_2$, it follows that

$$L_2^{-1} L_1 = D_2 U_2 U_1^{-1} D_1^{-1}. \tag{3.2}$$

Now the left-hand side of (3.2) is a unit lower triangular matrix and the right-hand side is an upper triangular matrix. Hence both sides are an identity matrix. This means $L_2^{-1} L_1 = I$ or $L_1 = L_2$. A similar argument shows that $U_1 = U_2$. Finally, since L_1 and U_1 are nonsingular, the equation $L_1 D_1 U_1 = L_1 D_2 U_1$ implies that $D_1 = D_2$.

It remains to show that A has an LDU decomposition if and only if $A^{[1]}, \ldots, A^{[n-1]}$ are nonsingular. First suppose that $A = LDU$ is an LDU decomposition of A. Then L, D, and U are nonsingular. Since L and U are triangular and D is diagonal, $L^{[k]}$, $D^{[k]}$, and $U^{[k]}$ are nonsingular. However, $A^{[k]} = L^{[k]} D^{[k]} U^{[k]}$, hence $A^{[k]}$ is nonsingular.

Conversely, suppose that $A^{[1]}, A^{[2]}, \ldots, A^{[n-1]}$ are nonsingular. Then by Theorem 2.5 the algorithm of Gaussian elimination can be carried to completion and we can write $A = LA_n$, where L is given by (3.1). Now the diagonal elements of A_n are the pivot elements $\alpha_{kk}^{(k)}$ $(k = 1, 2, \ldots, n)$, which by Theorem 2.5 are nonzero. Let $D = \operatorname{diag}(\alpha_{11}^{(1)}, \alpha_{22}^{(2)}, \ldots, \alpha_{nn}^{(n)})$ and let $U = D^{-1} A_n$. Then $A = LDU$ is an LDU decomposition of A. ∎

From the proof of Theorem 3.1 it is evident that when an LDU decomposition exists, its elements can be written in terms of the quantities calculated by Algorithm 2.4. Specifically,

$$\begin{aligned}
\lambda_{ij} &= \mu_{ij} && (i = 2, 3, \ldots, n;\ j = 1, 2, \ldots, i-1),\\
\delta_i &= \alpha_{ii}^{(i)} && (i = 1, 2, \ldots, n),\\
v_{ij} &= \frac{\alpha_{ij}^{(i)}}{\alpha_{ii}^{(i)}} && (i = 1, 2, \ldots, n;\ j = i, i+1, \ldots, n).
\end{aligned} \tag{3.3}$$

We now turn to the description of algorithms for computing LU decompositions of a nonsingular matrix. Different LU decompositions may be obtained by treating the diagonal in the LDU decomposition differently. There are three important variants. The first associates D with the lower triangular part to give the factorization

$$A = L'U = (LD)U.$$

This is known as the Crout decomposition. The second, called the Doolittle decomposition, associates D with the upper triangular part

$$A = LU' = L(DU).$$

When A is symmetric and has a unique LDU decomposition, the decomposition must have the form

$$A = LDL^{\mathrm{T}}.$$

If the diagonal elements of D are positive, then we can form the matrix $D^{1/2} = \operatorname{diag}(\delta_{11}^{1/2}, \ldots, \delta_{nn}^{1/2})$. Thus A can be written as

$$A = L'L'^{\mathrm{T}} = (LD^{1/2})(D^{1/2}L^{\mathrm{T}}).$$

This third variant is known as the Cholesky decomposition of A.

Any of these three decompositions can be calculated by performing the Gaussian reduction on A and using (3.3) to calculate the LDU decomposition (in fact, Gaussian elimination gives the Doolittle decomposition immediately). However, there are practical advantages to having algorithms that compute the decompositions directly. We shall confine our discussion to the Crout and Cholesky decompositions.

In the Crout reduction, we seek a factorization of the form

$$A = LU, \tag{3.4}$$

where L is lower triangular and U is unit upper triangular. This can be accomplished as follows. In terms of scalars, Equation (3.4) can be written

$$\alpha_{ij} = \sum_{l=1}^{\min\{i,j\}} \lambda_{il}v_{lj} \qquad (i, j = 1, 2, \ldots, n). \tag{3.5}$$

In particular, since $v_{11} = 1$,

$$\alpha_{i1} = \lambda_{i1}v_{11} = \lambda_{i1} \qquad (i = 1, 2, \ldots, n). \tag{3.6}$$

Moreover

$$\alpha_{1j} = \lambda_{11} v_{1j},$$

so that

$$v_{1j} = \frac{\alpha_{1j}}{\lambda_{11}} \qquad (j = 2, 3, \ldots, n). \tag{3.7}$$

Equations (3.6) and (3.7) determine the first column of L and the first row of U.

Now suppose that the first $k-1$ columns of L and the first $k-1$ rows of U have been computed. Then since $v_{kk} = 1$,

$$\alpha_{ik} = \lambda_{ik} + \sum_{l=1}^{k-1} \lambda_{il} v_{lk} \qquad (i = k, k+1, \ldots, n),$$

whence

$$\lambda_{ik} = \alpha_{ik} - \sum_{l=1}^{k-1} \lambda_{il} v_{lk} \qquad (i = k, k+1, \ldots, n). \tag{3.8}$$

Similarly

$$v_{kj} = \lambda_{kk}^{-1}\left(\alpha_{kj} - \sum_{l=1}^{k-1} \lambda_{kl} v_{lj}\right) \qquad (j = k+1, \ldots, n). \tag{3.9}$$

Equations (3.8) and (3.9) express the elements of the kth column of L and the kth row of U in terms of known quantities.

Concerning the organization of storage, note that after α_{ij} is used to compute λ_{ij} or v_{ij}, whichever, it is no longer used. Hence the nonzero elements of L and U can overwrite the corresponding elements of A as they are generated. Thus the Crout decomposition can be computed by the following algorithm.

ALGORITHM 3.2. (*Crout reduction*). Let A be of order n. This algorithm overwrites A with an LU decomposition, where U is unit upper triangular.

1) For $k = 1, 2, \ldots, n$

   1) $\alpha_{ik} \leftarrow \lambda_{ik} = \alpha_{ik} - \sum_{l=1}^{k-1} \lambda_{il} v_{lk}$ $\qquad (i = k, \ldots, n)$

   2) $\alpha_{kj} \leftarrow v_{kj} = \lambda_{kk}^{-1}\left(\alpha_{kj} - \sum_{l=1}^{k-1} \lambda_{kl} v_{lj}\right)$ $\qquad (j = k+1, \ldots, n)$

By (3.3) the elements λ_{ij} are the elements $\alpha_{ij}^{(j)}$ of the matrix A_j in the Gaussian reduction. In particular the elements λ_{kk} are the pivots in the Gaussian reduction and are nonzero provided the leading principal submatrices of A are nonsingular. Hence Algorithm 3.2 can be carried to completion if the leading principal submatrices of A are nonsingular.

The algorithm requires approximately $n^3/3$ multiplications for its completion, which is the same as for Gaussian elimination. Since Equations (3.3) allow us to obtain one decomposition from the other, there would seem to be no reason for preferring one over the other. However, in the Crout reduction the inner products in statements 1.1 and 1.2 can be accumulated in double precision, thereby reducing the effects of rounding errors. An equivalent modification for Gaussian elimination would require one to retain the elements $\alpha_{ij}^{(k)}$ in double precision, which doubles the storage requirements. Thus on a computer that can accumulate inner products in double precision, the Crout reduction is preferable to Gaussian elimination.

The following example illustrates the computations involved in the Crout reduction of a 3×3 matrix. It also shows the disasterous effects of a small diagonal element.

EXAMPLE 3.3. Let

$$A = \begin{pmatrix} 0.001 & 2.000 & 3.000 \\ -1.000 & 3.712 & 4.623 \\ -2.000 & 1.072 & 5.643 \end{pmatrix}$$

(A is the matrix of Example 2.8). In four-digit arithmetic, the Crout reduction goes as follows.

$$\begin{aligned}
\lambda_{11} &= 0.001, \\
\lambda_{21} &= -1.000, \\
\lambda_{31} &= -2.000, \\
v_{12} &= \mathrm{fl}\left(\frac{2.000}{0.001}\right) = 2000, \\
v_{13} &= \mathrm{fl}\left(\frac{3.000}{0.001}\right) = 3000, \\
\lambda_{22} &= \mathrm{fl}[3.712 + (1.000)(2000)] = 2004, \\
\lambda_{32} &= \mathrm{fl}[1.072 + (2.000)(2000)] = 4001, \\
v_{23} &= \mathrm{fl}\left(\frac{4.623 + (1.000)(3000)}{2004}\right) = 1.500, \\
\lambda_{33} &= \mathrm{fl}[5.643 + (2.000)(3000) - (4001)(1.500)] = 4.000.
\end{aligned}$$

Note that the calculation of λ_{33} entails the cancellation of three significant figures. Hence we can expect the computed value of λ_{33} to be very inaccurate, and indeed the true value of λ_{33}, rounded to four figures, is 5.922. The accumulation of inner products will not help, for an error in the fourth place of, say, λ_{32} generates an error in the first place of λ_{33}. In other words, the act of rounding λ_{32} completely destroys the possibility of computing λ_{33} accurately.

Example 3.3 indicates the necessity of avoiding small diagonal elements in the Crout reduction. As we did in Gaussian elimination, we shall seek a cure for the problem by permuting the rows and columns of the original matrix A. There is no convenient analog of complete pivoting for the Crout reduction; hence we shall confine ourselves to partial pivoting, that is to row interchanges.

The principal obstacle to incorporating interchanges into the Crout algorithm is that we cannot know that λ_{kk} is small until we have computed it. However, at this stage an interchange in the rows of A will change its Crout decomposition, which will then have to be recomputed. Fortunately there is a very simple relation between the Crout decompositions of a matrix and the matrix obtained by interchanging two of its rows.

This relation is best derived by considering what happens when Gaussian elimination is applied to A with the two of its rows are permuted. For definiteness suppose that $A_1 = A$ is of order 5 and the third and fifth rows of A are interchanged to give a new matrix

$$A_1' = \begin{pmatrix} \alpha_{11}^{(1)} & \alpha_{12}^{(1)} & \alpha_{13}^{(1)} & \alpha_{14}^{(1)} & \alpha_{15}^{(1)} \\ \alpha_{21}^{(1)} & \alpha_{22}^{(1)} & \alpha_{23}^{(1)} & \alpha_{24}^{(1)} & \alpha_{25}^{(1)} \\ \alpha_{51}^{(1)} & \alpha_{52}^{(1)} & \alpha_{53}^{(1)} & \alpha_{54}^{(1)} & \alpha_{55}^{(1)} \\ \alpha_{41}^{(1)} & \alpha_{42}^{(1)} & \alpha_{43}^{(1)} & \alpha_{44}^{(1)} & \alpha_{45}^{(1)} \\ \alpha_{31}^{(1)} & \alpha_{32}^{(1)} & \alpha_{33}^{(1)} & \alpha_{34}^{(1)} & \alpha_{35}^{(1)} \end{pmatrix}.$$

By examining Algorithm 2.4, it is easy to verify that after one step of Gaussian elimination, A_1' becomes

$$A_2' = \begin{pmatrix} \alpha_{11}^{(1)} & \alpha_{12}^{(1)} & \alpha_{13}^{(1)} & \alpha_{14}^{(1)} & \alpha_{15}^{(1)} \\ \mu_{21} & \alpha_{22}^{(2)} & \alpha_{23}^{(2)} & \alpha_{24}^{(2)} & \alpha_{25}^{(2)} \\ \mu_{51} & \alpha_{52}^{(2)} & \alpha_{53}^{(2)} & \alpha_{54}^{(2)} & \alpha_{55}^{(2)} \\ \mu_{41} & \alpha_{42}^{(2)} & \alpha_{43}^{(2)} & \alpha_{44}^{(2)} & \alpha_{45}^{(2)} \\ \mu_{31} & \alpha_{32}^{(2)} & \alpha_{33}^{(2)} & \alpha_{34}^{(2)} & \alpha_{35}^{(2)} \end{pmatrix},$$

where the α's and μ's are the same quantities that would be obtained by applying Algorithm 2.4 to A_1. Another step gives

$$A_3' = \begin{pmatrix} \alpha_{11}^{(1)} & \alpha_{12}^{(1)} & \alpha_{13}^{(1)} & \alpha_{14}^{(1)} & \alpha_{15}^{(1)} \\ \mu_{21} & \alpha_{22}^{(2)} & \alpha_{23}^{(2)} & \alpha_{24}^{(2)} & \alpha_{25}^{(2)} \\ \mu_{51} & \mu_{52} & \alpha_{53}^{(3)} & \alpha_{54}^{(3)} & \alpha_{55}^{(3)} \\ \mu_{41} & \mu_{42} & \alpha_{43}^{(3)} & \alpha_{44}^{(3)} & \alpha_{45}^{(3)} \\ \mu_{31} & \mu_{32} & \alpha_{33}^{(3)} & \alpha_{34}^{(3)} & \alpha_{35}^{(3)} \end{pmatrix}.$$

It follows from (3.3) that if we apply Algorithm 3.2 to A_1' and stop after statement 1.1 when $k = 3$, we shall have obtained the array

$$\begin{pmatrix} \lambda_{11} & v_{12} & v_{13} & v_{14} & v_{15} \\ \lambda_{21} & \lambda_{22} & v_{23} & v_{24} & v_{25} \\ \lambda_{51} & \lambda_{52} & \lambda_{53} & \alpha_{54} & \alpha_{55} \\ \lambda_{41} & \lambda_{42} & \lambda_{43} & \alpha_{44} & \alpha_{45} \\ \lambda_{31} & \lambda_{32} & \lambda_{33} & \alpha_{34} & \alpha_{35} \end{pmatrix}.$$

In other words, after statement 1.1 when $k = 3$ the only differences between the arrays obtained from A and A' is that the third and fifth row have been interchanged.

This is generally the case. If after statement 1.1 in Algorithm 3.2 we interchange rows k and l ($l > k$), the effect will be the same as if we had interchanged rows k and l in the original matrix before starting. This means that we can perform interchanges as we go along, as is done in the following algorithm.

ALGORITHM 3.4. Let A be of order n. This algorithm overwrites A with the Crout decomposition of $I_{n-1,\varrho_{n-1}} \cdots I_{1\varrho_1}A$.

1) For $k = 1, 2, \ldots, n$

   1) $\alpha_{ik} \leftarrow \lambda_{ik} = \alpha_{ik} - \sum_{l=1}^{k-1} \lambda_{il}v_{lk} \qquad (i = k, \ldots, n)$

   2) Find ϱ_k such that $|\lambda_{\varrho_k,k}| \geq |\lambda_{ik}| \qquad (i = k, \ldots, n)$

   3) $\alpha_{kj} \leftrightarrow \alpha_{\varrho_k j} \qquad (j = 1, 2, \ldots, n)$

   4) $\alpha_{kj} \leftarrow v_{kj} = \lambda_{kk}^{-1}\left(\alpha_{kj} - \sum_{l=1}^{k-1} \lambda_{kl}v_{lj}\right) \qquad (j = k+1, \ldots, n)$

The numbers $\lambda_{kk}, \lambda_{k+1,k}, \ldots, \lambda_{nk}$ are the numbers $\alpha_{kk}^{(k)}, \alpha_{k+1,k}^{(k)}, \ldots, \alpha_{n,k}^{(k)}$ obtained by applying Gaussian elimination to $I_{k-1,\varrho_{k-1}} \cdots I_{1\varrho_1}A$. Thus in choosing ϱ_k to maximize $|\lambda_{\varrho_k,k}|$ we are making the same choices as we would in Gaussian elimination with partial pivoting. In other words, if for each k there is a unique largest λ_{ik} $(i = k, \ldots, n)$, then the pivot indices are the same as those that would result from Gaussian elimination with partial pivoting.

We now turn to the Cholesky decomposition of a symmetric matrix A into the product LL^{T}, where L is lower triangular. Such a decomposition need not exist. For example, since

$$\alpha_{11} = \lambda_{11}^2 \geq 0,$$

no matrix with a negative (1, 1)-element can have a Cholesky decomposition. However, for a very important class of matrices, called positive definite matrices, the decomposition exists. Hence we shall begin our discussion of the Cholesky decomposition with an exposition of the properties of positive definite matrices.

DEFINITION 3.5. A symmetric matrix A is *positive definite* if

$$x \neq 0 \quad \Rightarrow \quad x^{\mathrm{T}}Ax > 0, \tag{3.10}$$

and *positive semidefinite* if

$$x \neq 0 \quad \Rightarrow \quad x^{\mathrm{T}}Ax \geq 0.$$

It might be felt that the criterion (3.10) is not very practical for establishing whether a given matrix is positive definite. In the following example we use (3.10) to show that the members of a very broad class of matrices are automatically positive definite.

EXAMPLE 3.6. Let A be an $m \times n$ matrix and let

$$B = A^{\mathrm{T}}A.$$

In Example 1.4.21 we showed that B is symmetric. We shall now show that B is positive semidefinite and that if $\operatorname{rank}(A) = n$, it is positive definite. To see this let $x \in \mathbb{R}^n$ be nonzero and $y = Ax$. Then

$$x^{\mathrm{T}}Bx = x^{\mathrm{T}}A^{\mathrm{T}}Ax = y^{\mathrm{T}}y = \sum_{i=1}^{n} \eta_i^2 \geq 0. \tag{3.11}$$

Hence B is positive semidefinite. If A is of rank n, then $x \neq 0$ implies $y = Ax \neq 0$. Hence from (3.11), $x^{\mathrm{T}}Bx > 0$, and B is positive definite.

Before establishing the existence of LL^{T} decomposition for positive definite matrices we prove the following lemma.

LEMMA 3.7. A principal submatrix of a positive definite matrix is positive definite.

PROOF. Let A' be the principal submatrix formed from the intersection of rows and columns $i_1 < i_2 < \cdots < i_r$. Let $x' \neq 0$ be an r-vector. Let x be the n-vector defined by

$$\xi_{i_k} = \xi_k' \qquad (k = 1, 2, \ldots, r),$$

$$\xi_j = 0 \qquad (j \neq i_1, i_2, \ldots, i_r).$$

Then $x \neq 0$, and it is easily verified that $x^{\mathrm{T}}Ax = x'^{\mathrm{T}}A'x'$. Since A is positive definite

$$0 < x^{\mathrm{T}}Ax = x'^{\mathrm{T}}A'x',$$

and A' is positive definite. ∎

THEOREM 3.8. If A is positive definite, then there is a unique lower triangular matrix L with positive diagonal elements such that $A = LL^{\mathrm{T}}$.

PROOF. The proof is by induction on the order of A. If A is of order unity and positive definite, then $\alpha_{11} > 0$, and L is uniquely defined by $\lambda_{11} = \sqrt{\alpha_{11}}$.

Suppose the assertion is true of matrices of order $n - 1$ and A' is positive definite of order n. Because A' is symmetric, it can be partitioned in the form

$$A' = \begin{pmatrix} A & a \\ a^{\mathrm{T}} & \alpha \end{pmatrix}.$$

By Lemma 3.7, A is positive definite. We seek a lower triangular matrix L' satisfying $L'L'^{\mathrm{T}} = A'$. Let L' be partitioned in the form

$$L' = \begin{pmatrix} L & 0 \\ l^{\mathrm{T}} & \lambda \end{pmatrix}.$$

Then we require that

$$A = LL^{\mathrm{T}}, \tag{3.12}$$

$$Ll = a, \tag{3.13}$$

$$l^{\mathrm{T}}L^{\mathrm{T}} = a^{\mathrm{T}}, \tag{3.14}$$

$$l^{\mathrm{T}}l + \lambda^2 = \alpha. \tag{3.15}$$

By the induction hypothesis, there is a unique lower triangular matrix L with positive diagonal elements satisfying (3.12). Since L is nonsingular, $l = L^{-1}a$ is the unique vector satisfying (3.13) and (3.14). Finally if $\alpha - l^{\mathrm{T}}l > 0$, then λ will be uniquely defined by $\lambda = \sqrt{\alpha - l^{\mathrm{T}}l}$.

To show that $\alpha - l^{\mathrm{T}}l > 0$, note that the nonsingularity of L implies that A is nonsingular. Let $b = A^{-1}a$. Then because A is positive definite

$$\begin{aligned}
0 &< (b^{\mathrm{T}}, -1)\begin{pmatrix} A & a \\ a^{\mathrm{T}} & \alpha \end{pmatrix}\begin{pmatrix} b \\ -1 \end{pmatrix} \\
&= b^{\mathrm{T}}Ab - 2b^{\mathrm{T}}a + \alpha \\
&= \alpha - b^{\mathrm{T}}a \\
&= \alpha - a^{\mathrm{T}}A^{-1}a \\
&= \alpha - a^{\mathrm{T}}(LL^{\mathrm{T}})^{-1}a \\
&= \alpha - (L^{-1}a)^{\mathrm{T}}(L^{-1}a) \\
&= \alpha - l^{\mathrm{T}}l. \quad \blacksquare
\end{aligned}$$

The proof of Theorem 3.8 allows us to construct the Cholesky decomposition of a matrix by computing successively the decompositions of its leading principal submatrices. Let A_k be the leading principal submatrix of order k and let $A_k = L_k L_k^{\mathrm{T}}$ be the Cholesky decomposition of A_k. Then if

$$A_k = \begin{pmatrix} A_{k-1} & a_k \\ a_k^{\mathrm{T}} & \alpha_{kk} \end{pmatrix},$$

it follows from the proof of the theorem that

$$L_k = \begin{pmatrix} L_{k-1} & 0 \\ l_k^{\mathrm{T}} & \lambda_{kk} \end{pmatrix},$$

where

$$l_k = L_{k-1}^{-1} a_k \tag{3.16}$$

and

$$\lambda_{kk} = \sqrt{\alpha_{kk} - l_k^{\mathrm{T}} l_k}\,.$$

In constructing an algorithm for computing the Cholesky decomposition of a positive definite matrix A, we note that, since A is symmetric, we need only work with, say, its lower half. Moreover, the elements of L can overwrite the corresponding elements of A. Of course, in computing l_k from (3.16), we do not form L_k^{-1}; rather we solve the system

$$L_k l_k = a_k\,.$$

ALGORITHM 3.9. (*Cholesky reduction*). Let A be positive definite of order n. This algorithm returns the matrix L of the Cholesky decomposition of A in the lower half of the array A.

1) For $k = 1, 2, \ldots, n$
   1) For $i = 1, 2, \ldots, k-1$
      1) $\alpha_{ki} \leftarrow \lambda_{ki} = \lambda_{ii}^{-1}\left(\alpha_{ki} - \sum_{j=1}^{i-1} \lambda_{ij}\lambda_{kj}\right)$
   2) $\alpha_{kk} \leftarrow \lambda_{kk} = \sqrt{\alpha_{kk} - \sum_{j=1}^{k-1} \lambda_{kj}^2}$

If A is positive definite, the algorithm can always be carried to completion. It requires about $n^3/6$ multiplications, one half the number required for Gaussian elimination or the Crout reduction. This is to be expected, since the algorithm takes advantage of the symmetry of A to reduce the calculations involved. As in the Crout reduction, inner products may be accumulated in double precision for additional accuracy.

EXERCISES

1. Let A be symmetric and nonsingular and let $A = LDU$ be an LDU decomposition of A. Show that $L = U^{\mathrm{T}}$.

2. Give an efficient INFL program to calculate the Crout decomposition of an upper Hessenberg matrix.

3. In the Crout reduction let

$$L_k = \begin{pmatrix} \lambda_{11} & & 0 \\ \vdots & \ddots & \\ \lambda_{k1} & \cdots & \lambda_{kk} \end{pmatrix}, \qquad M_k = \begin{pmatrix} \lambda_{k+1,1} & \cdots & \lambda_{k+1,k} \\ \vdots & & \vdots \\ \lambda_{n1} & \cdots & \lambda_{nk} \end{pmatrix},$$

$$U_k = \begin{pmatrix} 1 & \cdots & v_{1k} \\ & \ddots & \vdots \\ 0 & & 1 \end{pmatrix}, \qquad V_k = \begin{pmatrix} v_{1,k+1} & \cdots & v_{1n} \\ \vdots & & \\ v_{k,k+1} & \cdots & v_{kn} \end{pmatrix}.$$

Let A be partitioned in the form

$$A = \begin{pmatrix} A^{[k]} & B_k \\ C_k & D_k \end{pmatrix}.$$

Show that the equations $L_k U_k = A^{[k]}$, $L_k V_k = B_k$, $M_{k+1} U_{k+1} = C_{k+1}$, uniquely determine L_k, U_k, V_k, and M_{k+1} provided $A^{[1]}, A^{[2]}, \ldots, A^{[k]}$ are nonsingular. Use this result to give a formal justification of interchanges in Algorithm 3.4.

4. Prove that if A is positive definite, then A^{-1} is positive definite.

5. Prove that if A is positive definite, then A can be written uniquely in the form $A = L^{\mathrm{T}}L$, where L is a lower triangular matrix with positive diagonal elements. [*Hint*: Apply Theorem 3.8 to A^{-1}.]

6. Show that if A is symmetric, diagonally dominant, and has positive diagonal elements, then A is positive definite [*Hint*: By Exercise 2.11, Gaussian elimination may be performed on A without pivoting. Show that the pivot elements are positive and hence conclude $A = LL^{\mathrm{T}}$, where L is lower triangular with positive diagonal elements. For an easier way see Exercise 6.4.10.]

7. Prove that if A is positive definite, then Algorithm 2.4 (Gaussian elimination without pivoting) will not fail.

8. Show that the element of a positive definite matrix that is largest in absolute value lies on the diagonal.

9. Give an efficient INFL program to calculate the Cholesky decomposition of a positive definite tridiagonal matrix.

10. Give an INFL program to overwrite a positive definite matrix A with its inverse. Work only with the lower half of the array A. [*Hint*:

Use Algorithm 3.9 to compute L of the Cholesky decomposition. Then compute L^{-1} and $L^{-T}L^{-1}$.]

11. Let A be positive definite. If one step of Gaussian elimination is performed on A, the result is a matrix of the form

$$\begin{pmatrix} \alpha & a^{T} \\ 0 & A' \end{pmatrix}.$$

Show that A' is positive definite.

12. Devise a variant of Algorithm 3.4 that, in the spirit of Exercise 2.20, avoids interchanges.

NOTES AND REFERENCES

In view of the close connections between Gaussian elimination and methods of triangular decomposition, it is not surprising that many of the "new" methods that have appeared in the literature from time to time are simply rearrangements of Gaussian elimination. The term "*LDU* decomposition" is due to Turing (1948), who related it to Gaussian elimination. A complete discussion is given by Householder (1964). See also the book "Numerical Methods of Linear Algebra" by Faddeev and Faddeeva (1960, 1963), which although it is somewhat out of date, contains much other interesting material.

Programs implementing the Crout reduction have been published by Bowdler, Martin, Peters, and Wilkinson (1966, HACLA/I/7). The Cholesky decomposition has been implemented by Martin, Peters, and Wilkinson (1965, HACLA/I/1) for full positive definite matrices by Martin and Wilkinson (1965, HACLA/I/4) for positive definite band matrices.

4. THE SOLUTION OF LINEAR SYSTEMS

In the introductory material to Section 2 we indicated how the method of Gaussian elimination could be used to solve the linear system

$$Ax = b, \tag{4.1}$$

where A is of order n. In this section we shall describe in detail how the decompositions of Sections 2 and 3 can be used to solve (4.1). The essential feature of the resulting methods is that the reduction of A and the solution of (4.1) can be separated: once A has been reduced, the reduced form can be used to solve (4.1) for any number of right-hand sides b.

We consider first the use of Gaussian elimination with complete pivoting. Algorithm 2.10 defines elementary permutations P_i, Q_i $(i = 1, 2, \ldots, n-1)$, and elementary lower triangular matrices M_i of index i such that

$$A_n = M_{n-1}P_{n-1}M_{n-2} \cdots M_1P_1AQ_1Q_2 \cdots Q_{n-1}$$

is upper triangular. Equation (4.1) can thus be written in the form

$$P_1M_1^{-1}P_2M_2^{-1} \cdots P_{n-1}M_{n-1}^{-1}A_nQ_{n-1}Q_{n-2} \cdots Q_1x = b,$$

and the solution is given by

$$x = Q_1Q_2 \cdots Q_{n-1}A_n^{-1}M_{n-1}P_{n-1}M_{n-2}P_{n-2} \cdots M_1P_1b.$$

This suggests the following algorithm for solving (4.1).

ALGORITHM 4.1. Let P_i, Q_i, M_i $(i = 1, 2, \ldots, n-1)$ and A_n be the matrices defined by Gaussian elimination with complete pivoting applied to A. Given the vector b, this algorithm computes the solution x of (4.1).

1) $y_1 = b$
2) For $k = 1, 2, \ldots, n-1$
   1) $y_{k+1} = M_kP_ky_k$
3) Solve the upper triangular system $A_nz_n = y_n$
4) For $k = n-1, n-2, \ldots, 1$
   1) $z_k = Q_kz_{k+1}$
5) $x = z_1$

The only place where the algorithm can break down is in statement 3. However, if A is nonsingular, then so is A_n, and the solution of $A_nz_n = y_n$ always exists. Hence, if A is nonsingular, Algorithm 4.1 can be carried to completion and yields the solution to (4.1).

It is instructive to consider the practical details of the implementation of Algorithm 4.1. In the first place, it is not necessary to store the y_i and z_i separately. Rather we can initially set $x = b$, and store each new vector in x as it is generated. Second, if Algorithm 2.10 is used to accomplish the Gaussian reduction, the matrices P_k, Q_k, and M_k are given in terms of the numbers ϱ_k, γ_k, and μ_{ik}. Since these matrices are of very simple form, it would be wasteful of time and storage to generate them in full and perform the matrix–vector multiplications indicated by the algorithm. Instead we may compute one vector directly from its predecessor. For example, statement 2.1 can be accomplished by the sequence of computations

1) $\eta_k^{(k)} \leftrightarrow \eta_{\varrho_k}^{(k)}$

2) $\eta_i^{(k+1)} = \eta_i^{(k)} - \mu_{ik}\eta_k^{(k)} \qquad (i = k+1, k+2, \ldots, n)$

Finally, statement 3 can be accomplished by Algorithm 1.3.

We sum up these considerations by rewriting Algorithm 4.1. In the sequel, we shall describe the use of the other decompositions in the spirit of Algorithm 4.1, leaving the detailed recastings as exercises.

ALGORITHM 4.2. Given the output from Algorithm 2.10 and the n-vector b, this algorithm computes the solution of (4.1).

1) $x = b$
2) For $k = 1, 2, \ldots, n-1$
   1) $\xi_k \leftrightarrow \xi_{\varrho_k}$
   2) $\xi_i \leftarrow \xi_i - \mu_{ik}\xi_k \qquad (i = k+1, k+2, \ldots, n)$
3) $x \leftarrow A_n^{-1}x$
4) For $k = n-1, n-2, \ldots, 1$
   1) $\xi_k \leftrightarrow \xi_{\gamma_k}$

If Algorithm 1.3 is used to accomplish step 3, Algorithm 4.2 requires about n^2 multiplications for its execution. This should be compared with the $n^3/3$ multiplications required to reduce A to triangular form. In other words, if we wish to solve a single linear system, the bulk of the work will be concentrated in the initial reduction. Thereafter, additional systems with the same matrix but different right-hand sides can be solved at relatively little additional cost.

Turning now to the use of Gaussian elimination with partial pivoting, we note that the reduction with partial pivoting differs from the reduction with complete pivoting only in the absence of column interchanges. Hence, having performed Gaussian elimination with partial pivoting on A, we may obtain an algorithm for solving (4.1) by deleting step 4 from Algorithm 4.1.

ALGORITHM 4.3. Let P_i and M_i $(i = 1, 2, \ldots, n - 1)$, and A_n be the matrices obtained by applying Gaussian elimination with partial pivoting to A. Given the n-vector b, this algorithm computes the solution of (4.1).

1) $x = b$
2) For $k = 1, 2, \ldots, n - 1$
   1) $x \leftarrow M_k P_k x$
3) $x \leftarrow A_n^{-1} x$

The Crout reduction of A with pivoting yields elementary permutations P_i $(i = 1, 2, \ldots, n - 1)$ such that

$$P_{n-1} P_{n-2} \cdots P_1 A = LU,$$

where L is lower triangular and U is unit upper triangular. Hence the solution of (4.1) is given by

$$x = U^{-1} L^{-1} P_{n-1} P_{n-2} \cdots P_1 b.$$

This leads to the following algorithm for solving (4.1).

ALGORITHM 4.4. Let LU be the Crout decomposition of $P_{n-1} P_{n-2} \cdots P_1 A$, and let b be given. This algorithm computes the solution of (4.1).

1) $x = b$
2) For $k = 1, 2, \ldots, n - 1$
   1) $x \leftarrow P_k x$
3) $x \leftarrow L^{-1} x$
4) $x \leftarrow U^{-1} x$

The algorithm can be carried to completion if A is nonsingular and requires about n^2 multiplications. In statement 4 a minor saving can be effected by taking advantage of the fact that U is *unit* upper triangular.

When A is symmetric and positive definite, it has a Cholesky decomposition in the form $A = LL^{\mathrm{T}}$, where L is lower triangular. Thus the solution of (4.1) is given by

$$x = L^{-\mathrm{T}} L^{-1} b$$

and we obtain the following algorithm.

ALGORITHM 4.5. Let LL^{T} be the Cholesky decomposition of the positive definite matrix A, and let b be given. This algorithm computes the solution of (4.1).

$$1)\quad x = L^{-1}b$$

$$2)\quad x \leftarrow L^{-\mathrm{T}}x$$

Any of the above decompositions can be used to find the inverse of A, either by inverting and multiplying the matrices in the decomposition, or by solving the n linear equations

$$Ax_i = e_i$$

for the columns of the inverse. The details are left as exercises. Again it must be stressed that in applications the inverse of a matrix is seldom required (cf. Section 1).

EXERCISES

1. Devise an efficient algorithm for solving upper-Hessenberg systems (cf. Exercise 2.6).
2. Devise an efficient algorithm for solving tridiagonal systems (cf. Exercise 2.7).
3. Write an INFL program that takes the output of Algorithm 3.4 and overwrites A with A^{-1} by forming the product $U^{-1}L^{-1}$. Be careful of the interchanges.
4. Give detailed INFL code that uses the output of the algorithms of Exercises 2.19, 1.20, 2.21, and 3.12 to solve linear systems.

5. THE EFFECTS OF ROUNDING ERROR

In this section we shall discuss the effects of rounding errors on the algorithms described in this chapter. For example, we shall show that if the algorithms of Section 4 are used to compute a solution of the equation

$$Ax = b, \tag{5.1}$$

the computed solution $\bar{x}$ satisfies

$$(A + H)\bar{x} = b, \tag{5.2}$$

and we shall give rigorous bounds on the sizes of the elements of H. If H can be shown to be small, then the algorithm is stable in the sense of Section 2.1.

Such an error analysis has two important limitations. In the first place, the error bounds are often far larger than the observed error. There are two reasons for this. First, in order to obtain reasonably simple error bounds one must use estimates that are obviously not sharp. Second, no rigorous upper bound on the error, however sharp, can satisfactorily account for the statistical nature of rounding error. It should not be concluded from this that the error analyses are useless. Even if the upper bound on H in (5.2) is an overestimate, it none the less guarantees the stability of the algorithm, provided it is reasonably small. In addition, an error analysis can suggest how to arrange the details of an algorithm for greater stability. For example, the bound on H in (5.2) contains factors that depend on the pivoting strategy used in the decomposition of A and thereby provides a rationale for choosing a pivoting strategy.

The second limitation is that a stability result such as (5.2) cannot insure the accuracy of the solution, unless the problem is well conditioned, for if the system (5.1) is ill conditioned, even a small random H will correspond to a large deviation in $\bar{x}$. However, it may happen that the matrix H that results from rounding error has elements that are so correlated that $\bar{x}$ is accurate, even when (5.1) is ill conditioned. Our theorems will say nothing about this phenomenon, since they only bound the size of the elements of H. Such correlated errors actually occur in solving triangular systems, whose solutions are usually computed to high accuracy.

A rigorous rounding-error analysis proceeds by repeated applications of the rounding-error bounds of Section 2.1. Although the analyses are usually conceptually straightforward, they are fussy in detail. For this reason we shall only state the results of the rounding-error analyses in this section (however, some of the analyses are given in Appendix 3). We shall assume that all calculations have been performed in t-digit floating-point arithmetic satisfying the bounds of Section 2.1. We further assume that underflow and overflow have not occured in the calculation. Finally, we assume that the order of the problem considered, say n, is so restricted that $n\mu \cdot 10^{-t} < .1$. This restriction, which in practice is always satisfied, is necessary to obtain bounds in a reasonably simple form (cf. Exercises 2.1.4 and 2.1.5).

We turn first to the solution of triangular systems by Algorithm 1.3.

THEOREM 5.1. Let $T \in \mathbb{R}^{n \times n}$ and $b \in \mathbb{R}^n$. Let $\bar{x}$ denote the computed solution of the equation $Tx = b$ obtained from Algorithm 1.3. Then $\bar{x}$ satisfies the equation

$$(T + E)\bar{x} = b,$$

where

$$|\varepsilon_{ij}| \leq (n + 1)\pi |\tau_{ij}| 10^{-t} \qquad (i, j = 1, 2, \ldots, n). \tag{5.3}$$

Here π is a constant of order unity that depends on the details of the arithmetic.

This theorem is in many respects quite satisfactory. It says that the computed solution is an exact solution of a problem in which T has been perturbed slightly. In fact the elements of $T + E$ differ from the corresponding elements of T by small relative errors. For moderate n, these relative errors are comparable to the errors made in rounding T itself, and if the elements of T are themselves computed, the errors ε_{ij} may be considerably smaller than the uncertainty in T. If inner products are accumulated in double precision, the factor n may be dropped from (5.3), so that the error matrix E is quite comparable with the error made in rounding T.

None the less, the above results illustrate the limitations mentioned at the first of this section. In the first place, the details of the analysis show that

$$|\varepsilon_{ij}| \leq (j - i + 2)\pi |\tau_{ij}| 10^{-t},$$

which may be considerably smaller than (5.3). More important, though, is the fact that *the solutions of triangular systems are usually computed to high accuracy*. This fact, which is not a consequence of Theorem 5.1, cannot be proved in general, for counter examples exist. However, it is true of many special kinds of triangular matrices and the phenomenon has been observed in many others. The practical consequences of this fact cannot be overemphasized.

The question of whether Algorithm 1.4 for inverting an upper triangular matrix is stable is open. Since each column s_i of the inverse was formed by solving the system $Ts_i = e_i$, it follows from Theorem 5.1 that s_i is the ith column of the inverse of a slightly perturbed matrix $T + E_i$. However, since the perturbation is different for each i, it does not follow that $S = (s_1, s_2, \ldots, s_n)$ is the inverse of some perturbed matrix $T + E$ or even that S is near the inverse of such a matrix.

None the less, in practice the computed inverses of triangular matrices are usually found to be quite accurate. This is to be expected, since the inverse is obtained by solving a set of triangular systems, and we have already observed that the solutions of triangular systems are usually computed with high accuracy. Moreover, it can be shown that for some special classes of matrices the computed inverse must be accurate, although nothing can be proved in general.

We shall now consider the error analyses of Gaussian elimination and the algorithms for triangular decomposition. We have already seen that the strategy for selecting the pivots in these algorithms can have a marked effect on their numerical properties, and we should expect the error analysis to reflect the pivoting strategy. It turns out that the choice of pivots affects the outcome by limiting the growth of elements computed in the course of the reduction.

This point is most clearly illustrated by Gaussian elimination. By Theorem 2.9 we may assume that any interchanges required by the pivoting strategy have already been performed. Thus let the algorithm for Gaussian elimination without pivoting be performed in floating-point arithmetic on the matrix $A = A_1$ yielding matrices $M_1, M_2, \ldots, M_{n-1}$, and $A_2, A_3, \ldots, A_n$. Let

$$\beta_k = \max\{|\alpha_{ij}^{(k)}|\} \qquad (k = 1, 2, \ldots, n)$$

and

$$\gamma = \frac{\max\{\beta_k : k = 1, 2, \ldots, n\}}{\beta_1}.$$

Thus γ, the ratio of the largest element of $A_1, A_2, \ldots, A_n$ to the largest element of A_1, is a measure of the growth of the matrices generated by the algorithm. With these definitions, we have the following result.

THEOREM 5.2. The matrices $M_1, M_2, \ldots, M_{n-1}, A_n$ computed by Algorithm 2.4 satisfy

$$M_1^{-1}M_2^{-1} \cdots M_{n-1}^{-1}A_n = A + E, \tag{5.4}$$

where

$$|\varepsilon_{ij}| \leq n\pi\beta_1\gamma 10^{-t} \tag{5.5}$$

for some constant π of order unity that is independent of A.

By the theorem, the computed triangular decomposition obtained from Gaussian elimination is the exact triangular decomposition of a perturbed matrix. The perturbation will be small compared to β_1 (that is, compared to the largest element of A_1) provided the growth factor γ is small. Thus it is important to establish bounds on the growth of the elements in the course of the reduction. For partial and complete pivoting upper bounds can be given for γ that do not depend on A. The bound for complete pivoting is

$$\gamma \leq \{n \cdot 2^1 3^{1/2} 4^{1/3} \cdots n^{1/(n-1)}\}^{1/2}.$$

This bound increases rather slowly with n, and moreover the proof that establishes it shows that it cannot be attained. In fact, no matrix with real elements is known for which γ is greater than n. Thus Gaussian elimination with complete pivoting is a stable algorithm.

The bound for partial pivoting is

$$\gamma \leq 2^{n-1}. \tag{5.6}$$

This is a rather fast growing function, and its use in the bound (5.5) suggests that the order of those matrices that can be safely decomposed by Gaussian elimination with partial pivoting is severely limited. For example, on a four-digit machine, if A_1 is of order 12, the elements E may be of the same size as the elements of A. Unfortunately matrices are known for which this bound on γ is attained, so that we cannot assert that Gaussian elimination with partial pivoting is unconditionally stable.

In practice, however, the bound (5.6) is seldom attained. Usually the elements of the reduced matrices $A_1, A_2, \ldots, A_n$ remain of the same order of magnitude or even show a progressive decrease in size. Moreover, for many commonly occuring kinds of matrices, such as Hessenberg matrices, the growth is much more restricted. Thus *in practice Gaussian elimination with partial pivoting must be considered a stable algorithm.*

The error analysis of the Crout reduction yields similar results. The computed L and U satisfy

$$LU = A + E$$

where the elements of E satisfy (5.5). The growth factor γ in the bounds for the Crout reduction is the same as the growth factor for Gaussian elimination. We have already noted that with partial pivoting γ can increase swiftly with the order of the matrix. Since partial pivoting is the only convenient strategy for the Crout reduction, we cannot claim that this reduction is stable in general. However, the observation that large growth factors are almost never encountered applies here, and for practical purposes the Crout reduction can be considered a stable algorithm.

We have mentioned earlier that an important reason for preferring the Crout reduction to Gaussian elimination is that the Crout reduction permits the accumulation of inner products in double precision. As far as the rounding-error analysis is concerned, the effect of this is to remove the factor n from the bound (5.5). Thus if γ is also of order unity, the elements of the error matrix E are of the same size as the errors made in rounding A, and in many cases they will be considerably smaller. This is about as much as can be expected in the way of stability.

The error analysis of the Cholesky algorithm for positive definite matrices differs from the others in that there is no growth factor (this is also true of Gaussian elimination and Crout reduction applied to positive definite matrices; see Exercise 5.2). In fact, we can prove that the algorithm computes a lower triangular matrix L satisfying

$$LL^{\mathrm{T}} = A + E,$$

where the elements of E satisfy (5.5) with $\gamma = 1$. When inner products are accumulated in double precision, the factor n may be removed from the bounds. Thus the Cholesky algorithm for positive definite matrices is unconditionally stable.

It should be stressed that the results quoted here for Gaussian elimination and the algorithms for triangular decomposition are the final, simplified results of rather detailed error analyses. An examination of the details of the analyses shows that the bounds on the elements of E are, in many frequently occurring cases, severe overestimates. Nonetheless, the above results indicate the major features that determine the stability of the algorithms.

For example, these bounds have something to say about the delicate question of choosing between Gaussian elimination with complete pivoting, Gaussian elimination with partial pivoting, and the Crout reduction with partial pivoting. When inner products cannot be accumulated in double precision the Crout reduction and Gaussian elimination have the same numerical properties. When inner products can be accumulated in double precision, the Crout reduction is superior. Thus there seems to be little reason for using Gaussian elimination with partial pivoting. This conclusion is reinforced by the fact that a Crout reduction program that does not accumulate inner products in double precision can be converted into one that does with very little trouble.

The choice is then between Gaussian elimination with complete pivoting and the Crout reduction. A reason customarily advanced against the former is that searching the entire matrix for the largest element is inordinately time consuming. This will generally be true if the algorithm is coded in one of the standard languages such as FORTRAN. However, if the algorithm is coded in machine language, the search can be performed as the elements are computed, usually at little additional cost. Since the stability of Gaussian elimination with complete pivoting is guaranteed, a person of conservative temperament who is willing to put up with a machine dependent subroutine might prefer it over the Crout reduction. On the other hand, practically speaking the Crout reduction has the same stability as Gaussian elimination with complete pivoting and can be coded efficiently in, say, FORTRAN. If inner products can be accumulated in double precision, the Crout reduction is superior, except in the unlikely cases where there is a significant growth in the elements.

Bounds for the solution of the linear system $Ax = b$ may be obtained by combining Theorem 5.1 and 5.2. For definiteness we shall suppose that the matrix A has been decomposed by Gaussian elimination with partial pivoting into the product

$$A = (M_1^{-1} M_2^{-1} \cdots M_{n-1}^{-1}) A_n = LU.$$

The elements of L are the multipliers μ_{ij} of Algorithm 2.4, which are less than unity in magnitude. Hence

$$|\lambda_{ij}| \leq 1.$$

The elements of U are taken from the matrices $A_1, A_2, \ldots, A_n$ of the reduction and hence satisfy

$$|v_{ij}| \leq \gamma\beta_1,$$

where β_1 is the magnitude of the largest element of A_1 and γ is the growth factor. Finally from Theorem 5.2 we know that

$$LU = A + E, \tag{5.7}$$

where the elements of E satisfy (5.5).

Now suppose that Algorithm 4.3 is used to solve the system $Ax = b$. Statement 2 of the algorithm is numerically equivalent to solving the lower triangular system

$$Ly = b, \tag{5.8}$$

while statement 3 amounts to solving the upper triangular system

$$Ux = y.$$

We have observed that triangular systems are usually solved with high accuracy. Hence it is not surprising that the computed solution $\bar{x}$ is often very near the true solution $\tilde{x}$ of the system

$$(A + E)\tilde{x} = b. \tag{5.9}$$

In other words, *the computed solution* $\bar{x}$ *of a system of linear equations is often near a solution* $\tilde{x}$ *of* (5.9), *where the error matrix E is independent of the right-hand side b*. However, this is not a rigorous result and counter examples can be constructed. The following theorem shows that if we drop the requirement that the error matrix be independent of the right-hand side, we can obtain a rigorous stability result.

THEOREM 5.3. Let $\bar{x}$ denote the solution of the system $Ax = b$ computed by Algorithm 4.3. Then $\bar{x}$ satisfies

$$(A + H)\bar{x} = b, \tag{5.10}$$

where

$$|\eta_{ij}| \leq (n\pi + 2n^2\varrho + n^3\varrho^2 10^{-t})\gamma\beta_1 10^{-t}, \tag{5.11}$$

with π and ϱ constants of order unity.

PROOF. By Theorem 5.1, the computed solution $\bar{y}$ of (5.8) satisfies

$$(L + F)\bar{y} = b, \tag{5.12}$$

where

$$|\phi_{ij}| \leq n\varrho\,|\lambda_{ij}|\,10^{-t} \leq n\varrho 10^{-t}$$

with ϱ a constant of order unity. The last inequality follows from the fact that partial pivoting insures that $|\lambda_{ij}| \leq 1$. In statement 3 of Algorithm

4.3 we solve the system $Ux = \bar{y}$, yielding a computed solution $\bar{x}$ that satisfies

$$(U + G)\bar{x} = \bar{y} \tag{5.13}$$

where

$$|\gamma_{ij}| \leq n\varrho\,|v_{ij}|\,10^{-t} \leq n\varrho\beta_1\gamma 10^{-t}.$$

Combining (5.7), (5.12), and (5.13), we find that $\bar{x}$ satisfies (5.10) where

$$H = E + LG + FU + FG.$$

Now the largest element of the product of two matrices of order n is bounded by n times the product of the largest elements of each matrix. Hence, using the bounds for the elements of E, F, G, L, and U, we obtain (5.11). ∎

The error matrix H of course depends on the vector b, since the matrices F and G do. The bound (5.11) on its elements depends critically on the growth factor γ. If γ is of order unity and n is not unreasonably large, the elements of H will be not too much larger than the errors made in rounding A; often they will be a good deal smaller. The term containing n^2 comes from the errors made in solving triangular systems. Since, as we have noted, these errors are often negligible, the true error may be more accurately reflected by the term $n\pi$.

Analogs of Theorem 5.3 hold for the Crout and Cholesky algorithms. If inner products are accumulated in double precision throughout the calculations, the first factor in (5.11) may be replaced by

$$\pi + 2n\varrho + n\varrho^2 10^{-t}.$$

In many cases the term $2n\varrho$, which accounts for the errors introduced by solving the triangular systems, may be omitted, so that the bound becomes effectively independent of the size of the matrix.

Since no general theorems on the computed inverses of triangular equations have been established, it is not surprising that little can be said about the computed inverses of general matrices. Suppose, for example, that the columns of the inverse X, are computed as the solutions of the systems $Ax_i = e_i$ $(i = 1, 2, \ldots, n)$. Then from Theorem 5.3 we know that each computed column $\bar{x}_i$ satisfies $(A + H_i)\bar{x}_i = e_i$; however, the error matrices H_i may be different for each column, and we cannot conclude that there is a single small matrix H such that $(A + H)\bar{X} = I$.

None the less, it follows from the remarks preceding Theorem 5.3 that each $\bar{x}_i$ will often be near the solution $\tilde{x}_i$ of the equation $(A + E)\tilde{x}_i = e_i$, where E is the error matrix from the LU decomposition of A. Since E is independent of the right-hand side e_i, it follows that $\bar{X}$ is near an inverse of $A + E$. This means that in many cases *the error in the computed inverse of A will be almost entirely due to the errors made in computing the LU decomposition of A.*

All of the rigorous results established in this section concern the stability of algorithms. The only results on the accuracy of computed solutions were in the form of informal observations. In Section 5.4 we will apply the theory of norms to obtain rigorous bounds on the errors in computed solutions.

We conclude this section with a word of caution. The factor β_1, which appears in the error bound of Theorem 5.2, serves the purpose of a normalizing factor. In effect the bound says that the error will be small compared to β_1, that is compared to the largest element of A. If all the elements of A are of about the same size, then this is a perfectly satisfactory result. On the other hand, if there is a wide disparity in the sizes of the elements of A, there is a danger that the smaller elements will be overwhelmed by the errors. If these small elements are critical to the reduction, the results will be inaccurate.

For example, consider the matrix

$$A = \begin{pmatrix} 3.000 & 6000. & 9000. \\ -1.000 & 3.712 & 4.623 \\ -2.000 & 1.072 & 5.643 \end{pmatrix}, \tag{5.14}$$

which was obtained from the matrix of Example 2.8 by multiplying its first row by 3000. If Gaussian elimination with partial pivoting is performed in four-digit arithmetic on A, the (1, 1)-element is accepted as the first pivot, and the calculation proceeds exactly as in Example 2.8 with a disasterous cancellation occuring in the computation of the (3, 3)-element of the reduced matrix. Our error analysis guarantees that the computed decomposition is the exact decomposition of $A + E$, where the elements of E are of the order $\beta_1 \cdot 10^{-4}$; however, this is small consolation, for $\beta_1 = 9000$ and the elements of E can be, and are, as large as unity.

Since the disparity in the sizes of the elements of A is responsible for the problem, it is natural to attempt to scale the rows and columns of A so that the matrix is balanced. A frequently suggested method is to scale A so that the largest element in each row and column lies between, say, 1 and 10.

For example, the original matrix of Example 2.8, which can be reduced safely by Gaussian elimination with partial pivoting, satisfies this criterion. However, the matrix

$$A = \begin{pmatrix} 3.000 & 1.000 & 1.000 \\ -1.000 & 6.187\times 10^{-4} & 5.137\times 10^{-4} \\ -2.000 & 1.787\times 10^{-4} & 6.270\times 10^{-4} \end{pmatrix},$$

obtained from (5.14) by dividing the second column by 6000 and the third column by 9000 also satisfies the criterion. It is easily verified that, in four-digit arithmetic, even complete pivoting will not yield an accurate reduction. The error matrix associated with the reduction has elements of order 10^{-4}, but this is no help since four of the elements of A are also of order 10^{-4}.

The difficulties involved in scaling can be overstressed. For many problems the criterion suggested above is quite satisfactory and for many others the correct scaling is perfectly evident. However, the above example shows that pat scaling strategies are suspect. In spite of intensive theoretical investigation, there is no satisfactory algorithm for scaling a general matrix.

EXERCISES

1. Compute the last column of the inverse of

$$\begin{pmatrix} 10^{-4} & .9 & -.4 \\ 0 & .9 & -.4 \\ 0 & 0 & 10^{-4} \end{pmatrix}$$

in four-digit arithmetic. Interpret your results.

2. Show that if Gaussian elimination is applied to a positive definite matrix, then the growth factor γ of Theorem 5.2 is equal to unity. [*Hint*: Show that the diagonal elements must decrease and then apply Exercises 3.8 and 3.11.]

3. Show that if A is upper Hessenberg of order n, then the growth factor γ for Gaussian elimination with partial pivoting is bounded by n.

4. Show that if A is tridiagonal, then the growth factor γ for Gaussian elimination with partial pivoting is bounded by two.

5. Show that if A^{T} is diagonally dominant, then the growth factor γ for Gaussian elimination with partial pivoting is bounded by two.

NOTES AND REFERENCES

All the material in this section is contained in works by Wilkinson (1961; 1963; AEP, Chapter IV). The 1961 paper contains a wealth of detail: in particular, a derivation of bounds for the growth factors for matrices of special form and a discussion of the high accuracy of the computed solutions of triangular systems.

The conjecture that the growth factor for complete pivoting is bounded by the order of the matrix has neither been proved nor disproved. Cryer (1968) proves the conjecture for matrices of order four and surveys other attempts.

For more on the problem of scaling or *equilibration*, as it is sometimes called, see the works by Bauer (1963), Forsythe and Strauss (1955), and Wilkinson (1961; AEP, Chapter IV).

NORMS, LIMITS, AND CONDITION NUMBERS CHAPTER 4

With few exceptions we have, in the first three chapters of this book, confined ourselves to the algebraic properties of vectors and matrices; that is we have developed those areas that could be described in terms of algebraic operations without requiring the notion of limit. In this chapter we shall introduce the ideas of norm and limit and apply them to the problems of assessing and improving the accuracy of approximate solutions of linear systems.

A vector norm on $\mathbb{R}^n$ is a function from $\mathbb{R}^n$ into the nonnegative real numbers whose value in some sense measures the size of a vector in $\mathbb{R}^n$. Examples of norms are the absolute value in $\mathbb{R}$ and the Euclidean length of a vector in $\mathbb{R}^2$. The idea of norm is closely related to the idea of limit, as is suggested by the equivalence

$$\lim_{k\to\infty} \xi_k = \xi \quad \Leftrightarrow \quad \lim_{k\to\infty} | \xi - \xi_k | = 0.$$

In Section 1 we shall introduce the idea of a vector norm on $\mathbb{R}^n$ and explore its relation to limits in $\mathbb{R}^n$. In Section 2 we shall extend the concept of norm to matrices. In Section 3.5, we discussed the effects of rounding error on some algorithms for solving linear systems. However, our formal results gave no indication of the accuracy of the solution; rather they implied that the errors in the solution would be proportional to the degree of ill conditioning of the system. In Sections 3 and 4 we shall use the theory of norms to define a *condition number* for a linear system whose size measures the effects of perturbations in the matrix on the final solution. Finally in Section 5, we shall analyze a method for refining an approximate solution of a linear system to produce a more accurate one.

1. NORMS AND LIMITS

The idea of a vector norm arises from the attempt to generalize the idea of the length of a vector in $\mathbb{R}^1$ or $\mathbb{R}^2$. The simplest example of a norm is the absolute value function, which measures the distance of a scalar from the origin. The absolute value $|\xi|$ of a number ξ may be defined by the equation

$$|\xi| = \sqrt{\xi^2}$$

and has the three well-known properties

$$\begin{aligned} &1.\quad \xi \neq 0 \;\Rightarrow\; |\xi| > 0, \\ &2.\quad |\alpha\xi| = |\alpha|\,|\xi|, \\ &3.\quad |\xi + \eta| \leq |\xi| + |\eta|. \end{aligned} \tag{1.1}$$

Because of (1.1.1), the absolute value is said to be *positive definite*, or simply *definite*. Because of (1.1.2), it is said to be *homogeneous*. The third condition is called the *triangle inequality* for reasons that will become clear in a moment.

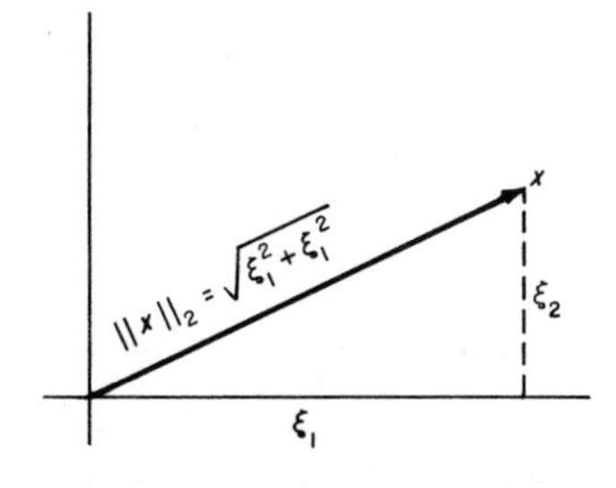

Fig. 1

A less trivial example of a norm is the Euclidean length of a vector in $\mathbb{R}^2$. For any $x \in \mathbb{R}^2$ it is denoted by $\| x \|_2$ and defined by

$$\| x \|_2 = \sqrt{\xi_1^2 + \xi_2^2}. \tag{1.2}$$

By the Pythagorean theorem, $\| x \|_2$ is the distance from x to the origin (Fig. 1).

The Euclidean norm $\| \cdot \|_2$ satisfies three conditions that correspond to the properties (1.1) satisfied by the absolute value. Specifically,

1. $x \neq 0 \;\Rightarrow\; \| x \|_2 > 0,$
2. $\| \alpha x \|_2 = | \alpha | \, \| x \|_2,$
3. $\| x + y \|_2 \leq \| x \|_2 + \| y \|_2.$

The first two conditions are clear from the definition of $\| \cdot \|_2$. The third condition may be interpreted geometrically as saying that the length of a side of a triangle is not greater than the sum of the lengths of the other two sides, whence the name triangle inequality (Fig. 2).

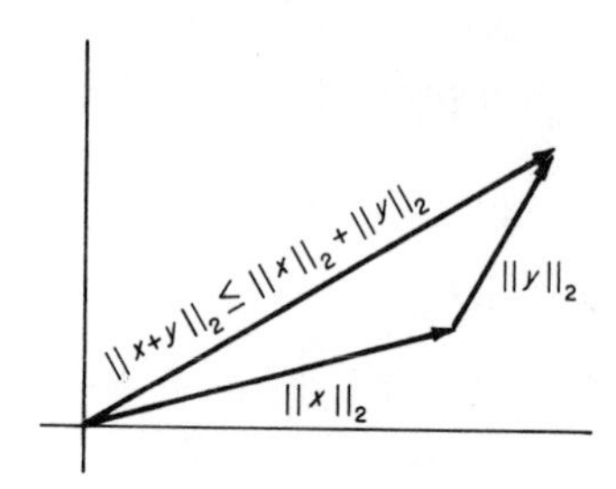

Fig. 2

One natural way of extending these ideas to $\mathbb{R}^n$ is to generalize formula (1.2) in the obvious way to obtain a Euclidean length of a vector in $\mathbb{R}^n$. However, this is somewhat restrictive, for there are many other functions that measure the size of a vector. For example, the number $\nu(x)$ defined for $x \in \mathbb{R}^2$ by

$$\nu(x) = \max\{| \xi_1 |, | \xi_2 |\} \tag{1.3}$$

is also a reasonable measure of the size of x and is computationally more convenient than (1.2). It turns out that in many applications, all that is required of a measure of size is that it be definite, homogeneous, and satisfy the triangle inequality. This motivates the following definition.

DEFINITION 1.1. A *vector norm* (or simply a *norm*) *on* $\mathbb{R}^n$ is a function $\nu : \mathbb{R}^n \to \mathbb{R}$ that satisfies the following conditions:

$$1. \quad x \neq 0 \quad \Rightarrow \quad \nu(x) > 0,$$

$$2. \quad \nu(\alpha x) = |\,\alpha\,|\, \nu(x),$$

$$3. \quad \nu(x + y) \leq \nu(x) + \nu(y).$$

The absolute value function is a norm on $\mathbb{R}^1$, and the function $\|\cdot\|_2$ defined by (1.2) is a norm on $\mathbb{R}^2$. The function ν defined by (1.3) is also a norm on $\mathbb{R}^2$.

The definition of norm has some elementary consequences. By property 1.1.1, the function ν is positive except at the zero vector. By property 1.1.2,

$$\nu(0) = \nu(0 \cdot x) = 0\nu(x) = 0,$$

so that the norm of the zero vector is zero. Also from 1.1.2, it follows that

$$\nu(-x) = \nu(-1 \cdot x) = |\,-1\,|\, \nu(x) = \nu(x).$$

An important inequality is the following.

$$|\, \nu(x) - \nu(y) \,| \leq \nu(x - y), \tag{1.4}$$

which in terms of the Euclidean norm says that the length of a side of a triangle is not less than the difference of the lengths of the other two sides (Fig. 3). To establish it, note that

$$\nu(x) = \nu[(x - y) + y] \leq \nu(x - y) + \nu(y).$$

Hence

$$\nu(x) - \nu(y) \leq \nu(x - y). \tag{1.5}$$

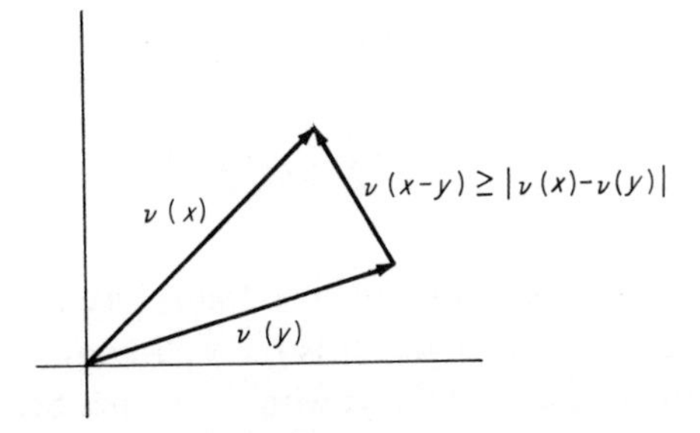

Fig. 3

If x and y are interchanged in (1.5), we obtain

$$\nu(y) - \nu(x) \leq \nu(y - x) = \nu(x - y). \tag{1.6}$$

Inequalities (1.5) and (1.6) together imply (1.4).

If y is replaced by $-y$ in (1.4), there results the related inequality

$$| \nu(x) - \nu(y) | \leq \nu(x + y).$$

We turn now to the construction of specific vector norms. We begin with three norms on $\mathbb{R}^n$ that are frequently used in analyzing matrix processes. They are the 1-, 2-, and ∞-norms defined by

$$\| x \|_1 = \sum_{i=1}^{n} | \xi_i |,$$

$$\| x \|_2 = \sqrt{\sum_{i=1}^{n} \xi_i^2} = \sqrt{x^{\mathrm{T}} x},$$

and

$$\| x \|_\infty = \max\{| \xi_i | : i = 1, 2, \ldots, n\}.$$

The 2-norm is the natural generalization of the Euclidean length of a 2- or 3-vector and is also called the Euclidean norm. The ∞-norm is sometimes called the maximum norm (max-norm) or the Chebyshev norm.

All three functions $\| \cdot \|_1$, $\| \cdot \|_2$, and $\| \cdot \|_\infty$ are obviously definite and homogeneous. Moreover, it is easy to show that $\| \cdot \|_1$ and $\| \cdot \|_\infty$ satisfy the triangle inequality. For example,

$$\begin{aligned} \| x + y \|_1 &= \sum_{i=1}^{n} | \xi_i + \eta_i | \\ &\leq \sum_{i=1}^{n} (| \xi_i | + | \eta_i |) \\ &= \sum_{i=1}^{n} | \xi_i | + \sum_{i=1}^{n} | \eta_i | = \| x \|_1 + \| y \|_1. \end{aligned}$$

To establish the triangle inequality for the 2-norm we require an inequality which is important in its own right. It is the well-known *Cauchy inequality* (which, incidentally, is also associated with the names Schwarz and Bunyakovski).

THEOREM 1.2. For all $x, y \in \mathbb{R}^n$,

$$| x^{\mathrm{T}}y | \leq \| x \|_2 \| y \|_2, \tag{1.7}$$

with equality if and only if x and y are linearly dependent.

PROOF. If either x or y is zero, the theorem is trivially true. Hence suppose that both x and y are nonzero. Then $\| x \|_2$ and $\| y \|_2$ are both nonzero. Let $x' = x/\| x \|_2$ and $y' = y/\| y \|_2$. Then $\| x' \|_2 = \| y' \|_2 = 1$, and the inequality (1.7) is equivalent to the inequality

$$| x'^{\mathrm{T}}y' | \leq 1.$$

Now suppose $x'^{\mathrm{T}}y' \geq 0$. Then

$$\begin{aligned} 0 \leq \| x' - y' \|_2^2 &= (x' - y')^{\mathrm{T}}(x' - y') \\ &= x'^{\mathrm{T}}y' + y'^{\mathrm{T}}y' - 2x'^{\mathrm{T}}y' \\ &= \| x' \|_2^2 + \| y' \|_2^2 - 2x'^{\mathrm{T}}y' \\ &= 2 - 2x'^{\mathrm{T}}y'. \end{aligned} \tag{1.8}$$

Hence $x'^{\mathrm{T}}y' \leq 1$. If $x'^{\mathrm{T}}y' < 0$, the inequality $-x'^{\mathrm{T}}y' \leq 1$ follows similarly, with $x' + y'$ replacing $x' - y'$ in (1.8).

Now suppose $| x^{\mathrm{T}}y | = \| x \|_2 \| y \|_2$ with say $x^{\mathrm{T}}y \geq 0$. This can only happen when the inequality in (1.8) is an equality. Hence $\| x' - y' \|_2 = 0$, and $x' = y'$. This implies that

$$x = \frac{\| x \|_2}{\| y \|_2} y$$

and hence that x and y are linearly dependent. The case $x^{\mathrm{T}}y < 0$ is treated similarly.

Finally, suppose that x and y are linearly dependent. Since $y \neq 0$, $x = \alpha y$ for some scalar α. Then $\| x \|_2 = | \alpha | \| y \|_2$, and

$$| x^{\mathrm{T}}y | = | \alpha | y^{\mathrm{T}}y = | \alpha | \| y \|_2 \| y \|_2 = \| x \|_2 \| y \|_2. \quad \blacksquare$$

The above proof makes heavy use of the identity $\| x \|_2^2 = x^{\mathrm{T}}x$. This characterization is also used in establishing that $\| \cdot \|_2$ is a norm.

THEOREM 1.3. The function $\| \cdot \|_2$ satisfies the triangle inequality.

PROOF. The proof is purely computational.

$$\begin{aligned}\| x + y \|_2^2 &= (x + y)^T(x + y) \\ &= x^T x + 2x^T y + y^T y \\ &\leq \| x \|_2^2 + 2 \| x \|_2 \| y \|_2 + \| y \|_2^2 \\ &= (\| x \|_2 + \| y \|_2)^2,\end{aligned}$$

where the inequality follows from the Cauchy inequality. ■

We have used the same symbol $\| \cdot \|_2$ to denote the 2-norm on $\mathbb{R}^1$, $\mathbb{R}^2$, $\mathbb{R}^3$, ... While this is not a serious abuse of notation, it can be avoided by introducing some terminology which will prove useful later in connection with matrix norms.

DEFINITION 1.4. Let $\nu : \bigcup_{n=1}^{\infty} \mathbb{R}^n \to \mathbb{R}$. Then ν is a *family of vector norms* if for each $n = 1, 2, 3, \ldots$ the restriction of ν to $\mathbb{R}^n$ is a norm.

Thus each of the functions $\| \cdot \|_1$, $\| \cdot \|_2$, $\| \cdot \|_\infty$ is a family of vector norms. Incidentally, these norms are special cases of the Hölder norms or p-norms defined by

$$\| x \|_p = \sqrt[p]{\sum_{i=1}^{n} | \xi_i |^p}, \qquad 1 \leq p < \infty.$$

($\| x \|_\infty$ is $\lim_{p \to \infty} \| x \|_p$.) A proof that these are indeed norms will be found in the exercises.

A useful technique for constructing new norms from old is contained in the following theorem.

THEOREM 1.5. Let ν be a norm on $\mathbb{R}^m$ and $A \in \mathbb{R}^{m \times n}$ have linearly independent columns. Then the function $\mu : \mathbb{R}^n \to \mathbb{R}$ defined by

$$\mu(x) = \nu(Ax)$$

is a norm on $\mathbb{R}^n$.

PROOF. Because A has linearly independent columns and ν is a norm,

$$x \neq 0 \Rightarrow Ax \neq 0 \Rightarrow \nu(Ax) > 0 \Rightarrow \mu(x) > 0.$$

Hence μ is definite. Since

$$\mu(\alpha x) = \nu(\alpha Ax) = |\alpha| \, \nu(Ax) = |\alpha| \, \mu(x),$$

μ is homogeneous. Finally,

$$\mu(x + y) = \nu[A(x + y)] = \nu(Ax + Ay) \leq \nu(Ax) + \nu(Ay) = \mu(x) + \mu(y),$$

so that μ satisfies the triangle inequality. ■

Theorem 1.5 may be applied to give an important class of norms.

THEOREM 1.6. Let $A \in \mathbb{R}^{n \times n}$ be positive definite. Then the function $\nu : \mathbb{R}^n \to \mathbb{R}$ defined by

$$\nu(x) = \sqrt{x^{\mathrm{T}} A x}$$

is a norm on $\mathbb{R}^n$.

PROOF. By Theorem 3.3.8, A can be written in the form $A = LL^{\mathrm{T}}$, where L is a nonsingular lower triangular matrix. Then

$$\nu(x) = \sqrt{(L^{\mathrm{T}}x)^{\mathrm{T}}(L^{\mathrm{T}}x)} = \| L^{\mathrm{T}}x \|_2 .$$

Hence by Theorem 1.5, ν is a norm. ■

Note that when $A = I$, the function ν of Theorem 1.6 reduces to the 2-norm.

The notions of norm and limit are closely connected, and in more abstract settings it is often convenient to define the notion of limit in terms of a norm. However, in $\mathbb{R}^n$ there is a very natural definition of limit in terms of limits of real numbers. We shall start with this definition and explore its connection with vector norms later (Theorem 1.12). We begin with a review of the properties of limits of real numbers.

We shall denote a sequence $\alpha_1, \alpha_2, \alpha_3, \ldots$ by $\langle \alpha_k \rangle$. The $\varepsilon - N$ definition of the limit of a sequence of real numbers may be found in almost any calculus text. For our purposes, we need only the elementary properties of the limit. Specifically, the sequence $\langle \alpha_k \rangle$ can have at most one limit. If $\langle \alpha_k \rangle$ has the limit α, we write

$$\alpha = \lim_{k \to \infty} \alpha_k,$$

or, when it will cause no confusion,

$$\alpha = \lim \alpha_k.$$

If $\alpha = \lim \alpha_k$ and $\beta = \lim \beta_k$, then

$$\lim(\alpha_k \pm \beta_k) = \alpha \pm \beta \tag{1.9}$$

and

$$\lim \alpha_k \beta_k = \alpha\beta.$$

Moreover, if $\alpha \neq 0$, then, for all sufficiently large k, $\alpha_k \neq 0$ and

$$\lim \frac{\beta_k}{\alpha_k} = \frac{\beta}{\alpha}.$$

Finally we have the characterization

$$\lim \alpha_k = \alpha \quad \Leftrightarrow \quad \lim |\alpha - \alpha_k| = 0. \tag{1.10}$$

In discussing limits of vectors, we shall need to refer to the components of a sequence of vectors. To do this we shall adopt the following convention. If $\langle x_k \rangle$ is a sequence of n-vectors, we shall denote the ith component of x_k by $\xi_i^{(k)}$. We have already used this convention in Section 3.2 in connection with the matrices A_k generated by Gaussian elimination.

In $\mathbb{R}^n$ it is natural to say that a sequence of vectors converges to a limit of each if its components converges to a limit. For example, in $\mathbb{R}^2$ we should say that the sequence of vectors

$$\begin{pmatrix} 2.1 \\ 0.9 \end{pmatrix}, \quad \begin{pmatrix} 2.01 \\ 0.99 \end{pmatrix}, \quad \begin{pmatrix} 2.001 \\ 0.999 \end{pmatrix}, \quad \ldots, \quad \begin{pmatrix} 2 + 10^{-k} \\ 1 - 10^{-k} \end{pmatrix}, \quad \ldots$$

is converging to the vector $(2, 1)^{\mathrm{T}}$. On the other hand, we should say that the sequence

$$\begin{pmatrix} 1 \\ 2 \end{pmatrix}, \quad \begin{pmatrix} 1 \\ -2 \end{pmatrix}, \quad \begin{pmatrix} 1 \\ 2 \end{pmatrix}, \quad \begin{pmatrix} 1 \\ -2 \end{pmatrix}, \quad \ldots, \quad \begin{pmatrix} 1 \\ (-1)^{k+1}2 \end{pmatrix}, \quad \ldots$$

has no limit, since the second component never settles down. This suggests the following definition.

DEFINITION 1.7. Let $\langle x_k \rangle$ be a sequence of n-vectors and let $x \in \mathbb{R}^n$. Then x is a *limit* of the sequence $\langle x_k \rangle$ (written $x = \lim_{k\to\infty} x_k$) if

$$\lim_{k\to\infty} \xi_i^{(k)} = \xi_i \qquad (i = 1, 2, \ldots, n).$$

Since the limit of a sequence of real numbers is unique, Definition 1.7 implies that the limit of a sequence of vectors must be unique. Moreover, the vector limit has some of the algebraic properties of the scalar limit, as the following theorem shows.

THEOREM 1.8. Let the sequences $\langle a_k \rangle$, $\langle b_k \rangle$, and $\langle \lambda_k \rangle$ have limits $a, b,$ and λ. Then

$$\lim(a_k \pm b_k) = a \pm b \tag{1.11}$$

and

$$\lim \lambda_k a_k = \lambda a. \tag{1.12}$$

PROOF. We prove (1.11), leaving (1.12) as an exercise. By Definition 1.7, we know that $\lim \alpha_i^{(k)} = \alpha_i$ and $\lim \beta_i^{(k)} = \beta_i$, and we must show that

$$\lim_{k\to\infty} (\alpha_i^{(k)} \pm \beta_i^{(k)}) = \alpha_i \pm \beta_i \qquad (i = 1, 2, \ldots, n).$$

However, for each i, this is simply (1.9). ∎

The relation between limits and norms is suggested by the equivalence (1.10). Namely, if ν is a norm on $\mathbb{R}^n$ and $\langle x_k \rangle$ is a sequence of n-vectors, then $\lim x_k = x$ if and only if $\lim \nu(x - x_k) = 0$. However, the proof of this fact requires a fairly deep result on the equivalence of norms. To illustrate its statement, consider the following scalar inequalities:

$$\max\{|\xi_i| : i = 1, 2, \ldots, n\} \leq \sum_{i=1}^{n} |\xi_i| \leq n \max\{|\xi_i| : i = 1, 2, \ldots, n\}.$$

In terms of norms, the inequalities state that in $\mathbb{R}^n$

$$1 \cdot \|x\|_\infty \leq \|x\|_1 \leq n \|x\|_\infty.$$

Thus whenever $\|\cdot\|_1$ appears in an expression as an upper bound, we may replace it by $n \cdot \|\cdot\|_\infty$. Whenever $\|\cdot\|_1$ appears as a lower bound we may replace it with $1 \cdot \|\cdot\|_\infty$. In other words, by multiplying the ∞-norm by suitable constants, we can make it equivalent to the 1-norm, at least for the purpose of stating upper and lower bounds. The following theorem, which we give without proof, states that such an equivalence obtains for any two norms.

THEOREM 1.9. Let μ and ν be norms on $\mathbb{R}^n$. Then there are positive constants $\sigma_{\nu\mu}$ and $\tau_{\nu\mu}$ such that for all $x \in \mathbb{R}^n$

$$\sigma_{\nu\mu}\mu(x) \leq \nu(x) \leq \tau_{\nu\mu}\mu(x).$$

Moreover, $\sigma_{\nu\mu}$ and $\tau_{\nu\mu}$ may be chosen so that for some nonzero $y \in \mathbb{R}^n$ we have $\sigma_{\nu\mu}\mu(y) = \nu(y)$ and for some nonzero $z \in \mathbb{R}^n$ we have $\tau_{\nu\mu}\mu(z) = \nu(z)$.

EXAMPLE 1.10. The following tables give the constants τ and σ for the p-norms ($p = 1, 2, \infty$) or $\mathbb{R}^n$.

σ_{pq}

$p \backslash q$	1	2	∞
1	1	1	1
2	$n^{-1/2}$	1	1
∞	n^{-1}	$n^{-1/2}$	1

,

τ_{pq}

$p \backslash q$	1	2	∞
1	1	$n^{1/2}$	n
2	1	1	$n^{1/2}$
∞	1	1	1

.

Note that in Example 1.10, $\sigma_{qp} = \tau_{pq}^{-1}$. This relation holds for any two norms μ and ν. In fact $\sigma_{\nu\mu}$ is the largest constant for which

$$\sigma_{\nu\mu}\mu(x) \leq \nu(x), \qquad x \in \mathbb{R}^n,$$

and $\tau_{\nu\mu}$ is the smallest constant for which

$$\mu(x) \leq \tau_{\mu\nu}\nu(x), \qquad x \in \mathbb{R}^n.$$

Hence

$$\sigma_{\nu\mu} = \tau_{\mu\nu}^{-1}.$$

We are now in a position to establish the relation between norms and limits. We first establish it for the ∞-norm.

LEMMA 1.11. Let $\langle x_k \rangle$ be a sequence of n-vectors and $x \in \mathbb{R}^n$. Then

$$\lim x_k = x \quad \Leftrightarrow \quad \lim \| x - x_k \|_\infty = 0.$$

PROOF. From Definition 1.7 and the equivalence (1.10) we have the following equivalences.

$$\begin{aligned} \lim x_k = x \quad &\Leftrightarrow \quad \lim \xi_i^{(k)} = \xi_i \qquad (i = 1, 2, \ldots, n) \\ &\Leftrightarrow \quad \lim | \xi_i - \xi_i^{(k)} | = 0 \qquad (i = 1, 2, \ldots, n) \\ &\Leftrightarrow \quad \lim \max \{ | \xi_i - \xi_i^{(k)} | : i = 1, 2, \ldots, n \} = 0 \\ &\Leftrightarrow \quad \lim \| x - x_k \|_\infty = 0. \quad \blacksquare \end{aligned}$$

THEOREM 1.12. Let ν be a norm on $\mathbb{R}^n$ and $x, x_1, x_2, x_3, \ldots \in \mathbb{R}^n$. Then

$$\lim_{k \to \infty} x_k = x \quad \Leftrightarrow \quad \lim_{k \to \infty} \nu(x - x_k) = 0. \tag{1.13}$$

PROOF. If $\lim x_k = x$, by Lemma 1.11 we have $\lim \| x - x_k \|_\infty = 0$. However, by Theorem 1.10, $\nu(x - x_k) \leq \tau_{\nu\infty} \| x - x_k \|_\infty$. Hence $\lim \nu(x - x_k) = 0$. Conversely, let $\lim \nu(x - x_k) = 0$. Then by Theorem 1.10, $\| x - x_k \|_\infty \leq \tau_{\infty\nu} \nu(x - x_k)$, and hence $\lim \| x - x_k \|_\infty = 0$. By Lemma 1.11, we have $\lim x_k = x$. $\blacksquare$

The importance of Theorem 1.12 is that it allows us to use the right-hand side of (1.13) as an alternate definition of limit. Since the norm ν is completely arbitrary, it may be chosen, say, to make it easy to show that $\lim \nu(x - x_k) = 0$. However it is chosen, though, we are assured that if $\lim \nu(x - x_k) = 0$, then $\lim x_k = x$ in the sense of Definition 1.7.

EXERCISES

1. Show that if $\nu : \mathbb{R}^1 \to \mathbb{R}$ is a norm, then $\nu[(\xi)] = \lambda | \xi |$, for some $\lambda > 0$.

2. Show that the function ν defined by Equation (1.3) is a vector norm.

3. Prove the *polarization identity*

$$x^{\mathrm{T}} y = \frac{\| x + y \|_2^2 - \| x - y \|_2^2}{4}.$$

4. Show that $\| \cdot \|_\infty$ satisfies the triangle inequality.

5. Show by example that the condition that A have linearly independent columns cannot be removed from Theorem 1.5.

6. Let $\alpha_1, \alpha_2, \ldots, \alpha_n$ be positive. Show that the function $\nu : R^n \to R$ defined by $\nu(x) = (\sum_{i=1}^n \alpha_i \xi_i^2)^{1/2}$ is a norm.

7. Find linearly independent vectors x and y such that $\| x + y \|_\infty = \| x \|_\infty + \| y \|_\infty$.

8. Show that $\| x + y \|_2 = \| x \|_2 + \| y \|_2$ if and only if x and y are linearly dependent and $x^T y \geq 0$.

9. Let ν be a norm on R^n and let $\mathcal{B}_\nu = \{x : \nu(x) \leq 1\}$. Prove that $\mathcal{B}_\nu$ satisfies the following conditions.

   1. There is an $\varepsilon > 0$ such that $\{x : \| x \|_2 \leq \varepsilon\} \subset \mathcal{B}_\nu$.
   2. (*Convexity*) If $x, y \in \mathcal{B}_\nu$ and α, β are nonnegative scalars with $\alpha + \beta = 1$, then $\alpha x + \beta y \in \mathcal{B}_\nu$.
   3. (*Equilibration*) If $x \in \mathcal{B}_\nu$ and $| \alpha | \leq 1$, then $\alpha x \in \mathcal{B}_\nu$.

   The set $\mathcal{B}_\nu$ is called the "unit ν-ball."

10. Describe the unit 1-, 2-, and ∞-balls in R^2; in R^3.

11. Establish the bounds of Example 1.10. For each σ and τ give the vectors y and z promised in Theorem 1.9.

12. Prove that $\| x \|_\infty = \lim_{p\to\infty} \| x \|_p$.

13. Show that if $p, q > 1$ with $p^{-1} + q^{-1} = 1$, then $| \alpha\beta | \leq | \alpha |^p/p + | \beta |^q/q$, for all scalars α and β. [*Hint*: Show that the function $\phi(\tau) = \tau^p/p + \tau^{-q}/q$ satisfies $\phi(\tau) \geq 1$ for all positive τ. Let $\tau = | \alpha |^{1/q} | \beta |^{-1/p}$.]

14. (*Hölder inequality*) Prove that if $p, q > 1$ with $p^{-1} + q^{-1} = 1$ and $x, y \in R^n$, then $| x^T y | \leq \| x \|_p \| y \|_q$. [*Hint*: Assume $\| x \|_p = \| y \|_q = 1$, and apply Exercise 3 to $| \xi_i | | \eta_i |$. Then sum.]

15. (*Minkowski inequality*) Let $x, y \in R^n$. Show that for $1 < p < \infty$ that $\| x + y \|_p \leq \| x \|_p + \| y \|_p$. [*Hint*: $\| x + y \|_p \leq (\| x \|_p + \| y \|_p)(\sum_{i=1}^n | \xi_i + \eta_i |^{q(p-1)})^{1/q}$, where $p^{-1} + q^{-1} = 1$.]

2. MATRIX NORMS

In this section we shall consider the problem of extending the idea of a vector norm to matrices. At first glance this does not seem to be a very difficult problem. We have seen (page 32) that the set of matrices $\mathbb{R}^{m\times n}$ is a vector space which is essentially identical with $\mathbb{R}^{mn}$. Consequently any vector norm on $\mathbb{R}^{mn}$ induces a definite, homogeneous function on $\mathbb{R}^{m\times n}$ that satisfies the triangle inequality, and it is natural to call such a function a matrix norm. For example, the 2-norm on $\mathbb{R}^{mn}$ induces the *Frobenius norm* $\|\cdot\|_F$ on $\mathbb{R}^{m\times n}$ defined by

$$\| A \|_F = \sqrt{\sum_{i=1}^{m} \sum_{j=1}^{n} \alpha_{ij}^2}. \tag{2.1}$$

The same proof that $\|\cdot\|_2$ is a vector norm shows that $\|\cdot\|_F : \mathbb{R}^{m\times n} \to \mathbb{R}$ is a definite, homogeneous function that satisfies the triangle inequality

$$\| A + B \|_F \leq \| A \|_F + \| B \|_F.$$

We shall adopt this general definition of matrix norm. However, most matrix norms defined in this way are not very useful. The reason is that this definition of matrix norm gives no relation between the norms of two matrices and the norm of their product. Consequently we shall restrict our attention to matrix norms satisfying a *consistency condition*, which for the Frobenius norm reads

$$\| AB \|_F \leq \| A \|_F \| B \|_F. \tag{2.2}$$

We begin our development with the general definition of matrix norm.

DEFINITION 2.1. A function $\nu : \mathbb{R}^{m\times n} \to \mathbb{R}$ is a *matrix norm on* $\mathbb{R}^{m\times n}$ if

(1) $A \neq 0 \ \Rightarrow\ \nu(A) > 0, \qquad A \in \mathbb{R}^{m\times n}$,

(2) $\nu(\alpha A) = |\alpha|\, \nu(A), \qquad A \in \mathbb{R}^{m\times n}, \quad \alpha \in \mathbb{R}$,

(3) $\nu(A + B) \leq \nu(A) + \nu(B), \qquad A, B \in \mathbb{R}^{m\times n}$.

This definition has some immediate consequences. Since a matrix norm on $\mathbb{R}^{m\times n}$ is essentially a vector norm on $\mathbb{R}^{mn}$, it follows that a matrix norm enjoys all the properties of a vector norm. In particular, any two matrix norms on $\mathbb{R}^{m\times n}$ are equivalent in the sense of Theorem 1.9. This means that if we define convergence in $\mathbb{R}^{m\times n}$, in analogy with Definition 1.7, as elementwise convergence, then $\lim A_k = A$ if and only if $\lim \nu(A - A_k) = 0$ for any matrix norm ν. We shall return to the subject of limits of matrices at the end of this section.

Since we have agreed to identify $n \times 1$ matrices with n-vectors, any matrix norm on $\mathbb{R}^{n\times 1}$ is also a vector norm on $\mathbb{R}^n$, and conversely. As with vector norms, we may speak of families of matrix norms.

DEFINITION 2.2. The function $\nu : \bigcup_{m,n=1}^{\infty} \mathbb{R}^{m\times n} \to \mathbb{R}$ is a *family of matrix norms* if for each $m, n \geq 1$ the restriction of ν to $\mathbb{R}^{m\times n}$ is a matrix norm.

It follows from this definition and the preceding comments that if ν is a family of matrix norms, then the restriction of ν to $\bigcup_{n=1}^{\infty} \mathbb{R}^{n\times 1}$ is a family of vector norms.

EXAMPLE 2.3. The Frobenius norms defined for any $A \in \mathbb{R}^{m\times n}$ by (2.1) form a family of matrix norms. The restriction of $\|\cdot\|_F$ to $\bigcup_{n=1}^{\infty} \mathbb{R}^{n\times 1}$ is the family of vector 2-norms; that is $\| x \|_F = \| x \|_2$ for all vectors x.

The Frobenius norm is a natural generalization of the Euclidean vector norm. Since it is relatively easy to compute and satisfies the inequality (2.2), it is used rather frequently in matrix computations. However, all matrix norms satisfying Definition 2.1 need not satisfy an inequality such as (2.2). For example, a natural generalization of the ∞-norm is the function ν defined for $A \in \mathbb{R}^{m\times n}$ by

$$\nu(A) = \max\{|\alpha_{ij}| : i = 1, 2, \ldots, m; \quad j = 1, 2, \ldots, n\}. \qquad (2.3)$$

It is easily verified that ν is a matrix norm in the sense of Definition 2.1. However, if

$$A = B = \begin{pmatrix} 1 & 1 \\ 1 & 1 \end{pmatrix},$$

then $\nu(A)\nu(B) = 1$ while $\nu(AB) = 2$, so that it is not true that $\nu(AB) \leq \nu(A)\nu(B)$. For this reason the matrix norm (2.3) is seldom encountered in the literature.

These considerations suggest that we consider the properties of matrix norms that satisfy a relation like (2.2). Since A, B, and AB will in general have different dimensions, we may use a different norm for each member of (2.2).

DEFINITION 2.4. Let $\mu : \mathbb{R}^{l\times m} \to \mathbb{R}$, $\nu : \mathbb{R}^{m\times n} \to R$, and $\varrho : \mathbb{R}^{l\times n} \to \mathbb{R}$ be matrix norms. Then μ, ν, and ϱ are *consistent* if

$$\varrho(AB) \leq \mu(A)\nu(B)$$

for all $A \in \mathbb{R}^{l\times m}$ and $B \in \mathbb{R}^{m\times n}$. A matrix norm on $\mathbb{R}^{n\times n}$ is consistent if it is consistent with itself. A family of matrix norms ν is consistent if

$$\nu(AB) \leq \nu(A)\nu(B)$$

whenever the product AB is defined.

As an example of a family of consistent matrix norms we show that the Frobenius norms of Example 2.3 are consistent.

THEOREM 2.5. The family $\|\cdot\|_F$ is consistent.

PROOF. We first show that

$$\| Ax \|_2 \leq \| A \|_F \, \| x \|_2 , \tag{2.4}$$

for any $A \in \mathbb{R}^{l\times m}$ and $x \in \mathbb{R}^m$. Let A be partitioned by rows: $A^T = (a_1, a_2, \ldots, a_l)$. Then

$$Ax = \begin{pmatrix} a_1^T x \\ a_2^T x \\ \vdots \\ a_l^T x \end{pmatrix}.$$

Hence

$$\| Ax \|_2^2 = \sum_{i=1}^{l} | a_i^T x |^2.$$

By the Cauchy inequality $| a_i^T x | \leq \| a_i \|_2 \, \| x \|_2$. Hence

$$\| Ax \|_2^2 \leq \| x \|_2^2 \sum_{i=1}^{l} \| a_i \|_2^2.$$

However, it is easily verified that $\| A \|_F^2 = \sum_{i=1}^{l} \| a_i \|_2^2$. Hence (2.4) is satisfied.

Now let $C = AB$, where $A \in \mathbb{R}^{l \times m}$ and $B \in \mathbb{R}^{m \times n}$. If $B = (b_1, b_2, \ldots, b_n)$ is partitioned by columns, then

$$\| C \|_F^2 = \| AB \|_F^2 = \| (Ab_1, Ab_2, \ldots, Ab_n) \|_F^2$$
$$= \sum_{j=1}^{n} \| Ab_j \|_2^2 \leq \| A \|_F^2 \sum_{j=1}^{n} \| b_j \|_2^2 = \| A \|_F^2 \| B \|_F^2,$$

which is the consistency relation. ■

By specialization, Definition 2.4 includes the notion of consistency of a vector norm and a matrix norm. For example, if $\| \cdot \| : \mathbb{R}^{n \times n} \to \mathbb{R}$ is a matrix norm and $\nu : \mathbb{R}^n \to \mathbb{R}$ is a vector norm, then ν and $\| \cdot \|$ are consistent if

$$\nu(Ax) \leq \| A \| \, \nu(x).$$

The inequality (2.4) in the proof of Theorem 2.5 shows that the Frobenius matrix norm is consistent with the Euclidean vector norm.

It sometimes happens that we are given a consistent matrix norm on $\mathbb{R}^{n \times n}$ and require a consistent vector norm. The proof of the following theorem shows how to construct such a norm.

THEOREM 2.6. Let $\| \cdot \| : \mathbb{R}^{n \times n} \to \mathbb{R}$ be a consistent matrix norm. Then there is a norm on $\mathbb{R}^n$ that is consistent with $\| \cdot \|$.

PROOF. Let $a \neq 0$ be an n-vector. Define the function $\nu : \mathbb{R}^n \to \mathbb{R}$ by

$$\nu(x) = \| x a^T \|, \qquad x \in \mathbb{R}^n. \tag{2.5}$$

It is easily verified that ν is a vector norm. Moreover,

$$\nu(Ax) = \| (Ax) a^T \| = \| A(x a^T) \| \leq \| A \| \, \| x a^T \| = \| A \| \, \nu(x),$$

so that ν is consistent with $\| \cdot \|$. ■

We turn now to an important process for constructing matrix norms consistent with given vector norms. The idea of the construction is the following. Let $A \in \mathbb{R}^{m \times n}$ and let $\nu : \mathbb{R}^n \to \mathbb{R}$ and $\mu : \mathbb{R}^m \to \mathbb{R}$ be norms. If $\nu(x) = 1$ and $y = \alpha x$, then $\nu(y) = | \alpha |$ and

$$\mu(Ay) = \mu(\alpha Ax) = | \alpha | \, \mu(Ax) = \nu(y) \mu(Ax).$$

In other words, the number $\mu(Ax)$ measures how much the linear transformation A magnifies (or diminishes) any vector that is a multiple of x. If there is a largest such number $\mu(Ax)$, this largest magnification constant is a natural candidate for a norm of A. In order to insure the existence of a vector x for which $\mu(Ax)$ is maximal, we need the following theorem, which we state without proof.

THEOREM 2.7. Let $\mu : \mathbb{R}^m \to \mathbb{R}$ and $\nu : \mathbb{R}^n \to \mathbb{R}$ be vector norms, and let $A \in \mathbb{R}^{m\times n}$. Then there are vectors $y, z \in \mathbb{R}^n$ such that $\nu(y) = \nu(z) = 1$ and for all $x \in \mathbb{R}^n$

$$\nu(x) = 1 \quad \Rightarrow \quad \mu(Az) \leq \mu(Ax) \leq \mu(Ay).$$

Otherwise put, Theorem 2.7 says that the numbers

$$\max_{\nu(x)=1} \mu(Ax) \qquad \text{and} \qquad \min_{\nu(x)=1} \mu(Ax)$$

are well defined. Moreover the maximum and minimum values can be attained for certain vectors y and z.

Now suppose we define the function $\| \cdot \| : \mathbb{R}^{m\times n} \to \mathbb{R}$ by

$$\| A \| = \max_{\nu(x)=1} \mu(Ax). \tag{2.6}$$

Then it follows that

$$\mu(Ax) \leq \| A \| \, \nu(x), \tag{2.7}$$

so that the function $\| \cdot \|$ is consistent with the vector norms μ and ν. We can also prove that $\| \cdot \|$ is a matrix norm. However, we can prove much more, as the following theorem shows.

THEOREM 2.8. Let ν be a family of vector norms. For $A \in \mathbb{R}^{m\times n}$ define

$$\| A \|_\nu = \max_{\nu(x)=1} \nu(Ax).$$

Then $\| \cdot \|_\nu : \bigcup_{m,n=1}^{\infty} \mathbb{R}^{m\times n} \to \mathbb{R}$ is a consistent family of matrix norms. Moreover, if on $\mathbb{R}^1$ we have

$$\nu[(\xi)] = | \, \xi \, |, \tag{2.8}$$

then

$$\| x \|_\nu = \nu(x), \qquad x \in \mathbb{R}^n.$$

PROOF. By Theorem 2.7, the function $\| \cdot \|_\nu$ is well defined. By (2.7) it satisfies

$$\nu(Ax) \leq \| A \|_\nu \nu(x).$$

To show that $\| \cdot \|_\nu$ is a family of matrix norms, we must show that for $m, n \geq 1$, the restriction of $\| \cdot \|_\nu$ to $\mathbb{R}^{m\times n}$ is a matrix norm.

1. *Definiteness*: Let $A \neq 0$. Then, say, the ith column of A is nonzero, so that $Ae_i \neq 0$. Then $0 < \nu(Ae_i) \leq \| A \|_\nu \nu(e_i)$. Since $\nu(e_i) \neq 0$, it follows that $\| A \|_\nu > 0$.

2. *Homogeneity*:

$$\| \alpha A \|_\nu = \max_{\nu(x)=1} \nu(\alpha Ax) = \max_{\nu(x)=1} | \alpha | \nu(Ax) = | \alpha | \max_{\nu(x)=1} \nu(Ax) = | \alpha | \| A \|_\nu .$$

3. *Triangle inequality*: Let x be such that $\nu(x) = 1$ and $\nu[(A + B)x] = \| A + B \|_\nu$. Then

$$\begin{aligned} \| A + B \|_\nu &= \nu[(A + B)x] \leq \nu(Ax) + \nu(Bx) \\ &\leq \| A \|_\nu \nu(x) + \| B \|_\nu \nu(x) = \| A \|_\nu + \| B \|_\nu . \end{aligned}$$

To show that $\| \cdot \|_\nu$ is consistent, let $\| AB \|_\nu = \nu(ABx)$, where $\nu(x) = 1$. Then

$$\begin{aligned} \| AB \|_\nu &= \nu[A(Bx)] \leq \| A \|_\nu \nu(Bx) \leq \| A \|_\nu \| B \|_\nu \nu(x) \\ &= \| A \|_\nu \| B \|_\nu . \end{aligned}$$

Finally if $\nu[(\xi)] = | \xi |$, it follows that for $y \in \mathbb{R}^n$

$$\| y \|_\nu = \max_{\substack{x \in \mathbb{R}^1 \\ \nu(x)=1}} \nu(yx) = \max_{|\xi|=1} \nu(y\xi) = \max_{|\xi|=1} | \xi | \nu(y) = \nu(y). \quad \blacksquare$$

The family of matrix norms $\| \cdot \|_\nu$ is said to be *subordinate* to the family of vector norms ν. The same proof shows that the function $\| \cdot \|$ defined on $\mathbb{R}^{m\times n}$ by (2.6) is a matrix norm. Such a norm is sometimes called an *operator norm* (subordinate to μ and ν). The most frequently occurring special case is when $m = n$ and $\mu = \nu$, in which case (2.6) defines a consistent norm on $\mathbb{R}^{n\times n}$. It should be noted that unless $\mu = \nu$, we cannot guarantee that the resulting norm is consistent.

The most natural family of vector norms to use in Theorem 2.8 is one of the p-norms ($p = 1, 2, \infty$). Since the p-norms satisfy (2.8), the subordinate matrix norms defined by the theorem give the same values as the

p-norms when applied to vectors. Hence there is no possibility of confusion in using the same symbol $\| \cdot \|_p$ for the matrix norm and the vector norm.

The 1-norm and the ∞-norm may be very easily computed.

THEOREM 2.9. Let $A \in \mathbb{R}^{m \times n}$. Then

$$\| A \|_1 = \max\left(\sum_{i=1}^{m} | \alpha_{ij} | : j = 1, 2, \ldots, n \right) \tag{2.9}$$

and

$$\| A \|_\infty = \max\left(\sum_{j=1}^{n} | \alpha_{ij} | : i = 1, 2, \ldots, m \right). \tag{2.10}$$

PROOF. The strategy of the proof is the same in both cases. Letting λ be the right-hand side of (2.9) or (2.10), as the case may be, we first show that for any vector x we have $\| Ax \|_p \leq \lambda \| x \|_p$ $(p = 1, \infty)$. This implies that

$$\| A \|_p \leq \lambda.$$

Next we find a particular vector x with $\| x \|_p = 1$ such that $\| Ax \|_p = \lambda$. This shows that

$$\| A \|_p \geq \lambda.$$

The two inequalities together imply that $\| A \|_p = \lambda$.

Specifically, for the 1-norm let $A = (a_1, a_2, \ldots, a_n)$ be partitioned by columns. Then $\lambda = \max\{\| a_j \|_1 : j = 1, 2, \ldots, n\}$ and

$$\begin{aligned} \| Ax \|_1 &= \| \xi_1 a_1 + \xi_2 a_2 + \cdots + \xi_n a_n \|_1 \\ &\leq | \xi_1 | \, \| a_1 \|_1 + | \xi_2 | \, \| a_2 \|_1 + \cdots + | \xi_n | \, \| a_n \|_1 \\ &\leq (| \xi_1 | + | \xi_2 | + \cdots + | \xi_n |) \max\{\| a_j \|_1\} = \lambda \| x \|_1. \end{aligned}$$

On the other hand if $\lambda = \| a_k \|_1$, then $\| e_k \|_1 = 1$ and

$$\| Ae_k \|_1 = \| a_k \|_1 = \lambda.$$

This establishes (2.9).

For the ∞-norm, $\lambda = \max\{\sum_{j=1}^{n} | \alpha_{ij} | : i = 1, 2, \ldots, n\}$. For any x we have

$$\begin{aligned} \| Ax \|_\infty &= \max_i \left\{ \left| \sum_j \alpha_{ij} \xi_j \right| \right\} \\ &\leq \max_i \left\{ \sum_j | \alpha_{ij} | \, | \xi_j | \right\} \\ &\leq \max_i \left\{ \sum_j | \alpha_{ij} | \right\} \max_j \{| \xi_j |\} = \lambda \| x \|_\infty. \end{aligned}$$

On the other hand if $\lambda = \sum_{j=1}^{n} | \alpha_{kj} |$ and $x = (\operatorname{sign}(\alpha_{k1}), \operatorname{sign}(\alpha_{k2}), \ldots, \operatorname{sign}(\alpha_{kn}))^{\mathrm{T}}$, then $\| x \|_\infty = 1$ and $\| Ax \|_\infty = \lambda$. ■

Thus the 1-norm of a matrix is equal to the maximum of the 1-norms of its columns. For this reason the matrix 1-norm is often called the *column sum norm*. Similarly the ∞-norm is called the *row sum norm*. Because these norms are so easy to compute they, along with the Frobenius norm, are often used in matrix algorithms.

There is no computationally convenient characterization of the matrix 2-norm (to anticipate a little, $\| A \|_2^2$ is the largest eigenvalue of $A^{\mathrm{T}}A$). However, the 2-norm does have a number of nice properties that make it useful for theoretical purposes. Some of these properties are contained in the following theorem.

THEOREM 2.10. Let $A \in \mathbb{R}^{m \times n}$. Then

1. $\| A \|_2 = \max_{\|x\|_2 = \|y\|_2 = 1} | y^{\mathrm{T}}Ax |$,
2. $\| A^{\mathrm{T}} \|_2 = \| A \|_2$,
3. $\| A^{\mathrm{T}}A \|_2 = \| A \|_2^2$.

PROOF. For part 1, let $\| x \|_2 = \| y \|_2 = 1$. Then by the Cauchy inequality,

$$| y^{\mathrm{T}}Ax | \leq \| y \|_2 \| Ax \|_2 \leq \| y \|_2 \| x \|_2 \| A \|_2 = \| A \|_2.$$

On the other hand let $\| x \|_2 = 1$ and $\| Ax \|_2 = \| A \|_2$. Set $y = Ax / \| Ax \|_2$. Then $\| y \|_2 = 1$ and

$$| y^{\mathrm{T}}Ax | = \frac{x^{\mathrm{T}}A^{\mathrm{T}}Ax}{\| Ax \|_2} = \frac{\| Ax \|_2^2}{\| Ax \|_2} = \| Ax \|_2 = \| A \|_2,$$

so that for this choice of x and y equality is attained.

For part 2, note that

$$\| A^{\mathrm{T}} \|_2 = \max_{\|x\|_2 = \|y\|_2 = 1} | y^{\mathrm{T}}A^{\mathrm{T}}x | = \max_{\|y\|_2 = \|x\|_2 = 1} | x^{\mathrm{T}}Ay | = \| A \|_2.$$

Finally, for part 3 we have

$$\| A^{\mathrm{T}}A \|_2 \leq \| A^{\mathrm{T}} \|_2 \| A \|_2 \leq \| A \|_2^2,$$

where the first inequality follows from the consistency of $\| \cdot \|_2$ and the second from part 2 of this theorem. For the reverse inequality, let $\| x \|_2 = 1$ and $\| Ax \|_2 = \| A \|_2$. Then by part 1 with $y = x$.

$$\| A^{\mathrm{T}}A \|_2 \geq | x^{\mathrm{T}}A^{\mathrm{T}}Ax | = \| Ax \|_2^2 = \| A \|_2^2. \quad \blacksquare$$

We conclude this section with a brief discussion of matrix limits. The results are analogous to the corresponding results for vectors. However, because matrices have a product defined among them, there are some additional results. We begin with the definition of the limit of a sequence of matrices.

DEFINITION 2.11. Let $A, A_2, A_3, \ldots \in \mathbb{R}^{m\times n}$. Then $\lim_{k\to\infty} A_k = A$ if

$$\lim_{k\to\infty} \alpha_{ij}^{(k)} = \alpha_{ij} \qquad (i = 1, 2, \ldots, m; \quad j = 1, 2, \ldots, n).$$

As was mentioned at the beginning of this section, the notion of limit of a sequence of matrices can be characterized in terms of any matrix norm ν. Namely

$$\lim_{k\to\infty} A_k = A \Leftrightarrow \lim_{k\to\infty} \nu(A - A_k) = 0.$$

Many of the algebraic properties of limits of scalars carry over to limits of matrices. A few are given in the following theorem, in which it is assumed that the dimensions of the matrices in the statements are consistent with the operators involved.

THEOREM 2.12. Let $\lim A_k = A$ and $\lim B_k = B$. Then

1. $\lim(A_k + B_k) = A + B$,
2. $\lim A_k B_k = AB$,
3. if A is nonsingular, then, for all sufficiently large k, A_k is nonsingular and $\lim A_k^{-1} = A^{-1}$.

The proofs of parts 1 and 2 are immediate from the definition of limit and are left as exercises. The proof of part 3 must be deferred to the end of the next section.

It should be noted that other properties of the limit are immediate consequences of this theorem. For example, since $\lambda A = \operatorname{diag}(\lambda, \lambda, \ldots, \lambda)A$, it follows from 2.12.2 that $\lim \lambda A_k = \lambda \lim A_k$. Other properties of the limit are given in the exercises.

EXERCISES

1. Verify that the function ν defined by (2.5) is a vector norm.

2. Establish the inequality (2.7) from the definition (2.6).

3. Prove that if $A = (a_1, a_2, \ldots, a_n)$ is partitioned by columns then $\| A \|_F^2 = \| a_1 \|_2^2 + \| a_2 \|_2^2 + \cdots + \| a_n \|_2^2$.

4. Show that $\| AB \|_F \leq \| A \|_2 \| B \|_F$ and $\| AB \|_F \leq \| A \|_F \| B \|_2$.

5. Show that $\| \cdot \|_2$ and $\| \cdot \|_F$ are not the same by calculating $\| I_n \|_2$ and $\| I_n \|_F$ for $n > 1$.

6. Show directly that for $x \in \mathbb{R}^n$, $\| x \|_2 = \| x^T \|_2$.

7. Let $\| \cdot \|_\nu : \mathbb{R}^{n \times n} \to \mathbb{R}$ be the operator norm subordinate to the vector norm $\nu : \mathbb{R}^n \to \mathbb{R}$. Show that $\| I \|_\nu = 1$.

8. Let $\nu : \mathbb{R}^{n \times n} \to \mathbb{R}$ be defined by

$$\nu(A) = n\{\max | \alpha_{ij} | : i, j = 1, 2, \ldots, n\}.$$

Show that ν is a consistent matrix norm.

9. Show that the function $\nu : \bigcup_{m,n=1}^{\infty} \mathbb{R}^{m \times n} \to \mathbb{R}$ defined by

$$\nu(A) = \sum_{i=1}^{m} \sum_{j=1}^{n} | \alpha_{ij} |, \qquad A \in \mathbb{R}^{m \times n},$$

is a consistent family of matrix norms.

10. Let $\nu : \mathbb{R}^{m \times n} \to \mathbb{R}$ be a matrix norm. Let $B \in \mathbb{R}^{m \times m}$ and $C \in \mathbb{R}^{n \times n}$ be nonsingular. Show that the function $\mu : \mathbb{R}^{m \times n} \to \mathbb{R}^n$ defined by $\mu(A) = \nu(BAC)$ is a matrix norm.

11. Let $\nu : \mathbb{R}^n \to \mathbb{R}$ be a norm and let $\| \cdot \|_\nu$ be the operator norm on $\mathbb{R}^{n \times n}$ subordinate to ν. Show that if A is nonsingular, then

$$\| A^{-1} \|_\nu^{-1} = \min_{\nu(x)=1} \nu(Ax).$$

12. Let ν be a vector norm on $\mathbb{R}^n$, let $B \in \mathbb{R}^{n \times n}$ be nonsingular, and let μ the norm defined by $\mu(x) = \nu(Bx)$. Let $\| \cdot \|_\nu$ and $\| \cdot \|_\mu$ denote the operator norms subordinate to ν and μ. Show that $\| A \|_\mu = \| BAB^{-1} \|_\nu$.

13. For $p, q = 1, 2, \infty, F$ establish the following table of constants τ_{pq} such that for all $A \in \mathbb{R}^{n\times n}$, $\| A \|_p \leq \tau_{pq} \| A \|_q$.

p \ q	1	2	∞	F
1	1	$n^{1/2}$	n	$n^{1/2}$
2	$n^{1/2}$	1	$n^{1/2}$	1
∞	n	$n^{1/2}$	1	$n^{1/2}$
F	$n^{1/2}$	$n^{1/2}$	$n^{1/2}$	1

14. Let $\nu : \mathbb{R}^{n\times n} \to \mathbb{R}$ be a matrix norm. Show that there is a constant λ such that the function μ defined by $\mu(A) = \lambda\nu(A)$ is a consistent matrix norm.

15. Let $A \in \mathbb{R}^{n\times n}$ and μ and ν be norms on $\mathbb{R}^n$. Show that there exists a vector x with $\nu(x) = 1$ such that for all y with $\nu(y) = 1$

$$\mu(Ax) \leq \mu(Ay).$$

16. Prove that if $\lim \lambda_k = \lambda$ and $\lim A_k = A$, then $\lim \lambda_k A_k = \lambda A$.

17. Prove that if A is nonsingular and $\lim AB_k = C$, then $\lim B_k$ exists and $A \lim B_k = C$. Show by example that the hypothesis of nonsingularity of A cannot be removed.

18. The sequence $\langle A_k \rangle$ is a Cauchy sequence with respect to the matrix norm ν if for every $\varepsilon > 0$ there is an integer N such that if $m, n \geq N$, then $\nu(A_m - A_n) < \varepsilon$. Show that if μ and ν are matrix norms and $\langle A_k \rangle$ is a Cauchy sequence with respect to ν, then $\langle A_k \rangle$ is a Cauchy sequence with respect to μ. This justifies speaking simply of Cauchy sequences of matrices.

19. Show that $\langle A_k \rangle$ is a Cauchy sequence if and only if for each i, j the sequence $\langle \alpha_{ij}^{(k)} \rangle$ is a Cauchy sequence. Conclude from the fact that Cauchy sequences of scalars converge, that Cauchy sequences of matrices converge.

20. The infinite series $\sum_{k=0}^{\infty} A_k$ of matrices converges if the sequence of partial sums $S_k = \sum_{i=1}^{k} A_i$ converges. Prove that if ν is a matrix

norm and $\sum_{k=0}^{\infty} \alpha_k$ is a convergent series of scalars such that $\nu(A_k) \leq \alpha_k$ ($k = 0, 1, 2, \ldots$), then $\sum_{k=0}^{\infty} A_k$ converges.

21. Let $\phi(\tau) = \sum_{k=0}^{\infty} \gamma_k \tau^k$ be a power series with radius of convergence ϱ. Let $A \in \mathbb{R}^{n\times n}$. Show that if, for any consistent matrix norm ν, $\nu(A) < \varrho$, then the series $\sum_{k=0}^{\infty} \gamma_k A^k$ converges. [The limit of the series is written $\phi(A)$.]

22. Let $A \in \mathbb{R}^{n\times n}$ and define $e^A = \sum_{k=0}^{\infty} A^k/k!$. Prove that if $AB = BA$, then $e^{A+B} = e^A e^B$.

NOTES AND REFERENCES

The concept of norm has long been used in functional analysis; its application to matrix theory is more recent. Details and further references are given by Householder (1964, Chapter II), who has done much to popularize the use of norms in analyzing matrix process. Mention should also be made of Bauer's "Theory of Norms" (1967), which unfortunately is available only as a report.

In many papers matrix norms are used only for square matrices of a fixed order. Since the product is always defined among such matrices, it is customary to add consistency as a property of a matrix norm, a usage which conflicts with our definition.

The notion of a family of norms appears to be new. It is merely a terminological device for saying in a precise way what everyone knows anyway.

Theorems 1.9 and 2.7 require compactness arguments for their proofs, but are not otherwise very difficult. A proof of Theorem 1.9 may be found in the book of Issacson and Keller (1966).

3. INVERSES OF PERTURBED MATRICES

In this section we shall consider the following problem. Let $A \in \mathbb{R}^{n\times n}$ be nonsingular and let E be an $n\times n$ matrix that is presumed to be small. How small must E be so that the perturbed matrix $A + E$ is also nonsingular, and by how much does $(A + E)^{-1}$ differ from A^{-1}? In answering these questions we shall use the theory of norms developed in the last two sections to make precise the phrases "how small" and "how much."

One application of our results is to the computation of matrix inverses. In Section 3.5 we indicated that the computed inverse of a matrix A would often be near the exact inverse of a slightly perturbed matrix

$A + E$. However, this result does not guarantee the accuracy of the computed inverse, for A^{-1} and $(A + E)^{-1}$ may differ greatly. In the terminology of Section 2.3 such a matrix would be called *ill conditioned with respect to inversion.* Our analysis not only gives conditions under which A is ill conditioned, but it also associates with A a *condition number* that measures the degree of its ill-conditioning.

In order to speak rigorously about the sizes of errors involving vectors and matrices, we introduce the following generalizations of absolute and relative error.

DEFINITION 3.1. Let $A, B \in \mathbb{R}^{m\times n}$ with B regarded as an approximation to A. The *residual* of B is the matrix

$$A - B.$$

If $\nu : \mathbb{R}^{m\times n} \to \mathbb{R}$ is a norm, the *error in B with respect to ν* is the number

$$\nu(A - B).$$

If $A \neq 0$, the *relative error in B with respect to ν* is the number

$$\frac{\nu(A - B)}{\nu(A)}.$$

Where it will cause no confusion, we drop the phrase "with respect to ν" and refer simply to the error or the relative error. The notions of residual, error, and relative error are of course defined for n-vectors regarded as $n\times 1$ matrices.

EXAMPLE 3.2. On a given computer suppose that the rounded value of a number β is $\text{fl}(\beta) = \beta(1 + \varrho)$, where $|\varrho| \leq 10^{-t}$. Let $A \in \mathbb{R}^{m\times n}$ and let α be the magnitude of the largest element of A. Let B be obtained from A by rounding the elements of A. Then $B = A + E$, where

$$\sum_{j=1}^{n} |\varepsilon_{ij}| \leq \sum_{j=1}^{n} |\varrho_{ij}|\,|\alpha_{ij}| \leq 10^{-t}\,\|A\|_\infty.$$

Hence

$$\|E\|_\infty \leq \|A\|_\infty\, 10^{-t},$$

and

$$\frac{\|A - B\|_\infty}{\|A\|_\infty} \leq 10^{-t};$$

that is, the relative error with respect to the ∞-norm in a matrix rounded to t figures is not greater than 10^{-t}.

The definition of error and relative error uses the notion of a norm to combine information about a set of numbers into a single number. It is not surprising that we should lose information in the process. For example, a small component of a vector with a low relative error may have a high relative error. Moreover, a change of norm may change the relative error markedly.

EXAMPLE 3.3. Let

$$A = \begin{pmatrix} 1.000 & .0050 \\ .0050 & .0001 \end{pmatrix},$$

$$B = \begin{pmatrix} 1.0001 & .0051 \\ .0051 & .0002 \end{pmatrix},$$

and

$$C = \begin{pmatrix} 1.0001 & .0050001 \\ .0050001 & .00010001 \end{pmatrix}.$$

Obviously C is a fairly good approximation to A, while B has a relative error of unity in its (2, 2)-element. In spite of this both matrices have a relative error of about 10^{-4} with respect to the ∞-norm. On the other hand, if we introduce the norm ν defined by

$$\nu(A) = \| \operatorname{diag}(1, 10^2)\, A \operatorname{diag}(1, 10^2) \|_\infty,$$

then

$$\frac{\nu(A - B)}{\nu(A)} \cong 1$$

while

$$\frac{\nu(A - C)}{\nu(A)} \cong 10^{-4}.$$

Thus the norm ν exposes the inaccuracy in the (2, 2)-element of B.

Example 3.3 indicates the necessity of choosing a norm to fit the problem, or alternatively, of scaling the problem to fit the norm. In applications to linear systems, the problem of choosing a norm is closely related to the scaling problems we discussed at the end of Section 3.5. Fortunately, in many cases the choice of norm is perfectly evident, as it was in Example 3.3.

We now turn to the question of the accuracy of perturbed inverses that was raised at the beginning of this section. We shall answer the question by exhibiting a bound on the relative error in $(A + E)^{-1}$. For the rest of this and the next section all matrices will be assumed square of order n. The symbol $\| \cdot \|$ will denote both a consistent matrix norm on $\mathbb{R}^{n\times n}$ satisfying

$$\| I \| = 1 \tag{3.1}$$

and a vector norm on $\mathbb{R}^n$ that is consistent with the matrix norm $\| \cdot \|$; that is, $\| Ax \| \leq \| A \| \| x \|$. It should be noted that the results of this section depend critically on (3.1). In particular, they hold for the 1-, 2-, and ∞-norms, but not for the Frobenius norm.

We begin by considering perturbations of the identity matrix.

THEOREM 3.4. If $\| P \| < 1$, then $I - P$ is nonsingular, and

$$\| (I - P)^{-1} \| \leq (1 - \| P \|)^{-1}. \tag{3.2}$$

PROOF. Let $x \neq 0$. Then

$$\begin{aligned} \| (I - P)x \| = \| x - Px \| &\geq \| x \| - \| Px \| \geq \| x \| - \| P \| \| x \| \\ &\geq (1 - \| P \|) \| x \| > 0, \end{aligned}$$

since $1 - \| P \| > 0$. Hence if $x \neq 0$, then $(I - P)x \neq 0$, and $I - P$ is nonsingular.

Now from the equation

$$(I - P)(I - P)^{-1} = I,$$

it follows that

$$(I - P)^{-1} = I + P(I - P)^{-1}. \tag{3.3}$$

Hence

$$\| (I - P)^{-1} \| \leq \| I \| + \| P \| \| (I - P)^{-1} \|. \tag{3.4}$$

Since $\| I \| = 1$, the inequalities (3.4) and (3.2) are equivalent. ■

For the identity matrix, Theorem 3.4 answers the first question raised at the beginning of this section; namely how small must P be so that $I - P$

is nonsingular? The answers to the second question of the proximity of $(I - P)^{-1}$ to $I^{-1} = I$ is contained in the following corollary.

COROLLARY 3.5. If $\| P \| < 1$, then

$$\| I - (I - P)^{-1} \| \leq \frac{\| P \|}{1 - \| P \|}. \tag{3.5}$$

PROOF. From (3.3) we have

$$I - (I - P)^{-1} = -P(I - P)^{-1}.$$

Hence

$$\| I - (I - P)^{-1} \| \leq \| P \| \, \| (I - P)^{-1} \| \leq \| P \| \, (1 - \| P \|)^{-1}. \quad \blacksquare$$

For very small P, the term $(1 - \| P \|)^{-1}$ in (3.5) does not differ significantly from unity and the right-hand side of (3.5) becomes effectively $\| P \|$. In other words, the error in the inverse of a slightly perturbed identity matrix is roughly of the same order as the perturbation.

Corollary 3.5 can be used to estimate the error in the inverse of a perturbed matrix. Actually, a slightly more elaborate result will prove useful later.

THEOREM 3.6. Let A be nonsingular and let $\| A^{-1}E \| < 1$. Then $A + E$ is nonsingular and $(A + E)^{-1}$ can be written in the form

$$(A + E)^{-1} = (I + F)A^{-1}, \tag{3.6}$$

where

$$\| F \| \leq \frac{\| A^{-1}E \|}{1 - \| A^{-1}E \|}. \tag{3.7}$$

Moreover

$$\frac{\| A^{-1} - (A + E)^{-1} \|}{\| A^{-1} \|} \leq \frac{\| A^{-1}E \|}{1 - \| A^{-1}E \|}. \tag{3.8}$$

PROOF. Since

$$A + E = A(I + A^{-1}E)$$

and $\| A^{-1}E \| < 1$, it follows from Corollary 3.4 that $(I + A^{-1}E)$, and hence $A + E$, is nonsingular. Moreover

$$(A + E)^{-1} = (I + A^{-1}E)^{-1}A^{-1}.$$

If we define F by $I + F = (I + A^{-1}E)^{-1}$, then this equation is equivalent to (3.6), and by Theorem 3.5 (with $P = -A^{-1}E$) we have

$$\| F \| = \| I - (I + A^{-1}E)^{-1} \| \leq \frac{\| A^{-1}E \|}{1 - \| A^{-1}E \|}.$$

Finally, from (3.6),

$$A^{-1} - (A + E)^{-1} = -FA^{-1},$$

so that

$$\| A^{-1} - (A + E)^{-1} \| \leq \| F \| \| A^{-1} \| \leq \frac{\| A^{-1} \| \| A^{-1}E \|}{1 - \| A^{-1}E \|},$$

which is equivalent to (3.8). ■

Recall that if the approximation β to α has a relative residual ϱ, then $\beta = (1 - \varrho)\alpha$. Equation (3.6) is quite analogous, and it is not out of place to call the matrix $-F$ the relative residual of $(A + E)^{-1}$ as an approximation to A^{-1}. Accordingly, we should expect that $\| F \|$ would be the relative error in $(A + F)^{-1}$; and indeed, comparing the inequalities (3.7) and (3.8), we see that $\| F \|$ and $\| A^{-1} - (A + E)^{-1} \| / \| A^{-1} \|$, which is the relative error in $(A + E)^{-1}$, have the same upper bounds.

If $\| A^{-1}E \|$ is significantly less than unity, then the term $(1 - \| A^{-1}E \|)^{-1}$ in (3.8) is essentially unity and the relative error in $(A + E)^{-1}$ is approximately bounded by $\| A^{-1}E \|$. This suggests that if A^{-1} is very large, the error E in A may be magnified considerably in the inverse of $A + E$. The following trivial corollary of Theorem 3.6 makes these considerations precise.

COROLLARY 3.7. In Theorem 3.6, let

$$\varkappa(A) = \| A \| \| A^{-1} \|. \tag{3.9}$$

If $\| A^{-1} \| \| E \| < 1$, then

$$\| F \| \leq \frac{\varkappa(A) \dfrac{\| E \|}{\| A \|}}{1 - \varkappa(A) \dfrac{\| E \|}{\| A \|}} \tag{3.10}$$

and

$$\frac{\| A^{-1} - (A + E)^{-1} \|}{\| A^{-1} \|} \leq \frac{\varkappa(A) \dfrac{\| E \|}{\| A \|}}{1 - \varkappa(A) \dfrac{\| E \|}{\| A \|}}. \tag{3.11}$$

PROOF. We have

$$\| A^{-1}E \| \leq \| A^{-1} \| \| E \| = \| A \| \| A^{-1} \| \frac{\| E \|}{\| A \|} = \varkappa(A) \frac{\| E \|}{\| A \|}.$$

If this upper bound is substituted for $\| A^{-1}E \|$ in (3.7) and (3.8), there result the inequalities (3.10) and (3.11). ■

The left-hand side of (3.11) is the relative error in $(A + E)^{-1}$. If E is sufficiently small, the right-hand side is effectively $\varkappa(A) \| E \|/\| A \|$. Since $\| E \|/\| A \|$ is the relative error in $A + E$, the inequality (3.11) states that *the relative error in $A + E$ may be magnified by as much as $\varkappa(A)$ in passing to $(A + E)^{-1}$*. For this reason, $\varkappa(A)$ is called *the condition number of A with respect to inversion* (and with respect to the norm $\| \cdot \|$). If $\varkappa(A)$ is large, then the inverse of A is sensitive to small perturbations in A, and the problem of computing the inverse is ill conditioned. Incidentally, note that $\varkappa(A)$ is indeed a magnification constant, for

$$1 \leq \| I \| = \| AA^{-1} \| \leq \| A \| \| A^{-1} \| = \varkappa(A).$$

EXAMPLE 3.8. Suppose that we are given a t-digit approximation B to A, in the sense that $\| A - B \|/\| A \| \cong 10^{-t}$. If $\varkappa(A) = 10^p$, then the relative error in B^{-1} will be approximately 10^{p-t}, and if the elements of A^{-1} are about equal in magnitude, they will be accurate to about $t - p$ digits. In other words, if $\varkappa(A) = 10^p$, we may expect to lose roughly p significant figures in inverting an approximation to A.

It should be stressed that Corollary 3.7 is weaker than Theorem 3.6, since it is based on the inequality

$$\| A^{-1}E \| \leq \| A^{-1} \| \| E \|. \tag{3.12}$$

For random perturbations E, the inequality (3.12) will usually be almost an equality. However, if E has special properties, then Theorem 3.6 may give much sharper bounds than Corollary 3.7. For example if $E = \varepsilon A$, where $| \varepsilon | < 1$, then Theorem 3.6 gives

$$\| F \| \leq \frac{\varepsilon}{1 - \varepsilon} \tag{3.13}$$

while Corollary 3.7 gives

$$\| F \| \leq \frac{\varkappa(A)\varepsilon}{1 - \varkappa(A)\varepsilon}.$$

Obviously (3.13) is the better bound.

We conclude this section by completing the proof of Theorem 2.12.

THEOREM 2.12.3. If A is nonsingular and $\lim A_k = A$, then, for all sufficiently large k, A_k is nonsingular and $\lim A_k^{-1} = A^{-1}$.

PROOF. Let $E_k = A_k - A$. Then $\lim \| E_k \| = 0$. Hence $\lim \| A^{-1}E_k \| = 0$. Thus for all sufficiently large k, $\| A^{-1}E_k \| < 1$, and by Theorem 3.6, $A_k = A + E_k$ is nonsingular. Moreover, $A_k^{-1} = (I + F_k)A^{-1}$, where $\| F_k \| \leq \| A^{-1}E_k \|/(1 - \| A^{-1}E_k \|)$. It follows that $\lim \| F_k \| = 0$ and hence

$$\lim A_k^{-1} = \lim(I + F_k)A^{-1} = A^{-1}. \quad \blacksquare$$

EXERCISES

1. Let A and B be of order n with A nonsingular, and let B be regarded as an approximation to A. Define the *left relative residual* of B as the matrix $F = (A - B)A^{-1}$ and the *right relative residual* of B as the matrix $G = A^{-1}(A - B)$. Show that $B = (I - F)A$ and $B = A(I - G)$.

2. Let $A, E \in \mathbb{R}^{n \times n}$ with $\| EA^{-1} \| < 1$. Prove that $A + E$ is nonsingular and that there is a matrix G satisfying $\| G \| \leq \| EA^{-1} \|/(1 - \| EA^{-1} \|)$ such that $(A + E)^{-1} = A^{-1}(I + G)$.

3. Let $A, B \in \mathbb{R}^{m \times n}$ and let A have linearly independent columns. Show that it is possible to define a *left relative residual* F of B as an approximation to A so that $B = (I - F)A$. [*Hint*: A has a left inverse $(A^{\mathrm{T}}A)^{-1}A^{\mathrm{T}}$ (cf. Exercise 1.6.8).] What can be said if A has linearly independent rows?

4. Prove that if $\| P \| < 1$, then the *Neumann series* $I + P + P^2 + \cdots$ converges to $(I - P)^{-1}$. Use this to give an alternate proof of Theorem 3.4 and Corollary 3.5.

5. Show that if P is sufficiently small, then $(I - P)^{-1} \cong (I + P)$ in the sense that $\| (I + P) - (I - P)^{-1} \| \leq 2 \| P \|^2$. Conclude that the bound (3.5) is sharp.

6. Prove that the matrix F of Theorem 3.6 is given by the infinite series

$$F = -A^{-1}E[I - A^{-1}E + (A^{-1}E)^2 - \cdots].$$

Hence derive the bound (3.7) and conclude that it is sharp.

7. Let

$$A = \begin{pmatrix} 1 & 1 \\ 1 & 1.01 \end{pmatrix}.$$

For

$$E = \begin{pmatrix} .0001 & -.0001 \\ -.0001 & .0001 \end{pmatrix}$$

compute the relative error in $(A + E)^{-1}$. Also compute the error predicted by (3.8) and (3.11). Do the same for

$$E = \begin{pmatrix} .0001 & .0001 \\ .0001 & .0001 \end{pmatrix}.$$

Explain any disparity in the results.

8. Explain under what conditions the inequality (3.12) may be expected to be nearly an equality when the elements of E are random.

NOTES AND REFERENCES

The perturbation theory of this section is classical and goes through for bounded operators in a Banach space. The theorems are usually established by means of the Neumann series

$$(I - P)^{-1} = I + P + P^2 + P^3 + \cdots.$$

The condition number $\| A \|_2 \| A^{-1} \|_2$ first appears in the rounding-error analysis of von Neumann and Goldstein (1947) in a somewhat disguised form. Turing (1948) seems to have been the first to use the term condition number, which he defines in terms of matrix norms as is done in the text.

4. THE ACCURACY OF SOLUTIONS OF LINEAR SYSTEMS

In Section 3.5, we saw that the computed solution $\bar{x}$ of the nonsingular system of linear equations

$$Ax = b \tag{4.1}$$

satisfies the equation

$$(A + E)\bar{x} = b, \tag{4.2}$$

where E is in some sense small. However, the computed solution $\bar{x}$ will be accurate only if the problem of solving (4.1) is well conditioned in the sense of Section 2.1. In this section we shall derive bounds for the error in $\bar{x}$ as a function of E. As in the last section, the bounds will depend on the condition number $\varkappa(A)$ defined by (3.9).

Since bounds on the matrix E in (4.2) are known, it is possible to combine them with the bounds of this section to bound the error in $\bar{x}$. Given a condition number for A, this bound can be computed without computing $\bar{x}$. Such an error bound is called an *a priori* bound. It is not unreasonable to suppose that if we compute $\bar{x}$, the additional information obtained in the course of the computations can be used to obtain a better bound. Such a bound that depends on the approximate solution itself is called an *a posteriori* bound. We shall discuss *a posteriori* bounds for the errors in approximate solutions of (4.1); however, the results are, in general, disappointing.

The first problem we shall consider is that of assessing the effects of a perturbation in the vector b on the solution of (4.1). So far as rounding-error analysis is concerned, this is not a very important problem, since the effects of rounding error can be accounted for by perturbations in A alone. In practice, however, the components of b may be contaminated with errors—perhaps they themselves have been computed with rounding error—and it is important to have a bound for the error induced in the solution.

Specifically suppose that $\bar{b}$ is an approximation to b and $\bar{x}$ is the solution of the system

$$A\bar{x} = \bar{b}.$$

Then

$$x - \bar{x} = A^{-1}(b - \bar{b}),$$

and hence

$$\| x - \bar{x} \| \leq \| A^{-1} \| \, \| b - \bar{b} \|. \tag{4.3}$$

The inequality (4.3) already suggests that if $\| A^{-1} \|$ is large, then the error in $\bar{x}$ may be significantly greater than the error in $\bar{b}$. We can make this more precise by computing the relative error in the solution. From (4.1) it follows that

$$\| x \| \geq \frac{\| b \|}{\| A \|}. \tag{4.4}$$

Hence

$$\frac{\| x - \bar{x} \|}{\| x \|} \leq \| A \| \| A^{-1} \| \frac{\| b - \bar{b} \|}{\| b \|}.$$

If, as in Section 3, we define $\varkappa(A) = \| A \| \| A^{-1} \|$, then our results may be summarized in the following theorem.

THEOREM 4.1. Let A be nonsingular, $Ax = b \neq 0$, and $A\bar{x} = \bar{b}$. Then

$$\frac{\| x - \bar{x} \|}{\| x \|} \leq \varkappa(A) \frac{\| b - \bar{b} \|}{\| b \|},$$

where $\varkappa(A) = \| A \| \| A^{-1} \|$.

This theorem requires two comments. In the first place, the condition number $\varkappa(A)$ plays much the same role as it did in Corollary 3.7; the relative error in b may be magnified by as much as a factor of $\varkappa(A)$ in the solution. The second comment is that when A is ill conditioned the result is frequently unrealistic, since the lower bound (4.4) on $\| x \|$ is often a severe underestimate. In fact for most right-hand sides b,

$$\| x \| = \| A^{-1}b \| = \gamma \| A^{-1} \| \| b \|, \tag{4.5}$$

where $\gamma < 1$ is a constant near unity. If this estimate is used in place of (4.4), the result is

$$\frac{\| x - \bar{x} \|}{\| x \|} \leq \gamma^{-1} \frac{\| b - \bar{b} \|}{\| b \|},$$

so that the relative error $\bar{x}$ is hardly greater than the relative error in $\bar{b}$. If x and b satisfy (4.5) with γ near unity, then they are said to *reflect the condition of the matrix A*. Our observation is, then, that a problem that reflects the condition of A is insensitive to perturbations in b, even if $\varkappa(A)$ is large.

Either Theorem 3.6 or Corollary 3.7 may be used to estimate the error in the solution of (4.2), with Corollary 3.7 giving a weaker result. We state both results.

THEOREM 4.2. Let A be nonsingular, let $x \neq 0$ satisfy (4.1), and let $\bar{x}$ satisfy (4.2). If $\| A^{-1}E \| < 1$, then

$$\frac{\| x - \bar{x} \|}{\| x \|} \leq \frac{\| A^{-1}E \|}{1 - \| A^{-1}E \|}. \tag{4.6}$$

If $\| A^{-1} \| \| E \| < 1$, then

$$\frac{\| x - \bar{x} \|}{\| x \|} \leq \frac{\varkappa(A) \dfrac{\| E \|}{\| A \|}}{1 - \varkappa(A) \dfrac{\| E \|}{\| A \|}}. \tag{4.7}$$

PROOF. Since $\| A^{-1}E \| < 1$, by Theorem 3.6 the matrix $A + E$ is nonsingular and $(A + E)^{-1} = (I + F)A^{-1}$, where F satisfies (3.7). Then

$$\bar{x} = (A + E)^{-1}b = (I + F)A^{-1}b = (I + F)x,$$

or

$$x - \bar{x} = -Fx.$$

Hence

$$\frac{\| x - \bar{x} \|}{\| x \|} \leq \| F \|.$$

If (3.7) is used to estimate F, then (4.6) results. If (3.10) is used, then (4.7) results. ∎

It is interesting to note that the bounds are independent of the right-hand side b. The number $\varkappa(A)$ again serves as a condition number in (4.7), telling how much the relative error in A is magnified in the solution. Unless $\| A^{-1}E \| \ll \| A^{-1} \| \| E \|$, which is unlikely for random perturbations, the bounds (4.6) and (4.7) will give comparable results. Moreover, these bounds will be sharp if $\| Fx \| \cong \| F \| \| x \|$, which will usually be true. This contrasts with the bound in Theorem 4.1 which is likely to be an overestimate when A is ill conditioned.

These bounds may be combined with the results of Section 3.5 to yield rigorous bounds on the accuracy of computed solutions of linear equations. For example, provided there is no growth in the elimination process, Theorem 3.5.3 may be interpreted as saying that the computed solution $\bar{x}$ of (4.1) satisfies

$$(A + H)\bar{x} = b, \tag{4.8}$$

where

$$\| H \|_\infty \leq \phi(n) \| A \|_\infty 10^{-t}. \tag{4.9}$$

Here $\phi(n)$ is a function of n whose form depends on the arithmetic details of the computation, but which in any event is not very big. If

$$\varkappa(A)\phi(n)10^{-t} < 1,$$

then the condition $\| A^{-1} \|_\infty \| H \|_\infty < 1$ is satisfied, and Theorem 4.2 applies to show that

$$\frac{\| x - \bar{x} \|_\infty}{\| x \|_\infty} \leq \frac{\varkappa(A)\phi(n)10^{-t}}{1 - \varkappa(A)\phi(n)10^{-t}}.$$

Recalling Example 3.8, we see that if $\phi(n)$ is not too large and $\varkappa(A) = 10^p$, then we may expect a solution computed in t-digit arithmetic to be accurate to about $t - p$ significant figures, at least when there is no severe imbalance in the sizes of the elements.

We turn now to the problem of calculating *a posteriori* bounds for an approximate solution $\bar{x}$ of (4.1). A natural check on $\bar{x}$ is to substitute it into Equation (4.1) and see to what extent it fails to satisfy the equation. This amounts to looking at the size of the *residual vector*

$$r = b - A\bar{x}.$$

If r is zero, then $\bar{x}$ is an exact solution of (4.1). If r is small, it might be expected that $\bar{x}$ is near a solution of (4.1). The following theorem states precisely in what sense this is true.

THEOREM 4.3. Let A be nonsingular and $Ax = b \neq 0$. Let $\bar{x}$ be given and set $r = b - A\bar{x}$. Then

$$\frac{\| x - \bar{x} \|}{\| x \|} \leq \varkappa(A) \frac{\| r \|}{\| b \|}. \tag{4.10}$$

PROOF. We have

$$A^{-1}r = A^{-1}b - \bar{x} = x - \bar{x}.$$

Hence

$$\| x - \bar{x} \| \leq \| A^{-1} \| \, \| r \|. \tag{4.11}$$

However,

$$\| x \| \geq \frac{\| b \|}{\| A \|}. \tag{4.12}$$

If (4.11) is divided by (4.12), the result is (4.10). ∎

As a practical means for assessing the accuracy of an approximate solution, Theorem 4.3 is somewhat disappointing; for it asserts that the accuracy of the solution depends not only on the size of the residual but also on the condition number of the matrix. If A is ill conditioned, even a very small residual cannot guarantee an accurate solution. Worse yet, it is possible for an accurate solution to have a large residual. These points are illustrated in the following example.

EXAMPLE 4.4. Let

$$A = \begin{pmatrix} 1.000 & 1.001 \\ 1.000 & 1.000 \end{pmatrix}$$

and $b = (2.001, 2.000)^{\mathrm{T}}$. Then the exact solution of the equation $Ax = b$ is $x = (1, 1)^{\mathrm{T}}$. However, the vector $\bar{x} = (2, 0)^{\mathrm{T}}$, which is in no sense near x, has the very small residual vector $r = (10^{-3}, 0)^{\mathrm{T}}$.

On the other hand, the solution of $Ax = b = (1, 0)^{\mathrm{T}}$ is $x = (-1000, 1000)^{\mathrm{T}}$, and the vector $\bar{x} = (-1001, 1000)^{\mathrm{T}}$ is very near x. However, the residual vector of $\bar{x}$ is $r = (0, -1)$ which is as large as b.

Further insight into the nature of the residual may be gained by considering the residual of a rounded solution $\tilde{x}$ of (4.1). From Example 3.2 we know that $\tilde{x}$ may be written in the form $\tilde{x} = x + e$, where $\| e \|_\infty \leq \| x \|_\infty 10^{-t}$. Hence

$$r = b - A\tilde{x} = b - Ax - Ae = -Ae,$$

and $\| r \|_\infty \leq \| A \|_\infty \| e \|_\infty \leq \| A \|_\infty \| x \|_\infty 10^{-t}$. Thus

$$\frac{\| r \|_\infty}{\| A \|_\infty} \leq \| x \|_\infty 10^{-t}. \tag{4.13}$$

Now if A is ill conditioned and x reflects the ill-conditioning of A, so that $\| x \|_\infty$ is large, then r will be large. In other words, *the rounded solution of an equation may have a large residual.* This phenomenon is illustrated by the second part of Example 4.4.

It is also instructive to compare the residual of the rounded solution $\tilde{x}$ of (4.1) with the residual of the solution $\bar{x}$ computed by, say, Gaussian elimination. We know that $\bar{x}$ satisfies (4.8), where H satisfies (4.9). Hence $r = b - A\bar{x} = H\bar{x}$, or

$$\frac{\| r \|_\infty}{\| A \|_\infty} \leq \phi(n)10^{-t} \| \bar{x} \|_\infty. \tag{4.14}$$

If $\phi(n)$ is not too large, the bound (4.14) is comparable to (4.13). Thus *the residual of the computed solution will be roughly of the same size as the residual of the exact solution rounded to t figures.* Note that if A is ill conditioned and $\bar{x}$ does *not* reflect the ill-conditioning of A, then $\| \bar{x} \|_\infty$ will be small, say of order unity, and $\| r \|_\infty$ will be of order 10^{-t}. This can happen even when $\bar{x}$ is inaccurate in all its significant figures.

So far all our bounds have the defect that they depend on the condition number of the matrix A and hence on A^{-1}. Since the calculation of inverses is relatively expensive, it is natural to ask how one can detect ill-conditioning without computing A^{-1}.

One sign of ill-conditioning is the emergence of a small pivot in the reduction of A. For example, the Crout reduction decomposes A into the product LU, where the diagonal elements of L are the pivots. If one of these diagonal elements is small, then L^{-1} will be large. In all probability $A^{-1} = U^{-1}L^{-1}$ will also be large, and hence A will be ill conditioned. For most matrices this is a fairly reliable indicator of ill-conditioning, especially if complete pivoting is used. However, there do exist ill-conditioned matrices for which no small pivots emerge.

A second sign of ill-conditioning is the emergence of a large solution. Suppose, for example, that $\| A \| = \| b \| = 1$. Then if $Ax = b$, we have

$$\| x \| \leq \| A^{-1} \| \| b \| = \| A^{-1} \| = \varkappa(A),$$

so that if $\| x \|$ is large, $\varkappa(A)$ must perforce be large. Unfortunately, an ill-conditioned system may have a small solution. Note that such a solution must also have a very small residual.

It would seem then that the only sure way of detecting ill-conditioning is to compute an approximation to A^{-1}. Actually, for full matrices this is not entirely out of the question, for if an LU decomposition of A has already been computed at a cost of $n^3/3$ multiplications, the decomposition $A^{-1} = U^{-1}L^{-1}$ may be computed for another $n^3/3$ multiplications. Thus for twice the work one can compute the upper bound $\| U^{-1} \| \| L^{-1} \|$ on $\| A^{-1} \|$. A cautious person might feel the additional work is not too high a price to pay for the security of knowing when his problems are ill conditioned. However, in the next section we shall consider a method for improving approximation solutions from which a fair estimate of the condition of the problem may be obtained.

EXERCISES

1. Let $A \in \mathbb{R}^{n \times n}$ be nonsingular and let X be an approximation to A^{-1}. Define the *residual matrix* R by $R = I - AX$. Show that

$$\frac{\| A^{-1} - X \|}{\| A^{-1} \|} \leq \| R \|.$$

Compare this result with the corresponding result for linear systems.

2. In Exercise 1 show that

$$\| I - XA \| \leq \varkappa(A) \| R \|.$$

Hence conclude that a matrix X can be a good approximate right inverse for A but a bad approximate left inverse.

3. Give an example of an ill-conditioned matrix A and a matrix X for which the bound in Exercise 2 is nearly attained.

4. Discuss the economics of computing A^{-1} to estimate the condition number of A in the special case when A is tridiagonal; upper Hessenberg.

NOTES AND REFERENCES

The material in this section is covered by Wilkinson (1963) and many others.

Given an approximate inverse X to A one can define, in analogy with the residual vector associated with a linear system, a residual matrix

$$R = I - AX.$$

Unlike the residual vector for linear systems, R does reflect the accuracy of the approximate inverse X; for

$$\| A^{-1} - X \| = \| A^{-1}R \| \leq \| A^{-1} \| \, \| R \|,$$

hence

$$\frac{\| A^{-1} - X \|}{\| A^{-1} \|} \leq \| R \|.$$

However, it should be noted that if A is ill conditioned, its rounded inverse may have a large residual.

Although it is possible to construct systems of equations that do not reveal any obvious signs of ill-conditioning, such systems do not often occur in practice. Noble (1969) gives a chart of various tests for ill-conditioning.

5. ITERATIVE REFINEMENT OF APPROXIMATE SOLUTIONS OF LINEAR SYSTEMS

In this section we shall analyze a method for improving an approximate solution of the equation

$$Ax = b, \tag{5.1}$$

where A is a matrix of order n which is not too ill conditioned, in a sense to be made precise later. The method starts with an approximate solution x_1 and produces a new approximate solution x_2 which is nearer x than x_1. Obviously one can again apply the method to the approximate solution x_2 to obtain an even better approximation x_3. Proceeding in this way, we obtain a sequence $\langle x_k \rangle$ of approximate solutions that converges to the true solution x. A process, such as we have just described, that generates successively a sequence of approximate solutions of a problem is called an *iteration*. The method to be described in this section is therefore called the method of iterative refinement of an approximate solution of (5.1).

The idea behind the method is simple. Let x_1 be an approximate solution of (5.1), presumably obtained by the techniques of Section 3.4. Let

$$r_1 = b - Ax_1$$

be the residual corresponding to x_1. Then if we solve the system

$$Ad_1 = r_1 \tag{5.2}$$

and calculate

$$x_2 = x_1 + d_1,$$

it follows that

$$x_2 = x_1 + A^{-1}r_1 = x_1 + A^{-1}(b - Ax_1) = x_1 + A^{-1}b - x_1 = A^{-1}b = x,$$

so that x_2 is the exact solution. Note that this procedure is quite cheap; for if the matrix A has been decomposed to compute x_1, say by Crout reduction, then it need not be decomposed again to solve (5.2). Since the computation of the residual requires only n^2 multiplications, the whole process can be carried out with $2n^2$ multiplications.

In practice the computations must be carried out with rounding error, so that x_2 will not be an exact solution. In fact, the errors made in solving (5.2) for d_1 are of the same kind made in solving initially for x_1, and it is not at all clear that x_2 will even be as good an approximate solution as x_1. It is therefore necessary to give a detailed analysis of the process to discover under what conditions x_2 is an improvement.

Since the analysis is lengthy, we begin by stating its final objective. We shall attempt to exhibit a constant $\eta > 0$, independent of x_1, such that

$$\| x - x_2 \|_\infty \leq \eta \| x - x_1 \|_\infty. \tag{5.3}$$

This being done, it follows by induction that

$$\| x - x_k \|_\infty \leq \eta^{k-1} \| x - x_1 \|_\infty. \tag{5.4}$$

If we can adjust the computations so that $\eta < 1$, then $\lim \eta^k = 0$, and it follows from (5.4) that $\lim x_k = x$. The result we shall actually obtain is a little weaker than this, but it will still suffice to establish the practical convergence of the method. For notational convenience we shall let the symbol $\| \cdot \|$ without a subscript denote the ∞-norm $\| \cdot \|_\infty$.

The process with rounding error may be described as follows.

$$\begin{aligned}
&1) \quad r_1 = \mathrm{fl}(b - Ax_1) \equiv b - Ax_1 + e_1 \\
&2) \quad d_1 = \mathrm{fl}(A^{-1}r_1) \equiv (A + H_1)^{-1}r_1 \\
&3) \quad x_2 = \mathrm{fl}(x_1 + d_1) \equiv x_1 + d_1 + g_1
\end{aligned} \tag{5.5}$$

Here the vectors g_1 and e_1 represent the difference of the true values from the computed values (n.b. e_1 is not the natural basis vector). The matrix H_1 is the error matrix of Theorem 3.5.3. We have seen (cf. page 195) that if computations are performed in t-digit arithmetic, then

$$\| H \| \leq \phi(n) \| A \| 10^{-t}.$$

It follows from Corollary 3.7 that if, say,

$$\phi(n)\varkappa(A)10^{-t} < \tfrac{1}{2}, \tag{5.6}$$

then

$$(A + H_1)^{-1} = (I + F_1)A^{-1}, \tag{5.7}$$

where

$$\| F_1 \| \leq \frac{\phi(n)\varkappa(A)10^{-t}}{1 - \phi(n)\varkappa(A)10^{-t}} < 1. \tag{5.8}$$

The first step in the analysis is to obtain a bound for $\| e_1 \|$ in terms of $\| x - x_1 \|$. To do this let

$$\varrho = \frac{\| e_1 \|}{\| r_1 \|}. \tag{5.9}$$

Note that ϱ_1 is the relative error in the residual. Now

$$r_1 - e_1 = b - Ax_1 = A(x - x_1).$$

Hence

$$\| r_1 \| - \| e_1 \| \leq \| A \| \| x - x_1 \|.$$

Substituting the value for $\| r_1 \|$ obtained from (5.9), we obtain

$$(\varrho^{-1} - 1) \| e_1 \| \leq \| A \| \| x - x_1 \|,$$

or

$$\| e_1 \| \leq \frac{\varrho \| A \| \| x - x_1 \|}{1 - \varrho}. \tag{5.10}$$

Now from statements 2 and 3 of (5.5), we have

$$x - x_2 = x - x_1 - (A + H_1)^{-1} r_1 - g_1.$$

In view of (5.7) and the definition of r_1, we have

$$\begin{aligned} x - x_2 &= x - x_1 - (I + F_1)A^{-1}(b - Ax_1 + e_1) - g_1 \\ &= x - x_1 - (I + F_1)(x - x_1 + A^{-1}e_1) - g_1 \\ &= -F_1(x - x_1) - (I + F_1)A^{-1}e_1 - g_1. \end{aligned} \tag{5.11}$$

Taking norms on both sides of (5.11) and remembering that $\| F_1 \| < 1$, we obtain

$$\| x - x_2 \| \leq \| F_1 \| \| x - x_1 \| + 2 \| A^{-1} \| \| e_1 \| + \| g_1 \|.$$

Finally if the bounds (5.8) and (5.10) are substituted for $\| F_1 \|$ and $\| e_1 \|$, the result is

$$\| x - x_2 \| \leq \varkappa(A)\left[\frac{\phi(n)10^{-t}}{1 - \phi(n)\varkappa(A)10^{-t}} + \frac{2\varrho}{1 - \varrho}\right] \| x - x_1 \| + \| g_1 \|. \tag{5.12}$$

If we define

$$\eta = \frac{\varkappa(A)\phi(n)10^{-t}}{1 - \varkappa(A)\phi(n)10^{-t}} + \frac{2\varkappa(A)\varrho}{1 - \varrho}, \tag{5.13}$$

then (5.12) becomes

$$\| x - x_2 \| \leq \eta \| x - x_1 \| + \| g_1 \|. \tag{5.14}$$

The bound (5.14) is not quite as good as the bound (5.3) since it contains the additional term $\| g_1 \|$. However, g_1 is the error made in adding the correction d_1 to x_1, and unless x_1 is very near the solution it will be negligible compared to $x - x_1$ and $x - x_2$. Thus the problem of convergence is still one of determining when η is less than unity.

The first term in the definition of η is easily disposed of. If $\varkappa(A)$ satisfies the restriction (5.6), then by (5.8) this first term is less than unity. Since the polynomial $\phi(n)$ from the rouding-error analysis is usually an overestimate, the actual bound will generally be a good deal smaller than unity. Thus if A is not too ill conditioned in the sense that it satisfies (5.6), we do not have to worry about the first term.

The second term is another story. Recall that the number ϱ is the relative error in the computed residual. Now we know from the discussion in Section that if x_1 was computed as an approximate solution of (5.1), then its residual satisfies

$$\| r_1 \| \leq \phi(n)10^{-t} \| A \| \| x_1 \|. \tag{5.15}$$

On the other hand, when the product Ax_1 is formed, one of the terms $\alpha_{ij}\xi_j^{(1)}$ will usually be of the same order of magnitude as $\| A \| \| x_1 \|$. Since by (5.15) the elements of r are almost 10^{-t} smaller than $\alpha_{ij}\xi_j^{(1)}$ one must cancel almost t significant figures when computing r_1. This means that if r_1 is calculated in t-digit arithmetic, the results will be inaccurate in almost their first figure and ϱ will be almost unity. Obviously, $2\varkappa(A)\varrho/(1 - \varrho)$, which is the second term in (5.13) may be much greater than unity.

On the other hand if r_1 is calculated in double precision ($2t$ digits), the cancellation of t figures will still leave t accurate figures. Thus ϱ will be about 10^{-t}, and if $\varkappa(A)$ satisfies (5.6), the term $2\varkappa(A)\varrho/(1 - \varrho)$ will be less than unity. Notice that double precision should be sufficient for all subsequent iterations of the process, since even the rounded exact solution will not have a residual much smaller than the right-hand side of (5.15). If inner products can be accumulated in double precision, the computation of the residual in double precision will require only single precision multiplications. Once the residual has been computed in double precision, it may be rounded to single precision and used in statement (2) of (5.5).

Thus if A is not too ill conditioned (this may still be very ill conditioned indeed) and if the residual r_1 is calculated in double precision, then the error $\| x - x_2 \|$ will be smaller than the error $\| x - x_1 \|$. If the process is applied iteratively to generate the sequence of approximate solutions $x_1, x_2, x_3, \ldots,$ each x_k will have a smaller error than its predecessor x_{k-1}. This decrease in error will continue to occur until the error is approximately equal to $\| g_k \|$ in (5.14). Practically, since g_k is the error made in adding the correction d_k to x_k, $\| g_k \|$ is approximately equal to $\| x_k \| \cdot 10^{-t}$. If $\| d_k \|$ is smaller than this, then d_k cannot be added accurately to x_k and we may as well terminate the iteration. The resulting approximation x_k will be very near the exact solution rounded to t figures.

We sum up all these considerations in the following algorithm.

ALGORITHM 5.1. Let x be an approximate solution of the equation $Ax = b$. If A is not too ill conditioned, this algorithm, when performed in t-digit arithmetic, returns in x an approximate solution that is nearly equal to the exact solution rounded to t figures.

1) Compute $r = b - Ax$ in double precision and round to single precision
2) Compute in single precision the solution of the equation $Ad = r$
3) If $\| d \|_\infty / \| x \|_\infty \leq 10^{-t}$, terminate the iteration
4) $x \leftarrow x + d$
5) Go to 1

Several comments should be made about this algorithm. In the first place, as it is constructed, it may loop indefinitely if A is too ill conditioned for the iteration to converge. In this case the corrections d_i will not show a steady decrease but will behave erratically, now decreasing—now increasing. Thus it is sufficient to terminate the algorithm with an error stop when $\| d_i \| / \| x_{i+1} \| \geq \| d_{i-1} \| / \| x_i \|$.

Secondly, we have not yet commented on the speed of convergence. However, from the approximate inequality

$$\frac{\| x - x_{k+1} \|}{\| x \|} \lesssim \eta \frac{\| x - x_k \|}{\| x \|}$$

it follows that if $\eta \cong 10^{-p}$, then the x_k will improve at the rate of about p

figures per iteration (cf. Example 2.1.2). This fact allows us to estimate the condition number of A, for if we are gaining p figures per iteration and x_1 is accurate to about q figures, then x_1 and x_2 will agree to q figures and x_2 and x_3 will agree to about $q+p$ figures. Thus

$$\frac{\| x_3 - x_2 \|}{\| x_2 - x_1 \|} = \frac{\| d_2 \|}{\| d_1 \|} \cong 10^{-p} \cong \eta.$$

We have already seen that $\eta \cong \varkappa(A)10^{-t}$. Hence $10^t \| d_2 \| / \| d_1 \|$ should be a fair estimate for $\varkappa(A)$ when A is very ill conditioned. Of course if the process converges in one iteration, as is likely to happen when $\varkappa(A)$ is significantly less than $10^{-t/2}$, this estimate is useless.

Finally, we note that, with a little cleaning up, our development amounts to a rigorous proof that Algorithm 5.1 must converge to something near the exact solution rounded to t digits, provided A is not too ill conditioned. However, we have not shown rigorously that if A is too ill conditioned, the iteration will diverge. This leaves open the possibility that, with a violently ill-conditioned matrix, the iteration may appear to converge to a false solution. It is generally conceded that the probability of this happening is negligible and that the results of Algorithm 5.1 can be taken at face value.

EXAMPLE 5.2. Let

$$A = \begin{pmatrix} 7.000 & 6.990 \\ 4.000 & 4.000 \end{pmatrix}$$

and $b = (34.97, 20.000)^{\mathrm{T}}$, so that $x = (2, 3)^{\mathrm{T}}$ is the solution of the equation $Ax = b$. The inverse of A to four figures is

$$A^{-1} = \begin{pmatrix} 100.0 & -174.8 \\ -100.0 & 175.0 \end{pmatrix},$$

and hence $\varkappa(A) = \| A \|_\infty \| A^{-1} \|_\infty \cong 3850$. The LU decomposition of A, computed in four-digit arithmetic is

$$\begin{pmatrix} 1.000 & 0.000 \\ .5714 & 1.000 \end{pmatrix} \begin{pmatrix} 7.000 & 6.990 \\ 0.000 & 0.006 \end{pmatrix}.$$

Let x_1 be the solution of $Ax = b$ computed using this decomposition. If Algorithm 5.1 in four-digit arithmetic is applied to refine x_1, there results the following table of values.

k	x	r	d
1	1.667 3.333	.00333 .00000	.3214 −.3172
2	1.988 3.016	−.02784 −.01600	.0160 −.0200
3	2.004 2.996	−.00004 .00000	−.00381 .00381
4	2.000 3.000	.0000 .0000	.0000 .0000

In the above example $\varkappa(A)10^{-4} \cong .4$, so that the A is almost too ill conditioned for the algorithm to work. On the basis of this estimate we should expect to gain about half a significant figure per iteration. The actual gain is more like one figure per iteration. The reason for this is that the convergence factor n is approximately equal to $\| F \|$ where $(I + F)A^{-1} = (A + E)^{-1}$ and E is the error made in solving the linear systems. Now most of this error E comes from the error in the LU decomposition, and it is easily verified that $\| A - LU \| \cong 3 \cdot 10^{-4}$. Thus $\| F \| \cong \| A^{-1} \| \| E \| \cong .8 \cdot 10^{-1}$, which is in accordance with the observed rate of convergence. For the same reason $10^4 \| d_2 \|/\| d_1 \| \cong 625$ is an underestimate for $\varkappa(A)$.

The behavior of the residuals is interesting. The computed solution x_1 is inaccurate in all its digits. Nonetheless, the residual is about as small as can be expected, as was predicted in the last section. It is mildly surprising that, although the refined solution x_2 is much more accurate, its residual is appreciably larger. This is typical of the process. Although the iterates become more accurate, the residuals do not necessarily decrease. This is to be expected since we have seen in Section 4 that the computed solution x_1 and the final iterate, which is essentially the rounded exact solution, should have residuals of about the same size. The only way in which the above example is atypical is that the residual corresponding to x_4 is zero. This is because x_4 is the exact solution whose components happen to be integers.

EXERCISE

1. The method of iterative refinement can be applied to calculate a solution to $Ax = b$ to any desired accuracy. The approximate solutions x_i

must, of course, be accumulated in higher and higher precision, and the residual vector, which will eventually get progressively smaller, must be calculated so that it is accurate. However, there is no point in calculating the residual so accurately that the second term in (5.13) is much less than the first. Assume that an upper bound for $\varkappa(A)$ is known and that the computer can operate in t, $2t$, $3t$, ... digit arithmetic. Write an algorithm for computing a solution of $Ax = b$ accurate to mt figures.

NOTES AND REFERENCES

The method of iterative refinement in the form described here is due to Wilkinson (1963), who analyzes the process for fixed-point arithmetic. Moler (1967) has given a detailed analysis for floating-point arithmetic, which essentially parallels the one given here.

The merits of iterative refinement are not unquestioned. Its critics point out that one must save the original matrix to compute the residual; and anyway it is not worthwhile to try to compute accurate solutions of ill-conditioned problems. In favor of the method it can be said that its slow convergence is a reliable test for ill-conditioning, and for many people that is sufficient justification.

Codes for the iterative refinement of positive definite systems have been published by Martin, Peters, and Wilkinson (1966, HACLA/I/2) and for general systems by Bowdler, Martin, Peters, and Wilkinson (1966, HACLA/I/7).

THE LINEAR LEAST SQUARES PROBLEM CHAPTER 5

An important problem in approximation theory is the following. Given a matrix $A \in \mathbb{R}^{m \times n}$ and an m-vector b, find a vector y in $\mathcal{R}(A)$ that is as near as possible to b. The vague term "near" may be made precise by introducing a norm ν on $\mathbb{R}^m$. The problem then becomes one of finding a vector $y \in \mathcal{R}(A)$ such that for all $z \in \mathcal{R}(A)$

$$\nu(b - y) \leq \nu(b - z).$$

It can be shown that this approximation problem always has a solution; however, the proof of this fact gives no hint of how to go about computing a solution. Most of the standard techniques for finding minima of functions are inapplicable to this problem, since, in general, a norm is not a differentiable function of its argument. Thus most methods for solving the problem are tailored to fit a special norm. In particular, algorithms exist for the 1-norm and the ∞-norm.

This chapter is devoted to the solution of the problem for the 2-norm. In this case the problem is called the linear least squares problem. The theory and practice of least squares problems is simplified by the fact that the square of the 2-norm is a differentiable function of its argument, although we shall not use this fact explicitly. In addition to its applications to the approximation of functions and data, the linear least squares problem is closely connected with the areas of statistics, such as regression analysis and the analysis of variance, that concern themselves with the normal distribution.

The theory of least squares is intimately related to the geometry of $\mathbb{R}^n$. Accordingly, Section 1 is devoted to exploring the consequences of a definition of angle in $\mathbb{R}^n$, with particular attention given to the properties of vectors that lie at right angles to one another. In Section 2 the theoretical problems surrounding the least squares problem are resolved and a computational method for computing solutions is described. In Section 3 a class of orthogonal matrices, analogous to the elementary lower triangular matrices of Section 3.2 are introduced and used to reduce a general matrix to trapezoidal form. This reduction is applied to the linear least squares problems. Finally, in Section 4, a method for refining least squares solutions is described.

1. ORTHOGONALITY

Let x and y be nonzero 3-vectors. It is a well-known fact from analytic geometry that the angle ϕ between x and y satisfies the relation

$$x^{\mathrm{T}}y = \| x \|_2 \| y \|_2 \cos \phi. \tag{1.1}$$

This relation can be used to define a notion of angle between vectors in $\mathbb{R}^n$. Specifically, if x and y are nonzero n-vectors, the Cauchy inequality implies that

$$-1 \leq \frac{x^{\mathrm{T}}y}{\| x \|_2 \| y \|_2} \leq 1.$$

Hence $x^{\mathrm{T}}y/(\| x \|_2 \| y \|_2)$ is the cosine of some angle ϕ lying between 0 and π, and we may call ϕ the angle between x and y; that is

$$\phi = \arccos \frac{x^{\mathrm{T}}y}{\| x \|_2 \| y \|_2}. \tag{1.2}$$

When x and y lie at right angles, the cosine of the angle $\pi/2$ between them is zero. In this case it follows from (1.1) that $x^{\mathrm{T}}y = 0$. Conversely, if $x^{\mathrm{T}}y = 0$, it follows from (1.2) that $\phi = \arccos(0) = \pi/2$, so that x and y lie at right angles. Thus the condition $x^{\mathrm{T}}y = 0$ may be taken as characterizing the perpendicularity or, as we shall call it, the orthogonality of the vectors x and y. This leads to the following definitions.

DEFINITION 1.1. Two n-vectors x and y are *orthogonal* if $x^{\mathrm{T}}y = 0$. The n-vectors $u_1, u_2, \ldots, u_r$ are orthogonal if they are pairwise orthogonal. The vector $x \in \mathbb{R}^n$ is orthogonal to the subspace $\mathcal{S} \subset \mathbb{R}^n$ if x is orthogonal to every vector in $\mathcal{S}$. Two subspaces $\mathcal{S}, \mathcal{T} \subset \mathbb{R}^n$ are orthogonal if each $s \in \mathcal{S}$ is orthogonal to each $t \in \mathcal{T}$.

Note that we do not require orthogonal vectors to be nonzero, even though strictly speaking it makes no sense to speak of the angle between a zero vector and another vector. However, the extended definition causes no difficulties and is notationally quite convenient. For example, it allows the concise statement "the only vector orthogonal to itself is the zero vector."

In practice a subspace is often given as the column space of a matrix. In this case we can determine directly when a vector is orthogonal to the subspace.

THEOREM 1.2. Let $A \in \mathbb{R}^{n \times r}$ and $x \in \mathbb{R}^n$. Then the following statements are equivalent.

1. x is orthogonal to $\mathcal{R}(A)$.
2. $A^{\mathrm{T}}x = 0$.
3. $x \in \mathcal{N}(A^{\mathrm{T}})$.

PROOF. Observe that by the definition of $\mathcal{N}(A^{\mathrm{T}})$, parts 2 and 3 are equivalent. To show that part 1 is equivalent to 2, first let x be orthogonal to $\mathcal{R}(A)$. Then if $A = (a_1, a_2, \ldots, a_r)$, each $a_i \in \mathcal{R}(A)$ and hence $a_i^{\mathrm{T}}x = 0$ $(i = 1, 2, \ldots, r)$. This is equivalent to part 2. On the other hand let $A^{\mathrm{T}}x = 0$ and let $y \in \mathcal{R}(A)$. Then $y = Az$ for some $z \in \mathbb{R}^r$. Hence

$$y^{\mathrm{T}}x = z^{\mathrm{T}}A^{\mathrm{T}}x = z^{\mathrm{T}}0 = 0,$$

which shows that x is orthogonal to y. ■

Since the Cauchy inequality was invoked to motivate the definition of orthogonality, it is not surprising that orthogonal vectors stand in special relation to the 2-norm. In particular, they satisfy the following generalization of the Pythagorean theorem.

THEOREM 1.3. If x and y are orthogonal, then

$$\| x + y \|_2^2 = \| x \|_2^2 + \| y \|_2^2.$$

PROOF.

$$\| x + y \|_2^2 = (x + y)^{\mathrm{T}}(x + y) = x^{\mathrm{T}}x + y^{\mathrm{T}}y + 2x^{\mathrm{T}}y = \| x \|_2^2 + \| y \|_2^2. \quad \blacksquare$$

Orthogonal vectors that have been scaled so that their 2-norm is unity occur often enough to have their own name.

DEFINITION 1.4. The vectors $u_1, u_2, \ldots, u_r$ are *orthonormal* if

1. $u_1, u_2, \ldots, u_r$ are orthogonal,
2. $\| u_i \|_2 = 1 \qquad (i = 1, 2, \ldots, r)$.

Observe that the first condition in the definition states that

$$u_i^{\mathrm{T}} u_j = 0 \qquad (i \neq j),$$

while the second states that

$$u_i^{\mathrm{T}} u_i = 1.$$

This means that if we form the matrix $U = (u_1, u_2, \ldots, u_r)$, then

$$U^{\mathrm{T}} U = I_r.$$

Thus the statement $U^{\mathrm{T}}U = I$ is a convenient way of saying that the matrix U has orthonormal columns.

If $U = (u_1, u_2, \ldots, u_r)$ has orthonormal columns, then

$$r = \operatorname{rank}(I_r) = \operatorname{rank}(U^{\mathrm{T}}U) \leq \operatorname{rank}(U) \leq r.$$

Hence $\operatorname{rank}(U) = r$ and the vectors $u_1, u_2, \ldots, u_r$ are linearly independent. In other words, *orthonormal vectors are linearly independent.*

Matrices with orthonormal columns interact nicely with the 2-norm and the Frobenius norm.

THEOREM 1.5. Let $U \in \mathbb{R}^{n \times r}$ have orthonormal columns. Then for any $A \in \mathbb{R}^{r \times s}$ we have

$$\| UA \|_p = \| A \|_p \qquad (p = 2, \mathrm{F}).$$

Moreover

$$\| U \|_2 = 1 \qquad \text{and} \qquad \| U \|_F = \sqrt{r}.$$

PROOF. By Theorem 4.2.10

$$\| UA \|_2^2 = \| (UA)^T UA \|_2 = \| A^T U^T UA \|_2 = \| A^T A \|_2 = \| A \|_2^2.$$

Note that this implies that for any $a \in \mathbb{R}^r$ we have $\| Ua \|_2 = \| a \|_2$.

Now let $A = (a_1, a_2, \ldots, a_s)$ be partitioned by columns. Then

$$\| UA \|_F^2 = \sum_{i=1}^{s} \| Ua_i \|_2^2 = \sum_{i=1}^{s} \| a_i \|_2^2 = \| A \|_F^2.$$

The fact that $\| U \|_2 = 1$ is established as follows:

$$\| U \|_2^2 = \| U^T U \|_2 = \| I \| = 1.$$

Finally, let $U = (u_1, u_2, \ldots, u_r)$. Then $\| u_i \|_2 = 1$, and

$$\| U \|_F^2 = \sum_{i=1}^{r} \| u_i \|_2^2 = r. \quad \blacksquare$$

It is important to note that $U^T U = I$ does not imply that $\| AU \|_2 = \| A \|_2$ or $\| AU \|_F = \| A \|_F$. In fact it is possible for AU to be zero when A is not zero. However, when U is square the equalities hold, as we shall see a little later (Theorem 1.7).

A linear transformation $f : \mathbb{R}^r \to \mathbb{R}^n$ is an *isometry* if

$$\| f(x) \|_2 = \| x \|_2,$$

for all $x \in \mathbb{R}^r$. In this terminology, Theorem 1.5 states that the linear transformation induced by a matrix with orthonormal columns is an isometry. The converse is also true; the matrix of an isometry has orthonormal columns. The proof of this fact is left as an exercise.

Square matrices with orthonormal columns occur very frequently in matrix work.

DEFINITION 1.6. An *orthogonal matrix* is a square matrix with orthonormal columns.

The important properties of orthogonal matrices are summed up in the following theorem.

THEOREM 1.7. Let U be orthogonal. Then

1. $U^{\mathrm{T}}U = UU^{\mathrm{T}} = I$,
2. $\| AU \|_2 = \| A \|_2$,
3. $\| AU \|_{\mathrm{F}} = \| A \|_{\mathrm{F}}$.

PROOF. The fact that $U^{\mathrm{T}}U = I$ follows from the orthonormality of the columns of U. It follows that U^{T} is the inverse of U. Since a matrix commutes with its inverse, we have $U^{\mathrm{T}}U = UU^{\mathrm{T}} = I$.

Now the equation $UU^{\mathrm{T}} = I$ implies that U^{T} has orthonormal columns. Hence by Theorem 4.2.10 and Theorem 1.5, for $p = 2$, F we have

$$\| AU \|_p = \| (AU)^{\mathrm{T}} \|_p = \| U^{\mathrm{T}}A^{\mathrm{T}} \|_p = \| A^{\mathrm{T}} \|_p = \| A \|_p. \qquad \blacksquare$$

We emphasize again that, by property 1 of the above theorem, *an orthogonal matrix is nonsingular, its transpose is its inverse, and it has orthonormal rows.*

EXAMPLE 1.8. Let $\{u_1, u_2, \ldots, u_n\}$ be a basis of orthonormal vectors for $\mathbb{R}^n$. Then any $x \in \mathbb{R}^n$ can be expressed in the form

$$x = \gamma_1 u_1 + \gamma_2 u_2 + \cdots + \gamma_n u_n, \tag{1.3}$$

where the γ_i are the components of the vector x with respect to the basis $\{u_1, u_2, \ldots, u_n\}$. Because the u_i are orthonormal, the γ_i can be computed rather simply. To do this, let $U = (u_1, u_2, \ldots, u_n)$ and $c = (\gamma_1, \gamma_2, \ldots, \gamma_n)^{\mathrm{T}}$. Then U is an orthogonal matrix, and (1.3) is the same as

$$x = Uc.$$

Since U is orthogonal

$$c = U^{\mathrm{T}}Uc = U^{\mathrm{T}}x,$$

or

$$\gamma_i = u_i^{\mathrm{T}}x \qquad (i = 1, 2, \ldots, n).$$

Thus the γ_i may be computed as inner products.

Example 1.8 shows that the use of an orthonormal basis may simplify calculations. In $\mathbb{R}^n$ the vectors $e_1, e_2, \ldots, e_n$ form a natural orthonormal basis; however, it is not at all obvious that one can find an orthonormal basis for an arbitrary subspace of $\mathbb{R}^n$. In order to attack this problem, we prove a factorization theorem that is important in its own right.

THEOREM 1.9. Let A have linearly independent columns. Then A can be written uniquely in the form

$$A = QR,$$

where Q has orthonormal columns and R is upper triangular with positive diagonal elements.

PROOF. By Example 3.3.6, $A^{\mathrm{T}}A$ is positive definite. Hence by Theorem 3.3.8, $A^{\mathrm{T}}A = R^{\mathrm{T}}R$, where R is upper triangular with positive diagonal elements. Let $Q = AR^{-1}$. Then $A = QR$, and

$$Q^{\mathrm{T}}Q = R^{-\mathrm{T}}A^{\mathrm{T}}AR^{-1} = R^{-\mathrm{T}}R^{\mathrm{T}}RR^{-1} = I,$$

so that Q has orthonormal columns.

To show the uniqueness of the factorization, let $A = Q'R'$, where $Q'^{\mathrm{T}}Q' = I$ and R' is upper triangular with positive diagonal elements. Then

$$A^{\mathrm{T}}A = R'^{\mathrm{T}}Q'^{\mathrm{T}}Q'R' = R'^{\mathrm{T}}R'.$$

However, the factorization of Theorem 3.3.8 is unique. Hence $R = R'$ and $Q' = AR'^{-1} = AR^{-1} = Q$. ■

The factorization in Theorem 1.9 is often called the QR factorization of the matrix A. The proof of the theorem is constructive in that it actually suggests an algorithm for computing the factorization. Namely,

1) Form $A^{\mathrm{T}}A$
2) Use Algorithm 3.3.9 to find an upper triangular R satisfying $R^{\mathrm{T}}R = A^{\mathrm{T}}A$
3) $Q = AR^{-1}$

However, this algorithm is not to be recommended. In the first place, it is by no means easy to compute R accurately, a point which will be treated later in Section 2. In the second place, even if R is computed accurately, the columns of the computed Q may be far from orthogonal. To see how this may come about, suppose that the elements of A are of order unity and that $\varkappa(A^{\mathrm{T}}A) \cong 10^6$. Then R^{-1} must have elements as large as 10^3. Since $Q = AR^{-1}$ has elements of order unity, its computation must involve the cancellation of three significant figures. Thus, if R is accurate to, say four figures, the computed Q will be accurate to only one. A stable way of computing the QR factorization will be discussed in the next section.

Theorem 1.9 may be used to show that every subspace has an orthonormal basis.

COROLLARY 1.10. Every nontrivial subspace of $\mathbb{R}^n$ has an orthonormal basis.

PROOF. Let the columns of A form a basis for the subspace in question, and let $A = QR$ as in Theorem 1.9. Then the columns of Q are orthonormal, hence linearly independent, and since R is nonsingular,

$$\mathcal{R}(Q) = \mathcal{R}(A). \quad \blacksquare$$

Two complementary subspaces $\mathcal{S}$ and $\mathcal{T}$ in $\mathbb{R}^n$ are *orthogonal complements* if they are orthogonal to one another. If $\mathcal{S}$ and $\mathcal{T}$ are orthogonal complements, then any vector x may be written uniquely in the form $x = s + t$ where $s \in \mathcal{S}$ is orthogonal to $t \in \mathcal{T}$. We have already seen that any subspace has a complement, although this complement need not be unique. However, a subspace can have only one orthogonal complement.

THEOREM 1.10. Let $\mathcal{S} \subset \mathbb{R}^n$ be a subspace. Then $\mathcal{S}$ has a unique orthogonal complement. In fact, if $\mathcal{R}(A) = \mathcal{S}$, then the orthogonal complement of $\mathcal{S}$ is $\mathcal{N}(A^{\mathrm{T}})$.

PROOF. Let $A \in \mathbb{R}^{n \times r}$ be such that $\mathcal{R}(A) = \mathcal{S}$. By Theorem 1.2, $\mathcal{N}(A^{\mathrm{T}})$ is the set of all vectors orthogonal to $\mathcal{S}$. Thus $\mathcal{N}(A^{\mathrm{T}})$ contains any orthogonal complement of $\mathcal{S}$, and the proof will be complete if we can show that $\mathcal{N}(A^{\mathrm{T}})$ is a complement of $\mathcal{S}$.

First suppose that $x \in \mathcal{S} \cap \mathcal{N}(A^{\mathrm{T}})$. Then, since $x \in \mathcal{N}(A^{\mathrm{T}})$, x is orthogonal to any vector in $\mathcal{S}$. However, $x \in \mathcal{S}$, hence $x^{\mathrm{T}}x = 0$, and $x = 0$. Thus $\mathcal{S} \cap \mathcal{N}(A^{\mathrm{T}}) = \{0\}$.

Second, $\mathrm{rank}(A) = \dim(\mathcal{S})$, but $\mathrm{null}(A^{\mathrm{T}}) = n - \mathrm{rank}(A^{\mathrm{T}}) = n - \mathrm{rank}(A) = n - \dim(\mathcal{S})$. Hence $\dim[\mathcal{N}(A^{\mathrm{T}})] + \dim(\mathcal{S}) = n$, and it follows that $\mathcal{N}(A^{\mathrm{T}}) \oplus \mathcal{S} = \mathbb{R}^n$. $\blacksquare$

EXERCISES

1. Construct vectors $u_1, u_2, u_3 \in \mathbb{R}^3$ such that u_2 is orthogonal to u_1 and u_3 is orthogonal to u_2, but u_3 is not orthogonal to u_1.

2. Show that if $\gamma^2 + \sigma^2 = 1$, then the matrix

$$\begin{pmatrix} \gamma & \sigma \\ -\sigma & \gamma \end{pmatrix}$$

is orthogonal.

3. Prove that if U has orthonormal columns and V is orthogonal, then UV has orthonormal columns. Conclude that the product of orthogonal matrices is orthogonal.

4. Show that if a matrix is orthogonal and triangular, then it is diagonal. What are its diagonal elements?

5. Let the linear transformation $f: \mathbb{R}^n \to \mathbb{R}^m$ be an isometry. Show that the matrix of f has orthonormal columns.

6. Show that if $n > m$, there is no isometry from $\mathbb{R}^n$ into $\mathbb{R}^m$.

7. Show that if U is orthogonal and x is a nonzero vector such that $Ux = \lambda x$ for some λ, then $|\lambda| = 1$.

8. Let $A \in \mathbb{R}^{n\times n}$ be positive definite. Two vectors u_1 and u_2 are A-orthogonal if $u_1{}^{\mathrm{T}} A u_2 = 0$. If $U \in \mathbb{R}^{n\times r}$ and $U^{\mathrm{T}}AU = I$, then the columns of U are said to be A-orthonormal. Show that every subspace has an A-orthonormal basis. [*Hint*: Let the linearly independent columns of B span the subspace. Show that $B^{\mathrm{T}}AB$ is positive definite. Write $B^{\mathrm{T}}AB = T^{\mathrm{T}}T$, where T is upper triangular and consider the matrix $U = BT^{-1}$.]

9. In $\mathbb{R}^3$ compute an orthonormal basis for the space spanned by the two vectors $(1, 1, 1)^{\mathrm{T}}$ and $(1.05, 1, 1)^{\mathrm{T}}$. Carry four figures and note anything unusual about the computations.

10. Let $A \in \mathbb{R}^{m\times n}$ have linearly independent columns. Show that the following *Gram–Schmidt* algorithm computes the QR factorization of A.

1) For $k = 1, 2, \ldots, n$
   1) $\varrho_{ik} = q_i{}^{\mathrm{T}} a_k \qquad (i = 1, 2, \ldots, k-1)$
   2) $q_k = a_k - \sum_{i=1}^{k-1} \varrho_{ik} q_i$
   3) $\varrho_{kk} = \| q_k \|_2$
   4) $q_k \leftarrow \dfrac{q_k}{\varrho_{kk}}$

Numerically this algorithm has nothing to recommend it. However, it is of considerable historical interest. [*Hint*: Show inductively that at statement 1.2 the vector q_k satisfies $q_j^{\mathrm{T}} q_k = 0$ $(j = 1, 2, \ldots, k-1)$. Show also that all the q_k satisfy $A - QR = 0$.]

11. Let $A \in \mathbb{R}^{m \times n}$ have linearly independent columns. Show that the following *modified Gram–Schmidt* algorithm computes the QR factorization of A.

    1) $q_k = a_k \qquad (k = 1, 2, \ldots, n)$

    2) For $k = 1, 2, \ldots, n$

        1) $\varrho_{kk} = \| q_k \|_2$

        2) $q_k \leftarrow \dfrac{q_k}{\varrho_{kk}}$

        3) $\varrho_{kj} = q_k^{\mathrm{T}} q_j \qquad (j = k+1, k+2, \ldots, n)$

        4) $q_j \leftarrow q_j - \varrho_{kj} q_k \qquad (j = k+1, k+2, \ldots, n)$

    While numerically this algorithm may produce q's that are far from orthogonal, the QR factorization obtained in this way may be used to solve least squares problem.

2. THE LINEAR LEAST SQUARES PROBLEM

In this section we shall be concerned with the problem of determining a vector x that minimizes the function

$$\varrho^2(x) = \| b - Ax \|_2^2, \tag{2.1}$$

where A is a fixed $m \times n$ matrix and b is an n-vector. The problem is called the *linear least squares* problem, since it requires that the sum of squares of the residual vector

$$r = b - Ax$$

be minimized. The scalar ϱ^2 is called the *residual sum of squares.* The problem is said to be linear because r depends linearly on x.

The linear least squares problem arises in a number of contexts. In particular, it is intimately connected with the approximation of data and with those parts of statistics that are concerned with the normal distribution. The following example indicates, in a general way, how such problems arise.

EXAMPLE 2.1. Suppose we are given points $p_1, p_2, \ldots, p_m$ and data $\phi_1, \phi_2, \ldots, \phi_m$ associated with these points. Suppose further we are given n functions $\psi_1, \psi_2, \ldots, \psi_n$ that can be evaluated at the points p_i. We desire to approximate the data ϕ_i by a linear combination of the ψ_i; that is we wish to determine constants $\gamma_1, \gamma_2, \ldots, \gamma_n$ such that

$$\phi_i \cong \gamma_1\psi_1(p_i) + \gamma_2\psi_2(p_i) + \cdots + \gamma_n\psi_n(p_i) \qquad (i = 1, 2, \ldots, m). \tag{2.2}$$

This amounts to saying that we wish to choose the γ_j to make the residuals

$$\varrho_i = \phi_i - \sum_{j=1}^{n} \gamma_j\psi_j(p_i) \qquad (i = 1, 2, \ldots, m)$$

as small as possible.

Now if $m > n$, there are more residuals than unknowns, and it will not generally be possible to make all of them zero. One approach is to make some of them zero and hope that the others are small. The least squares approach is to choose the γ_j so that

$$\varrho^2 = \sum_{i=1}^{m} \varrho_i^2 \tag{2.3}$$

is as small as possible. If this is done, no residual need be zero. But on the other hand, no residual can be greater than ϱ, so that if ϱ is sufficiently small, the right-hand side of (2.2) is a good approximation to ϕ_i.

The problem of determining the γ_j can be cast as a linear least squares problem as defined at the beginning of this section. Specifically let

$$A = \begin{pmatrix} \psi_1(p_1) & \psi_2(p_1) & \cdots & \psi_n(p_1) \\ \psi_1(p_2) & \psi_2(p_2) & \cdots & \psi_n(p_2) \\ \psi_1(p_3) & \psi_2(p_3) & \cdots & \psi_n(p_3) \\ \vdots & \vdots & & \vdots \\ \psi_1(p_m) & \psi_2(p_m) & \cdots & \psi_n(p_m) \end{pmatrix},$$

$b = (\phi_1, \phi_2, \ldots, \phi_m)^{\mathrm{T}}$, and $x = (\gamma_1, \gamma_2, \ldots, \gamma_n)^{r}$. Then it is easily verified that the number ϱ^2 of (2.3) is $\| b - Ax \|_2^2$, and the problem of choosing the γ_j to minimize ϱ^2 is the same as the problem of choosing x to minimize $\| b - Ax \|_2^2$.

It should not be supposed that the points p_i in the above example are necessarily members of a vector space. For example, with $m = 5$, the p_i may be the days Monday, Tuesday, Wednesday, Thursday, and Friday.

The ϕ_i may be the blood pressure of a patient on the day in question, and ψ_1 and ψ_2 may be his pulse rate and respiration rate. The approximation problem then becomes one of trying to predict the blood pressure of a patient as a linear combination of his pulse rate and his respiration rate.

We turn now to the theoretical and practical problems associated with the linear least squares problem. We shall consider the following questions:

1. When does the linear least squares problem have a solution?
2. When is it unique?
3. What are the effects of perturbations on the solution?
4. How may a solution be computed?

Some heuristic comments may make the answer to the first two questions clearer. First note that as x varies over $\mathbb{R}^n$ the vector $y = Ax$ varies over $\mathcal{R}(A)$. Thus instead of considering the problem of minimizing $\| b - Ax \|_2^2$ we may consider the problem of finding a vector $y_{\min}$ in $\mathcal{R}(A)$ that minimizes $\| b - y \|_2^2$. If $m = 3$ and $\operatorname{rank}(A) = 2$, it is obvious from geometric considerations (Fig. 1) that the minimum is attained when the residual $b - y$ is perpendicular to the plane $\mathcal{R}(A)$. Thus if we write $b = b_1 + b_2$, where b_1 is in $\mathcal{R}(A)$ and b_2 is in the orthogonal complement of $\mathcal{R}(A)$, then b_1 is the required vector $y_{\min}$ and b_2 is the residual vector $b - Ax$ at the minimum.

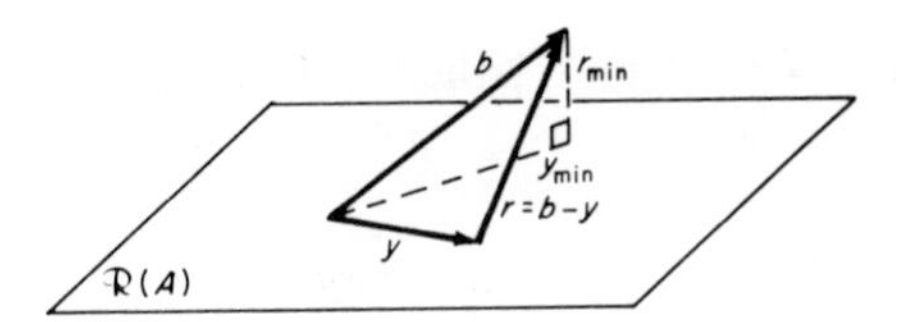

Fig. 1

Having found b_1, to find x we must solve the equation $b_1 = Ax$, but since $b_1 \in \mathcal{R}(A)$ this can always be done. Moreover the solution will be unique if and only if $\operatorname{null}(A) = 0$ (see Corollary 1.6.4).

The above is an outline of the existence and uniqueness proof of Theorem 2.2 below. Before stating the theorem, however, we shall introduce some terminology that will be used throughout this and the next section.

Let $\mathcal{S}$ be a subspace and let $\mathcal{T}$ be its orthogonal complement. Then any vector x can be written uniquely in the form $x = s + t$, where $s \in \mathcal{S}$ and $t \in \mathcal{T}$. The vector s is called the *projection of x onto* $\mathcal{S}$. In Fig. 5.1 the vector $y_{\min}$ is the projection of b onto $\mathcal{R}(A)$ and the vector $r_{\min} = b - y_{\min}$ is the projection of b onto the orthogonal complement of $\mathcal{R}(A)$. Note that if r

is the projection of any vector onto the orthogonal complement of $\mathcal{R}(A)$, then r is orthogonal to $\mathcal{R}(A)$, and by Theorem 1.2, $A^{\mathrm{T}}r = 0$.

We are now in a position to establish the existence of a least squares solution and determine when it is unique.

THEOREM 2.2. The linear least squares problem of minimizing (2.1) always has a solution. The solution is unique if and only if $\text{null}(A) = 0$.

PROOF. Let b_1 be the projection of b onto $\mathcal{R}(A)$ and b_2 be the projection of b onto the orthogonal complement of $\mathcal{R}(A)$. Then $b_1 - Ax$ belongs to $\mathcal{R}(A)$ and is orthogonal to b_2. Hence, by Theorem 1.3,

$$\varrho^2(x) = \| b - Ax \|_2^2 = \| b_1 - Ax \|_2^2 + \| b_2 \|_2^2.$$

Thus $\varrho^2(x)$ will be a minimum when $\| b_1 - Ax \|_2^2$ is a minimum. However, $b_1 \in \mathcal{R}(A)$, hence we can always find an x satisfying

$$Ax = b_1. \tag{2.4}$$

For this x, $\| b_1 - Ax \|_2^2 = 0$, which is certainly a minimum. By Corollary 1.6.4, the solution of (2.4) is unique if and only if $\text{null}(A) = 0$. ■

From the proof of Theorem 2.2 there follow two facts with important computational consequences.

COROLLARY 2.3. Let x be a solution of the least squares problem of minimizing (2.1). Then the residual vector $r = b - Ax$ satisfies

$$A^{\mathrm{T}}r = 0. \tag{2.5}$$

Equivalently,

$$A^{\mathrm{T}}Ax = A^{\mathrm{T}}b. \tag{2.6}$$

PROOF. Since x is a solution, $Ax = b_1$. Hence $r = b - Ax = b - b_1 = b_2$, but b_2 belongs to the orthogonal complement of $\mathcal{R}(A)$. Hence $0 = A^{\mathrm{T}}b_2 = A^{\mathrm{T}}r$. Equation (2.6) is obtained from (2.5) by substituting $b - Ax$ for r. ■

Theorem 2.2 shows that we can expect a unique solution to the linear least squares only when A has linearly independent columns. The theory and practice of least squares when A has linearly dependent columns is quite involved. To indicate just one problem, one has to decide which of an infinite number of solutions one will work with. For this reason we shall defer the case $\text{null}(A) \neq 0$ to Chapters 6 and 7 where the singular value decomposition is discussed. For the rest of this chapter we shall assume that A has linearly independent columns. Note that this implies that A has at least as many rows as columns.

The square system of Equation (2.6) is called the system of *normal equations* for the linear least squares problem. Since A has linearly independent columns, $A^{\mathrm{T}}A$ is nonsingular, and the solution x can be written in the form

$$x = (A^{\mathrm{T}}A)^{-1}A^{\mathrm{T}}b. \tag{2.7}$$

If we define

$$A^{\dagger} = (A^{\mathrm{T}}A)^{-1}A^{\mathrm{T}},$$

then (2.7) becomes

$$x = A^{\dagger}b.$$

The matrix $A^{\dagger}$ is called the *pseudo-inverse* or the *Moore–Penrose generalized inverse* of A. Just as the statement $x = A^{-1}b$ is a convenient way of saying that x is a solution of the equation $Ax = b$, so the statement $x = A^{\dagger}b$ is a convenient way of saying that x solves the linear least squares problem of minimizing (2.1). In Chapter 6 we shall consider a general definition of a pseudo-inverse that solves the linear least squares problem, even when $\text{null}(A) \neq 0$.

Regarding the third question of the sensitivity of least squares solutions to perturbations, we shall consider the effects of perturbations in the vector b and perturbations in the matrix A. The theory of regression in statistics examines the statistical effects on x of a random perturbation in b. We shall not pursue this line here. Rather we shall attempt to find a bound on the relative error in x due to a perturbation in b.

THEOREM 2.4. Let $A \in \mathbb{R}^{m\times n}$ have linearly independent columns and let $b, \bar{b} \in \mathbb{R}^m$. Let b_1 and $\bar{b}_1$ be the projections of b and $\bar{b}$ onto $\mathcal{R}(A)$. Then if $b_1 \neq 0$,

$$\frac{\| A^{\dagger}b - A^{\dagger}\bar{b} \|_2}{\| A^{\dagger}b \|_2} \leq \varkappa(A) \frac{\| b_1 - \bar{b}_1 \|_2}{\| b_1 \|_2}, \tag{2.8}$$

where

$$\varkappa(A) = \| A \|_2 \| A^{\dagger} \|_2 . \tag{2.9}$$

PROOF. Let b_2 be the projection of b onto the orthogonal complement of $\mathcal{R}(A)$. Then $A^{\mathrm{T}}b_2 = 0$. Since $b = b_1 + b_2$, it follows that

$$A^{\dagger}b = A^{\dagger}b_1 + A^{\dagger}b_2 = A^{\dagger}b_1 + (A^{\mathrm{T}}A)^{-1}A^{\mathrm{T}}b_2 = A^{\dagger}b_1.$$

Likewise, $A^{\dagger}\bar{b} = A^{\dagger}\bar{b}_1$. Hence

$$\| A^{\dagger}b - A^{\dagger}\bar{b} \|_2 = \| A^{\dagger}(b_1 - \bar{b}_1) \|_2 \leq \| A^{\dagger} \|_2 \| b_1 - \bar{b}_1 \|_2. \tag{2.10}$$

Now since $A^{\dagger}b$ is the solution of the least squares problem of minimizing (2.1), it follows from (2.4) that $AA^{\dagger}b = b_1$. Hence

$$\| A^{\dagger}b \|_2 \geq \frac{\| b_1 \|_2}{\| A \|_2}. \tag{2.11}$$

Dividing (2.10) by (2.11) gives (2.8). ■

The bound (2.8) of Theorem 2.4 is analogous to the bound of Theorem 4.4.1. The number $\varkappa(A)$ defined by (2.9) generalizes the previous definition. It serves as a condition number by telling how an error in b may be magnified in the least squares solution. However, it is important to note that only the part of the relative error lying in the column space of A counts. If b has a small projection onto $\mathcal{R}(A)$, that is if b_1 is small, a small perturbation in b can make a large change in b_1 and hence have a great effect on the least squares solution.

EXAMPLE 2.5. Let

$$A = \begin{pmatrix} 1 & 0 \\ 0 & 1 \\ 0 & 0 \end{pmatrix},$$

$$b = (.01, 0, 1.000)^{\mathrm{T}},$$

and

$$\bar{b} = (.0101, 0, 1.000)^{\mathrm{T}}.$$

Then the least squares solutions corresponding to b and $\bar{b}$ are $x = (.0100, 0)$ and $\bar{x} = (.0101, 0)$. Thus $\bar{x}$ has a relative error of about 10^{-2}, even though $\bar{b}$ has a relative error of only 10^{-4}. The bounds of Theorem 2.4 predict this phenomenon; for $\varkappa(A) = 1$, $b_1 = (.0100, 0, 0)^{\mathrm{T}}$, and $\bar{b}_1 = (.0101, 0, 0)^{\mathrm{T}}$. Thus the relative error in $\bar{b}_1$ is about 10^{-2}, the same size as the relative error in $\bar{x}$.

The condition number of A is nicely related to the condition number of the square matrix $A^{\mathrm{T}}A$.

THEOREM 2.6. Let A have linearly independent columns. Then

$$\varkappa^2(A) = \varkappa(A^{\mathrm{T}}A).$$

PROOF. From Theorem 4.2.10 we have

$$\| A \|_2^2 = \| A^{\mathrm{T}}A \|_2.$$

From the definition of $A^\dagger$ we have

$$\| A^\dagger \|_2^2 = \| A^\dagger A^{\dagger\mathrm{T}} \|_2 = \| (A^{\mathrm{T}}A)^{-1}A^{\mathrm{T}}A(A^{\mathrm{T}}A)^{-\mathrm{T}} \|_2 = \| (A^{\mathrm{T}}A)^{-1} \|_2.$$

Hence

$$\varkappa^2(A) = \| A \|_2^2 \| A^\dagger \|_2^2 = \| A^{\mathrm{T}}A \|_2 \| (A^{\mathrm{T}}A)^{-1} \|_2 = \varkappa(A^{\mathrm{T}}A). \quad \blacksquare$$

The analysis of the effects of perturbations in A on a least squares solution is quite difficult. Even to state a realistic theorem requires some new terminology. Let $\mathcal{S}$ be a subspace of $\mathbb{R}^m$ and $E \in \mathbb{R}^{m\times n}$. Then each column of E is an m-vector and may be projected onto $\mathcal{S}$. The matrix obtained by projecting the columns of E onto $\mathcal{S}$ is called *the projection of E onto* $\mathcal{S}$.

THEOREM 2.7. Let $A \in \mathbb{R}^{m\times n}$ have linearly independent columns. Let $E \in \mathbb{R}^{m\times n}$ and $b \in \mathbb{R}^n$. Let $E_1(b_1)$ and $E_2(b_2)$ be the projections of $E(b)$ onto $\mathcal{R}(A)$ and onto the orthogonal complement of $\mathcal{R}(A)$. Then if

$$\| A^\dagger \|_2 \| E_1 \|_2 < \tfrac{1}{2},$$

the columns of $A + E$ are linearly independent. Moreover, if $x = A^\dagger b$ and $\bar{x} = (A + E)^\dagger b$, then

$$\frac{\| x - \bar{x} \|_2}{\| x \|_2} \le 2\varkappa \frac{\| E_1 \|_2}{\| A \|_2} + 4\varkappa^2 \frac{\| E_2 \|_2}{\| A \|_2} \frac{\| b_2 \|_2}{\| b_1 \|_2} + 8\varkappa^3 \frac{\| E_2 \|_2^2}{\| A \|_2^2}, \tag{2.12}$$

where $\varkappa = \| A \|_2 \| A^\dagger \|_2$.

The bound (2.12) is quite complicated and merits some explanation. The third term in the bound depends on the square of $\| E_2 \|_2^2$ and will usually be unimportant compared to the first two terms which depend linearly on

$\| E_1 \|_2$ and $\| E_2 \|_2$. The first term is analogous to the bound (4.4.7) for linear systems. It says that the part of the relative error lying in $\mathcal{R}(A)$ is magnified by the constant $2\varkappa$.

The second term says that the part of the relative error lying in the orthogonal complement of $\mathcal{R}(A)$ is magnified by a factor of $4\varkappa^2 \| b_2 \|_2 / \| b_1 \|_2$. Now the ratio $\| b_2 \|_2 / \| b_1 \|_2$ measures how nearly b lies in $\mathcal{R}(A)$; if b is very nearly in $\mathcal{R}(A)$ the ratio will be very small. Thus the significance of the second term depends on how b is situated with respect to $\mathcal{R}(A)$. If $\| E_1 \|_2$ and $\| E_2 \|_2$ are comparable, then the first term dominates when b_2 is small while the second term dominates when b_2 is significant. Loosely stated, *if b very nearly lies in $\mathcal{R}(A)$, then $\varkappa(A)$ is the condition number for the least squares problem, otherwise $\varkappa^2(A)$ is the condition number.* This effect can be quite dramatic, as the following example shows.

EXAMPLE 2.8. The matrix

$$A = \begin{pmatrix} 1.000 & 1.050 \\ 1.000 & 1.000 \\ 1.000 & 1.000 \end{pmatrix}$$

is moderately ill conditioned, having a condition number of about 10^2. Let

$$A + E = \begin{pmatrix} 1.000 & 1.051 \\ 1.000 & 1.001 \\ 1.000 & 1.000 \end{pmatrix}.$$

Then E_1 and E_2 are comparable (what are they?). The following table gives least squares solutions for two different vectors b.

b	2.050 2.000 2.000	2.050 1.500 2.500
$A^\dagger b$	1.000 1.000	1.000 1.000
$(A + E)^\dagger b$	1.0097 0.9898	1.3088 0.6958

The first vector b lies in $\mathcal{R}(A)$. Hence the first term of the bound dominates and the least squares solution is not perturbed too badly. The second vector b has projections

$$b_1 = \begin{pmatrix} 2.050 \\ 2.000 \\ 2.000 \end{pmatrix}, \qquad b_2 = \begin{pmatrix} 0.000 \\ -0.500 \\ 0.500 \end{pmatrix}.$$

Hence $\| b_2 \|_2 / \| b_1 \|_2$ is significant, and the least squares solution is much more sensitive to the perturbation E.

The remainder of this section, as well as the next section, is devoted to the problem of computing least squares solutions. One way that immediately suggests itself is to solve the normal equations (2.6) for the solution x. Since we are assuming that A has linearly independent columns, $A^{\mathrm{T}}A$ is positive definite and can be reduced by the Cholesky algorithm. Thus we are led to the following tentative algorithm.

1) $C = A^{\mathrm{T}}A$
2) $d = A^{\mathrm{T}}b$
3) Solve the system $Cx = d$ using Algorithms 3.3.9 and 3.4.5

However, this algorithm requires some modification before it will work.

The first difficulty is that the condition of the system of normal equations may be worse than that of the original least squares problem. For example, suppose that A has elements of order unity and that the calculations are performed in t-digit arithmetic. Then the error made in calculating $A^{\mathrm{T}}A$ will be of the order 10^{-t}. If no further errors are made in the calculations, the computed solution will have an error on the order of $\varkappa(A^{\mathrm{T}}A)10^{-t}$. On the other hand, if b_2 is sufficiently small, then from Theorem 2.7 it follows that we should expect an error in x from rounding the elements of A on the order of $\varkappa(A)10^{-t}$. Since $\varkappa(A^{\mathrm{T}}A) = \varkappa^2(A)$, the error induced by solving the normal equations may be much larger than the error induced by rounding A. It is true that if b_2 is not small, then the two errors will be comparable. It is further true that in many cases we have no reason for believing that b_2 is small. However, an algorithm that may introduce unnecessary errors, even if only in exceptional cases, is unsatisfactory.

A more subtle problem is that the computed $A^{\mathrm{T}}A$ may not be positive definite, so that we cannot solve the normal equation by Cholesky's method. Suppose for example that $A^{\mathrm{T}}A$ is computed exactly and then rounded to t figures to give a matrix X. Then

$$X = A^{\mathrm{T}}A + E,$$

where $\| E \|_2 \cong \| A^T A \|_2 10^{-t}$. Hence

$$\begin{aligned} \| (A^T A)^{-1} E \|_2 &\le \| (A^T A)^{-1} \|_2 \| A^T A \|_2 10^{-t} \\ &= \varkappa(A^T A) 10^{-t} = \varkappa^2(A) 10^{-t}. \end{aligned} \tag{2.13}$$

Now we cannot guarantee that X is nonsingular unless $\| (A^T A)^{-1} E \|_2 < 1$. Thus the bound (2.13) suggests that unless $\varkappa(A) < 10^{t/2}$ the computed $A^T A$ may be singular or fail to be positive definite. It should be stressed that the condition $\varkappa(A) < 10^{t/2}$ is quite restrictive: a matrix with patently independent columns may fail it.

EXAMPLE 2.9. Let

$$A = \begin{pmatrix} 1.000 & 1.020 \\ 1.000 & 1.000 \\ 1.000 & 1.000 \end{pmatrix}.$$

Then $A^T A$ rounded to four figures is

$$X = \begin{pmatrix} 3.000 & 3.020 \\ 3.020 & 3.040 \end{pmatrix}.$$

However, X is not positive definite, since $(1, -1)\, X(1, -1)^T = 0$.

The above considerations suggests that if $\varkappa(A) < 10^t$, we will be safe computing in $2t$-digit arithmetic. In practice, we cannot know $\varkappa(A)$ before the calculations start. However, it the elements of A are known to t figures and $\varkappa(A) > 10^t$, we cannot guarantee that the approximate A has linearly independent columns. In other words if our problem is numerically well posed in t-digit arithmetic, we must have $\varkappa(A) < 10^t$. Using this criterion, we obtain the following algorithm.

ALGORITHM 2.10. Let A be an $m \times n$ matrix whose elements are known to t figures and let b be an m-vector. Perform the following calculations in $2t$-digit arithmetic.

1) $C = A^T A$
2) $d = A^T b$
3) Solve the system $Cx = d$

The statement that A is known to t figures is imprecise, since some elements will be more accurate than others. In practice one must estimate the overall accuracy of the elements of A and then work with the most convenient degree of precision that exceeds twice that accuracy. For example, if the elements of A had been calculated in single precision, one would work in double precision.

The requirement that the calculations be performed in double precision is not as onerous as one might suppose. The calculation of the lower half of $A^{\mathrm{T}}A$ (the upper half need not be calculated, since $A^{\mathrm{T}}A$ is symmetric) requires about $\frac{1}{2}mn^2$ additions and $\frac{1}{2}mn^2$ multiplications. If the computer being used can accumulate inner products in double precision, then only the additions need be performed in double precision. Moreover, in many problems $m \gg n$ so that the calculation of $A^{\mathrm{T}}A$ comprises the bulk of the work.

When m is very large, it may be impossible to retain the elements of the matrix A in the high speed storage of a computer. In this case it is possible to calculate $A^{\mathrm{T}}A$ and $A^{\mathrm{T}}b$ by bringing A in by blocks from some auxiliary storage. Specifically let A be partitioned in the form,

$$A = \begin{pmatrix} A_1 \\ A_2 \\ \vdots \\ A_k \end{pmatrix},$$

where each submatrix A_i can be contained in the high speed memory of the computer. Then

$$A^{\mathrm{T}}A = \sum_{i=1}^{k} A_i^{\mathrm{T}}A_i.$$

Hence $A^{\mathrm{T}}A$ may be calculated by the algorithm

1) $C = 0$

2) For $i = 1, 2, \ldots, k$
   1) $C \leftarrow C + A_i^{\mathrm{T}}A_i$

which at the ith stage requires only the submatrix A_i. Similarly if $b = (b_1^{\mathrm{T}}, b_2^{\mathrm{T}}, \ldots, b_k^{\mathrm{T}})^{\mathrm{T}}$ is partitioned conformally with A, then $A^{\mathrm{T}}b$ may be calculated by the algorithm

1) $d = 0$

2) | For $i = 1, 2, \ldots, k$
   | 1) $d \leftarrow d + A_i^{\mathrm{T}} b_i$

Of course the accumulation of C and d should be done at the same time.

EXERCISES

1. Let A have linearly independent columns. Show that
   (a) $AA^{\dagger}A = A$,
   (b) $A^{\dagger}AA^{\dagger} = A^{\dagger}$,
   (c) $A^{\dagger}A = (A^{\dagger}A)^{\mathrm{T}}$,
   (d) $AA^{\dagger} = (AA^{\dagger})^{\mathrm{T}}$.

2. Prove that if A has orthonormal columns, then $A^{\dagger} = A^{\mathrm{T}}$.

3. Let $A \in \mathbb{R}^{m \times n}$ have orthonormal columns. Give explicitly the solution of the linear least squares problem of minimizing $\| b - Ax \|_2$.

4. Let

$$A_\varepsilon = \begin{pmatrix} 1 & 1+\varepsilon \\ 1 & 1-\varepsilon \\ 1 & 1 \end{pmatrix}.$$

   Compute an upper bound on $\varkappa(A)$ as a function of ε.

5. Rework Example 2.8 with

$$A + E = \begin{pmatrix} 1.000 & 1.051 \\ 1.000 & 1.000 \\ 1.000 & 1.000 \end{pmatrix}.$$

   Explain your results. [*Hint*: What is E_2?]

6. Let $\mathcal{S} \subset \mathbb{R}^n$ be a subspace and let the columns of U form an orthonormal basis for $\mathcal{S}$. Let $P = UU^{\mathrm{T}}$. Show that for any $x \in \mathbb{R}^n$, Px is the projection of x onto $\mathcal{S}$. Conclude that the function that takes the projection of x onto $\mathcal{S}$ is a linear transformation on $\mathbb{R}^n$. The matrix of this linear transformation is called a *projector*. [*Hint*: Write $x = s + t$, where $s \in \mathcal{S}$ and t belongs to the orthogonal complement of $\mathcal{S}$. Show that $Ps = s$ and $Pt = 0$.]

7. Let P be a projector. Show that for any x,

$$\| x \|_2^2 = \| Px \|_2^2 + \| (I - P)x \|_2^2.$$

8. Let P be a projector. Show that P is symmetric ($P^{\mathrm{T}} = P$) and idempotent ($P^2 = P$).

9. Let P be the projector onto the subspace $\mathcal{S}$. Show that $I - P$ is the projector onto the orthogonal complement of $\mathcal{S}$.

10. Let P be a symmetric idempotent matrix. Show that P is the projector onto $\mathcal{R}(P)$.

11. Prove that if A has linearly independent columns, then $A(A^{\mathrm{T}}A)^{-1}A^{\mathrm{T}}$ is the projector onto $\mathcal{R}(A)$.

12. Let $v \in \mathbb{R}^n$ be nonzero. Construct the projector onto the space spanned by v.

13. Construct the projectors onto the spaces

$$\mathcal{R}\left[\begin{pmatrix} 1.000 & 1.001 \\ 1.000 & 1.000 \\ 1.000 & 1.000 \end{pmatrix}\right] \quad \text{and} \quad \mathcal{R}\left[\begin{pmatrix} 1.000 & 1.001 \\ 1.000 & 1.001 \\ 1.000 & 1.000 \end{pmatrix}\right].$$

[*Hint*: First construct the projectors onto the orthogonal complements.]

14. Let P and Q be projectors. Show that

$$PQ = QP = 0$$

if and only if $\mathcal{R}(P) \cap \mathcal{R}(Q) = \{0\}$ and $\mathcal{R}(P)$ and $\mathcal{R}(Q)$ are orthogonal. In this case $P + Q$ is the projector onto $\mathcal{R}(P) + \mathcal{R}(Q)$.

15. Show that if P is a projector and x is a nonzero vector such that $Px = \lambda x$, then $\lambda = 0$ or $\lambda = 1$.

NOTES AND REFERENCES

The geometric approach to the linear least squares problem taken here is standard. The operation of taking the projection of a vector onto a subspace is a linear transformation called the (orthogonal) projector onto the subspace. The theory of projectors is developed in the exercises.

Theorem 2.7 is due to Stewart (1969a). Golub and Wilkinson (1966) have given first-order estimates for the perturbed solution that amount to the same thing. A third approach is taken by Björk (1967a,b).

Until recently linear least squares problems were solved exclusively by forming and solving the normal equations. A piece of folklore from this period, presumably due to the problems associated with the squaring of the condition number, is that the normal equations must be solved in double precision. The analysis in the text shows this is only half the truth; if the matrix $A^{\mathrm{T}}A$ is not computed to high precision, even the exact solution of the resulting system may be inaccurate.

The emergence of good alternative techniques for solving linear least squares problems has caused the pendulum to swing in the other direction, and one frequently hears the warning that one should never use the normal equations. Although the normal equations are indeed tricky to use, the simplicity of the technique may make them preferable when A has special structure or when it is expected that A will be modified slightly from time to time. And if it is known *a priori* that A is reasonably well conditioned, single-precision calculations should suffice. In short, although the normal equations method cannot be recommended for general purpose use, it is a useful tool to have around.

The pseudo-inverse was first introduced by Moore (1919–1920) and later rediscovered by Penrose (1955, 1956), whose papers initiated a vogue in the subject. Bibliographies may be found in the books of Boullion and Odel (1971) and Rao and Mitra (1971). A survey of computational techniques for computing the pseudo-inverse is given by Peters and Wilkinson (1970b).

3. ORTHOGONAL TRIANGULARIZATION

The algorithm of Gaussian elimination as described in Chapter 3 reduces a matrix to upper triangular form by premultiplying it by a sequence of elementary lower triangular matrices. Each elementary lower triangular matrix is chosen to introduce zeros below a diagonal element. What makes it possible to choose such elementary lower triangular matrices is the fact that given a vector x, there is an elementary lower triangular matrix M such that Mx is a multiple of e_1. It is this property that is essential to the reduction; any other class of matrices having it can be used in place of elementary lower triangular matrices to reduce a matrix to lower triangular form. In this section we shall derive a triangularization algorithm based on a class of orthogonal matrices that have this crucial property. The algorithm will be applied to the solution of least squares problems and the computation of projections.

DEFINITION 3.1. An *elementary reflector* is a matrix of the form

$$I - 2uu^{\mathrm{T}},$$

where $u^{\mathrm{T}}u = 1$.

Elementary reflectors are also known as *elementary Hermitian matrices* and as *Householder transformations.* In fact either of the last two names is more common than the name elementary reflector. For the geometric consideration behind this terminology see Exercise 3.2.

Some simple consequences of the definition of elementary reflectors are contained in the following theorem.

THEOREM 3.2. Let U be an elementary reflector. Then

1. U is symmetric $(U^{\mathrm{T}} = U)$,
2. U is orthogonal $(U^{\mathrm{T}}U = I)$,
3. U is involutory $(U^2 = I)$.

PROOF. Let $U = I - 2uu^{\mathrm{T}}$, where $u^{\mathrm{T}}u = 1$. The matrix U is obviously symmetric. Moreover

$$\begin{aligned} U^{\mathrm{T}}U = U^2 &= (I - 2uu^{\mathrm{T}})(I - 2uu^{\mathrm{T}}) \\ &= I - 4uu^{\mathrm{T}} + 4uu^{\mathrm{T}}uu^{\mathrm{T}} = I, \end{aligned}$$

since $u^{\mathrm{T}}u = 1$. This establishes properties 2 and 3. ■

In application it frequently happens that the vector u that defines an elementary reflector is given in an unnormalized form. Since the computation of the normalized vector $u/\|u\|_2$ requires that a square root be taken to compute $\|u\|_2$, it is desirable to work with the unnormalized u. This may be done as follows.

THEOREM 3.3. Let $u \neq 0$ and let

$$\pi = \tfrac{1}{2}\|u\|_2^2.$$

Then

$$U = I - \pi^{-1}uu^{\mathrm{T}}$$

is an elementary reflector.

PROOF. Let $u' = u/\| u \|_2$. Then $\| u' \|_2 = 1$, and

$$U = I - 2\frac{uu^{\mathrm{T}}}{\| u \|_2^2} = I - 2u'u'^{\mathrm{T}}. \quad \blacksquare$$

The crucial property of elementary reflectors is that they can be used to introduce zeros into a vector. Specifically, given a vector x one can find an elementary reflector U such that $Ux = -\sigma e_1$ (the minus sign simplifies the notation in the sequel). The constant σ is determined up to its sign by the fact that U is orthogonal; for

$$| \sigma | = \| -\sigma e_1 \|_2 = \| Ux \|_2 = \| x \|_2,$$

so that $\sigma = \pm \| x \|_2$. The recipe for determining u is given in the following theorem.

THEOREM 3.4. Let $x \in \mathbb{R}^n$, $\sigma = \pm \| x \|_2$, and suppose that $x \neq -\sigma e_1$. Let

$$u = x + \sigma e_1$$

and

$$\pi = \tfrac{1}{2} \| u \|_2^2.$$

Then $U = I - \pi^{-1}uu^{\mathrm{T}}$ is an elementary reflector and $Ux = -\sigma e_1$.

PROOF. Since $x \neq -\sigma e_1$, $u \neq 0$, and by Theorem 3.3 U is an elementary reflector. Now since $x^{\mathrm{T}}x = \sigma^2$,

$$\begin{aligned} \pi &= \tfrac{1}{2} (x + \sigma e_1)^{\mathrm{T}}(x + \sigma e_1) \\ &= \tfrac{1}{2} (x^{\mathrm{T}}x + 2\sigma\xi_1 + \sigma^2) \\ &= \sigma^2 + \sigma\xi_1. \end{aligned}$$

Hence

$$\begin{aligned} Ux = x - \frac{uu^{\mathrm{T}}x}{\pi} &= x - \frac{(x + \sigma e_1)(x + \sigma e_1)^{\mathrm{T}}x}{\sigma^2 + \sigma\xi_1} \\ &= x - \frac{(x + \sigma e_1)(x^{\mathrm{T}}x + \sigma\xi_1)}{\sigma^2 + \sigma\xi_1} \\ &= x - (x + \sigma e_1) = -\sigma e_1. \quad \blacksquare \end{aligned}$$

The statement of Theorem 3.4 is almost an algorithm for computing u and π. Once the sign of σ has been decided upon, we can compute

$$\begin{aligned} v_1 &= \xi_1 + \sigma, \\ v_i &= \xi_i \qquad (i = 2, 3, \ldots, n), \end{aligned} \tag{3.1}$$

and

$$\pi = \sigma(\sigma + \xi_1) = \sigma v_1.$$

If σ and ξ_1 have different signs, then cancellation can occur in the computation of v_1 via (3.1). Hence we take σ to have the same sign as ξ_1; that is,

$$\sigma = \mathrm{sign}(\xi_1) \, \| x \|_2 .$$

Note that with this choice of σ we never have $x = -\sigma e_1$, unless $x = 0$.

In applications it is often the case that having formed u we no longer need x (or more accurately we need only $Ux = -\sigma e_1$, which is determined by the constant σ). Hence the components of u may overwrite the components of x. But the last $n - 1$ components of u are identical with the last $n - 1$ components of x; hence only the first component of x need be altered in computing u.

These considerations may be summed up in the following algorithm.

ALGORITHM 3.5. Given the nonzero n-vector x, this algorithm returns σ, π, and u such that $(I - \pi^{-1}uu^{\mathrm{T}})x = -\sigma e_1$. The components of u overwrite those of x.

1) $\sigma = \mathrm{sign}(\xi_1)\sqrt{\xi_1^2 + \xi_2^2 + \cdots + \xi_n^2}$
2) $\xi_1 \leftarrow v_1 = \xi_1 + \sigma$
3) $\pi = \sigma v_1$

Algorithm 3.5 is satisfactory for most purposes; however, it can fail because of overflows and underflows. For example, if the largest floating-point number is 10^{50} and some component of x is of order 10^{30}, the algorithm will overflow at statement 1 when that component is squared. An underflow may be even more serious, for many computer systems set underflows to zero without telling the user. If, for example, all the components of x underflow when they are squared, then the computed value of σ will be zero. This error will not be detected until later when one attempts to form the reciprocal of π, whose computed value σv_1 is also zero.

The cure for this problem lies in observing that the matrix U does not change when x is multiplied by a scalar (although π and u will change). Thus we may scale x so that $\| x \|_2^2$ can be computed accurately. This leads to the following algorithm.

ALGORITHM 3.6. Given the nonzero n-vector x, this algorithm returns σ, π, and u such that $(I - \pi^{-1}uu^{\mathrm{T}})x = -\sigma e_1$. The components of u overwrite those of x. The algorithm will work even when the squares of the components of x overflow or underflow.

1) $\eta = \max\{| \xi_i | : i = 1, 2, \ldots, n\}$
2) $\xi_i \leftarrow v_i = \xi_i/\eta \qquad (i = 1, 2, \ldots, n)$
3) $\sigma = \mathrm{sign}(v_1)\sqrt{v_1{}^2 + v_2{}^2 + \cdots + v_n{}^2}$
4) $v_1 \leftarrow v_1 + \sigma$
5) $\pi = \sigma v_1$
6) $\sigma \leftarrow \eta\sigma$

It should be stressed that the representation $U = I - \pi^{-1}uu^{\mathrm{T}}$ is not necessarily the best one for all purposes. When n is 2 or 3, for example, the representation $U = I - uv^{\mathrm{T}}$ $(v = \pi^{-1}u)$ is sometimes used (cf. Algorithm 7.4.4).

Once U has been determined, say by Algorithm 3.6, it is usually required to calculate UA, where A is some given matrix. If we write $A = (a_1, a_2, \ldots, a_r)$, then $UA = (Ua_1, Ua_2, \ldots, Ua_r)$, so that the problem of calculating UA is reduced to the problem of calculating the product Ua, where a is a vector. Needless to say, we do not actually form the matrix $U = I - \pi^{-1}uu^{\mathrm{T}}$; rather we write

$$Ua = (I - \pi^{-1}uu^{\mathrm{T}})a = a - (\pi^{-1}u^{\mathrm{T}}a)u.$$

This gives the following algorithm.

ALGORITHM 3.7. Let $u, a \in \mathbb{R}^n$ and $\pi \neq 0$. This algorithm overwrites a with the product $(I - \pi^{-1}uu^{\mathrm{T}})a$.

1) $\tau = \pi^{-1} \sum_{i=1}^{n} v_i\alpha_i$
2) $\alpha_i \leftarrow \alpha_i - \tau v_i \qquad (i = 1, 2, \ldots, n)$

Some accuracy may be gained in Algorithm 3.7 by accumulating the inner product in statement 1 in double precision. The algorithm requires $2n$ multiplications. Hence if $A \in \mathbb{R}^{n\times r}$ and UA is formed as suggested above, the cost will be $2nr$ multiplications.

We are now in a position to describe the algorithm of orthogonal triangularization, sometimes called Householder's reduction to upper triangular form. Given $A \in \mathbb{R}^{m\times n}$, the algorithm produces elementary reflectors $U_1, U_2, \ldots, U_r$ such that the product $A_{r+1} = U_r U_{r-1} \cdots U_1 A$ is upper trapezoidal. The algorithm is quite analogous to Gaussian elimination without pivoting. At the kth step, the matrix $A_k = U_{k-1} U_{k-2} \cdots U_1 A$ has the form illustrated below ($m = 6$, $n = 5$, $k = 3$)

$$\left(\begin{array}{cc:c:cc} x & x & x & x & x \\ 0 & x & x & x & x \\ \hdashline 0 & 0 & x & x & x \\ 0 & 0 & \textcircled{x} & x & x \\ 0 & 0 & \textcircled{x} & x & x \\ 0 & 0 & \textcircled{x} & x & x \end{array}\right).$$

We choose U_k to introduce zeros below the kth diagonal element as indicated.

In more detail, at the kth step, A_k has the form

$$A_k = \begin{pmatrix} R_k & r_k & B_k \\ 0 & c_k & D_k \end{pmatrix}, \tag{3.2}$$

where R_k is an upper triangular matrix of order $k - 1$ and $c_k \in \mathbb{R}^{m-k+1}$. We use Algorithm 3.5 or Algorithm 3.6 to find $u_k \in \mathbb{R}^{m-k+1}$ and π_k so that if $U_k' = I - \pi^{-1} u_k u_k^{\mathrm{T}}$, then $U_k' c_k = \varrho_{kk} e_1$. If we set

$$U_k = \begin{pmatrix} I_k & 0 \\ 0 & U_k' \end{pmatrix},$$

then

$$A_{k+1} = U_k A_k = \begin{pmatrix} R_k & r_k & B_k \\ 0 & \varrho_{kk} e_1 & U_k' D_k \end{pmatrix},$$

which carries the triangularization one step further. The upper triangular matrix R_{k+1} is given by

$$R_{k+1} = \begin{pmatrix} R_k & r_k \\ 0 & \varrho_{kk} \end{pmatrix}.$$

So far as storage is concerned we may overwrite c_k with $u_k = (v_{kk}, v_{k+1,k}, \ldots, v_{mk})^{\mathrm{T}}$. The number π_k and the diagonal element ϱ_{kk} must be stored elsewhere. One convenient artifice is to require that the array A have two additional rows and store π_k and ϱ_{kk} there. Thus for $m = 6$, $n = 5$, and $k = 3$ the matrix A will have the form [cf. (3.2)]

$$\begin{pmatrix} v & \varrho & \varrho & \beta & \beta \\ v & v & \varrho & \beta & \beta \\ v & v & \gamma & \delta & \delta \\ v & v & \gamma & \delta & \delta \\ v & v & \gamma & \delta & \delta \\ v & v & \gamma & \delta & \delta \\ \pi & \pi & & & \\ \varrho & \varrho & & & \end{pmatrix}.$$

To calculate A_{k+1} from A_k it is only necessary to overwrite D_k with $U_k' D_k$. This can be done by columns via Algorithm 3.7. All the above considerations are summed up in the following algorithm.

ALGORITHM 3.8. Let $A \in \mathbb{R}^{m \times n}$ and $r = \min\{m-1, n\}$. This algorithm returns elementary reflectors $U_1, U_2, \ldots, U_r$ such that $A_{r+1} = U_r U_{r-1} \cdots U_1 A$ is upper trapezoidal. The reflectors U_k are in the form $U_k = I - \pi_k^{-1} u_k u_k^{\mathrm{T}}$, where $u_k = (0, \ldots, 0, v_{kk}, v_{k+1,k}, \ldots, v_{mk})^{\mathrm{T}}$. The nonzero elements v_{ik} overwrite α_{ik}. The scalars π_k overwrite $\alpha_{m+1,k}$. The nonzero elements of A_{r+1}, that is the elements of R_{r+1}, overwrite the corresponding elements of A, except for the diagonal elements ϱ_{kk} which are stored in $\alpha_{m+2,k}$.

1) For $k = 1, 2, \ldots, r$
   1) $\eta = \max\{|\alpha_{ik}| : i = k, k+1, \ldots, m\}$
   2) If $\eta = 0$
      1) $\alpha_{m+1,k} = 0$
      2) Step k
   3) $\alpha_{ik} \leftarrow v_{ik} = \alpha_{ik}/\eta \qquad (i = k, k+1, \ldots, m)$
   4) $\sigma = \operatorname{sign}(v_{kk}) \sqrt{v_{kk}^2 + \cdots + v_{mk}^2}$
   5) $v_{kk} = v_{kk} + \sigma$
   6) $\alpha_{m+1,k} \leftarrow \pi_k = \sigma v_{kk}$
   7) $\alpha_{m+2,k} \leftarrow \varrho_{kk} = -\eta\sigma$

8) For $j = k + 1, k + 2, \ldots, n$

1) $\tau = \pi_k^{-1} \sum_{i=k}^{m} v_{ik}\alpha_{ij}$

2) $\alpha_{ij} \leftarrow \alpha_{ij} - \tau v_{ik} \qquad (i = k, k + 1, \ldots, m)$

2) If $m \leq n$, $\alpha_{m+2,m} \leftarrow \varrho_{mm} = \alpha_{mm}$

We have chosen to use the safer Algorithm 3.6 to determine the vectors u_k. The product $U_k'D_k$ is computed in statement 1.8 using Algorithm 3.7. Some accuracy may be gained by accumulating inner products in double precision in statements 1.4 and 1.8.1.

If, at statement 1.2, η is zero, then the elements below the kth diagonal element are already zero. In this case no transformation is performed and this is signaled by setting π_k to zero.

If $m \geq n$, the most frequently occurring case, then the algorithm requires about $mn^2 - \frac{1}{3}n^3$ multiplications.

Algorithm 3.8 is quite stable. If the computations were carried out exactly, Algorithm 3.8 would produce an orthogonal matrix

$$Q = U_r U_{r-1} \cdots U_1 \tag{3.3}$$

such that $A_{r+1} = QA$ is upper trapezoidal. The following theorem, which we state without proof, shows that Q in effect reduces a slightly perturbed matrix.

THEOREM 3.9. Let A be reduced by Algorithm 3.8 in t-digit arithmetic, producing the upper trapezoidal matrix A_{r+1}. Then there is an orthogonal matrix Q and a matrix E such that

$$Q(A + E) = A_{r+1}$$

where

$$\| E \|_{\mathrm{F}} \leq \phi(m, r) \| A \|_{\mathrm{F}} 10^{-t}. \tag{3.4}$$

Moreover, if c is the numerical result of applying transformations $U_1, U_2, \ldots, U_r$ defined by Algo· hm 3.8 to the vector b, then there is a vector e such that

$$Q(b + e) = c$$

and

$$\| e \|_2 \leq \phi(m, r) \| b \|_2 \cdot 10^{-t}. \tag{3.5}$$

In both (3.4) and (3.5), ϕ is a low degree polynomial in m and r. If inner products are accumulated in double precision, then ϕ is independent of m.

We turn now to applications of Algorithm 3.8. Perhaps the most immediate application of the algorithm is the computation of the QR factorization of A. Assume that A has linearly independent columns (hence $m \geq n$). With Q defined by (3.3) set

$$A_{k+1} = QA = \begin{pmatrix} R \\ 0 \end{pmatrix}, \tag{3.6}$$

where R is upper triangular of order n. Then if we partition Q^{T} in the form

$$Q^{\mathrm{T}} = (Q_1, Q_2), \tag{3.7}$$

where Q_1 has n columns, we have

$$A = Q^{\mathrm{T}} A_{k+1} = (Q_1, Q_2)\begin{pmatrix} R \\ 0 \end{pmatrix} = Q_1 R. \tag{3.8}$$

Since Q_1 has orthonormal columns and R is upper triangular, Equation (3.8) gives the QR factorization of A up to the signs of the diagonal elements of R.

The matrix R of (3.8) is produced explicitly in the course of Algorithm 3.8. The matrix Q_1 may be computed as the first n columns of the product

$$Q^{\mathrm{T}} = U_1 U_2 \cdots U_r.$$

However, it should be stressed that in most applications the matrix Q or the matrix Q_1 is not needed; rather it is sufficient to know the factored form (3.3). We shall illustrate this point by applying the output of Algorithm 3.8 to two problems: the computation of least squares solutions and the computation of projections. For the rest of this section we shall assume that A has linearly independent columns, so that R, defined by (3.6), is a nonsingular upper triangular matrix of order $n \leq m$.

We first consider the linear least squares problem of minimizing the 2-norm of

$$r = b - Ax. \tag{3.9}$$

If we set

$$Qb = \begin{pmatrix} c \\ d \end{pmatrix}, \tag{3.10}$$

where $c \in \mathbb{R}^n$, then, upon multiplying (3.9) by Q, we obtain

$$Qr = \begin{pmatrix} c \\ d \end{pmatrix} - \begin{pmatrix} R \\ 0 \end{pmatrix} x = \begin{pmatrix} c - Rx \\ d \end{pmatrix}. \tag{3.11}$$

Since Q is an orthogonal matrix,

$$\| r \|_2^2 = \| Qr \|_2^2 = \| c - Rx \|_2^2 + \| d \|_2^2. \tag{3.12}$$

Thus $\| r \|_2^2$ will be minimized if x is chosen so that

$$c - Rx = 0. \tag{3.13}$$

In this case it follows from (3.12) that $\| r \|_2^2 = \| d \|_2^2$. Moreover, from (3.11),

$$Qr = \begin{pmatrix} 0 \\ d \end{pmatrix}.$$

Hence

$$r = Q^{\mathrm{T}} \begin{pmatrix} 0 \\ d \end{pmatrix}. \tag{3.14}$$

Equations (3.10), (3.13), and (3.14) effectively give an algorithm for using the output of Algorithm 3.8 to solve the linear least squares problem. Note that we may calculate the product Qb in the form $U_r U_{r-1} \cdots U_1 b$ without ever having to calculate Q. Likewise we may calculate r in (3.14) from the factored form of Q^{T}.

ALGORITHM 3.10. Let the matrix A have linearly independent columns, let the array A contain the output of Algorithm 3.8, and let $b \in \mathbb{R}^m$. This algorithm solves the linear least squares problem of minimizing the 2-norm of $r = b - Ax$. The residual vector r at the minimum overwrites b.

1) For $i = 1, 2, \ldots, n$
   1) $b \leftarrow U_i b$
2) $\gamma_i = \beta_i \quad (i = 1, 2, \ldots, n)$
3) Solve the system $Rx = c$
4) $\beta_i = 0 \quad (i = 1, 2, \ldots, n)$
5) For $i = n, n - 1, \ldots, 1$
   1) $b \leftarrow U_i b$

The products $U_i b$ in statements 1.1 and 5.1 may be computed using Algorithm 3.7. In this connection it should be noted that only the last $m - i + 1$ components of b are altered.

Algorithm 3.10 is a very stable way of solving the linear least squares problem. To see this let Q, E, and e be the quantities whose existence is insured by Theorem 3.9. Then

$$Q(A + E) = \begin{pmatrix} R \\ 0 \end{pmatrix}$$

and

$$Q(b + e) = \begin{pmatrix} c \\ d \end{pmatrix}. \tag{3.15}$$

Now by Theorem 3.5.1 the solution x computed at statement 3 satisfies

$$(R + F)x = c, \tag{3.16}$$

where F is small. If we set

$$G = E + Q^{\mathrm{T}}\begin{pmatrix} F \\ 0 \end{pmatrix},$$

then, because Q is orthogonal, G is small along with E and F. Moreover

$$Q(A + G) = \begin{pmatrix} R + F \\ 0 \end{pmatrix}. \tag{3.17}$$

Hence if the reasoning used to derive Algorithm 3.10 is applied to (3.17), (3.15), and (3.16), we see that the computed solution x is the solution to the least squares problem of minimizing

$$\| (A + G)x - (b + e) \|_2^2. \tag{3.18}$$

When $m = n$, Algorithm 3.10, with statements 4 and 5 deleted, becomes an algorithm for solving the linear system $Ax = b$. The considerations of the last paragraph show that the computed solution satisfies $(A + G)x = b + e$. The bounds on G and e are quite comparable with the corresponding bounds for the methods of Section 3.4. Indeed, they are in some respects better, since they contain no growth factors. Thus orthogonal triangularization is an unconditionally stable way of solving linear systems. Nonetheless, the methods of Section 3.4 are usually preferred in practice. There are two reasons for this. In the first place, the reduction of Algorithm 3.8 requires about $\frac{2}{3}n^3$ multiplications as compared with $\frac{1}{3}n^3$ multiplications for Gaussian elimination or Crout reduction. In the second place, as we have already noted, the growth of elements that could render the methods of Chapter 3 unstable does not occur in practice.

As a final example of an application of Algorithm 3.8, we shall show how to compute the projection of a vector b onto $\mathcal{R}(A)$ and onto the orthogonal complement of $\mathcal{R}(A)$. Actually we have seen how to compute the latter projection, since we know from the theory of least squares that the residual vector computed in Algorithm 3.10 is just this projection. To give a more systematic derivation, let Q be partitioned as in (3.7) and Qb be partitioned as in (3.10). Then

$$b = Q^{\mathrm{T}}Qb = (Q_1, Q_2)\begin{pmatrix} c \\ d \end{pmatrix} = Q_1c + Q_2d.$$

Now $\mathcal{R}(Q_1) = \mathcal{R}(A)$ and since Q^{T} is orthogonal, $\mathcal{R}(Q_2)$ is the orthogonal complement of $\mathcal{R}(A)$. Hence $b_1 = Q_1c$ is the projection onto $\mathcal{R}(A)$ and $b = Q_2d$ is the projection onto the orthogonal complement of $\mathcal{R}(A)$.

For computational purposes it is convenient to take b_1 and b_2 in the form

$$b_1 = Q^{\mathrm{T}}\begin{pmatrix} c \\ 0 \end{pmatrix} \quad \text{and} \quad b_2 = Q^{\mathrm{T}}\begin{pmatrix} 0 \\ d \end{pmatrix}.$$

Again it is not necessary to form the matrix Q^{T} explicitly. Instead we may work with the factored form $Q^{\mathrm{T}} = U_rU_{r-1}\cdots U_1$.

ALGORITHM 3.11. Let A have linearly independent columns, let the array A contain the output of Algorithm 3.8, and let $b \in \mathbb{R}^m$. This algorithm computes the projection b_1 of b onto $\mathcal{R}(A)$ and the projection b_2 of b onto the orthogonal complement of $\mathcal{R}(A)$.

1) For $i = 1, 2, \ldots, r$
   1) $b \leftarrow U_ib$
2) $\beta_i^{(1)} = \beta_i \quad (i = 1, 2, \ldots, n)$
3) $\beta_i^{(1)} = 0 \quad (i = n+1, n+2, \ldots, m)$
4) $\beta_i^{(2)} = 0 \quad (i = 1, 2, \ldots, n)$
5) $\beta_i^{(2)} = \beta_i \quad (i = n+1, n+2, \ldots, m)$
6) For $i = r, r-1, \ldots, 1$
   1) $b_1 \leftarrow U_ib_1$
   2) $b_2 \leftarrow U_ib_2$

In theory it is only necessary to calculate, say, b_1, since b_2 can then be computed from the relation

$$b_2 = b - b_1. \tag{3.19}$$

However if b_2 is very small compared with b and b_1, cancellation will occur in (3.19) and the computed b_2 will not be orthogonal to b_1. The vectors b_1 and b_2 produced by Algorithm 3.11 are always very nearly orthogonal.

EXERCISES

1. In $\mathbb{R}^1$ show that if $x = y$, then there is no elementary reflector U such that $Ux = y$.

2. Let $U = I - 2uu^{\mathrm{T}}$ be an elementary reflector. Let x be given and let $x = v + w$ where v lies along u and w is orthogonal to u. Show that $Ux = -v + w$. Interpret this result geometrically in $\mathbb{R}^n$.

3. Let $x, y \in \mathbb{R}^n$ $(n > 1)$. Show that if $\| x \|_2 = \| y \|_2$, then there is an elementary reflector U such that $Ux = y$. [*Hint*: If $x \neq y$, take $u = (x - y)/\| x - y \|_2$.]

4. Show that if $x \in \mathbb{R}^n$, there is an elementary reflector U such that $Ux = -\sigma e_n$. Derive algorithms for computing U.

5. Let $A \in \mathbb{R}^{m \times n}$ with $m \geq n$. Derive an algorithm that produces elementary reflectors $U_1, U_2, \ldots, U_r$ such that $A_{r+1} = U_r U_{r-1} \cdots U_1 A$ is lower trapezoidal. [*Hint*: Starting with column n and working backward, introduce zeros above the diagonal elements. Use Exercise 3.4.]

6. Let A have the form

$$A = \begin{pmatrix} R \\ S \end{pmatrix},$$

where R is upper triangular. Describe an algorithm that has elementary reflectors to reduce A to upper trapezoidal form taking advantage of the zero elements in R.

7. Expand statements 1.1 and 5.1 of Algorithm 3.10 to compute the products $U_i b$ from Algorithm 3.7.

8. Let $A \in \mathbb{R}^{m \times n}$ have linearly independent columns and $C \in \mathbb{R}^{l \times n}$ have linearly independent rows. Let $d \in \mathbb{R}^l$. Derive an algorithm for solving the least squares problem of minimizing $\| b - Ax \|_2$ subject to the constraint $Cx = d$. [*Hint*: Determine orthogonal U so that UA is upper trapezoidal and orthogonal V so that CV is lower trapezoidal. Set $y = V^{\mathrm{T}}x$ and consider the problem of minimizing $\| Ub - UAVy \|_2$ subject to the constraint $CVy = d$.]

9. Let $x \in \mathbb{R}^2$ be nonzero. Let

$$\gamma = \xi_1/\sqrt{\xi_1^2 + \xi_2^2} \quad \text{and} \quad \sigma = \xi_2/\sqrt{\xi_1^2 + \xi_2^2}.$$

Show that

$$\begin{pmatrix} \gamma & \sigma \\ -\sigma & \gamma \end{pmatrix} x = \begin{pmatrix} \sqrt{\xi_1^2 + \xi_2^2} \\ 0 \end{pmatrix}.$$

10. A matrix of the form

$$P_{ij} = \begin{matrix} \\ \\ \\ \\ i \\ \\ \\ j \\ \\ \\ \\ \end{matrix} \begin{bmatrix} 1 & 0 & \cdots & 0 & \cdots & 0 & \cdots & 0 \\ 0 & 1 & \cdots & 0 & \cdots & 0 & \cdots & 0 \\ \vdots & \vdots & & \vdots & & \vdots & & \vdots \\ 0 & 0 & \cdots & \gamma & \cdots & \sigma & \cdots & 0 \\ \vdots & \vdots & & \vdots & & \vdots & & \vdots \\ 0 & 0 & \cdots & -\sigma & \cdots & \gamma & \cdots & 0 \\ \vdots & \vdots & & \vdots & & \vdots & & \vdots \\ 0 & 0 & \cdots & 0 & \cdots & 0 & \cdots & 1 \end{bmatrix},$$

(with column i marked above the $\gamma, -\sigma$ column and column j above the σ, γ column)

where $\gamma^2 + \sigma^2 = 1$, is called a plane rotation [in the (i, j)-plane]. Show that if $x \in \mathbb{R}^n$, there is a plane rotation P_{ij} such that if $y = P_{ij}x$, then

$$\eta_k = \xi_k, \qquad k \neq i, j,$$

$$\eta_i = \sqrt{\xi_i^2 + \xi_j^2},$$

$$\eta_j = 0.$$

Give an algorithm for computing P_{ij} that will not fail when ξ_i^2 or ξ_j^2 overflows or underflows.

11. Let P_{ij} be a plane rotation in the (i, j)-plane. Describe an efficient algorithm for forming the product $P_{ij}A$.

12. Let $A \in \mathbb{R}^{m\times n}$. Use plane rotations to introduce zeros into A in the order illustrated below

$$\begin{matrix} x & x & x \\ x^1 & x & x \\ x^2 & x^5 & x \\ x^3 & x^6 & x^8 \\ x^4 & x^7 & x^9 \end{matrix}$$

without disturbing the previously introduced zeros. Give a detailed INFL algorithm for this reduction to upper trapezoidal form. Compare the operation count for this algorithm with the one for Algorithm 3.8.

NOTES AND REFERENCES

Although elementary reflectors appear in a paper by Feller and Forsythe (1951) as a special case of more general transformations, Householder (1958a) was the first to use them in a systematic way to introduce zeros into a matrix. In his 1958 paper, Householder proves the following generalization of Theorem 3.4. *Let x and y be n-vectors with $\|x\|_2 = \|y\|_2$ and $x \neq y$. Then there is an elementary reflector R such that $Rx = y$.* The proof is trivial once it is realized that the vector u determining R is $(x - y)/\|x - y\|_2$.

Although Householder mentions least squares applications in his 1958 paper, Golub (1965) was the first to work out the details and in conjunction with Businger (1965, HACLA/I/8) publish an algorithm. In many statistical applications it is necessary to compute the quadratic form $x^{\mathrm{T}}(A^{\mathrm{T}}A)^{-1}x$. Since $A^{\mathrm{T}}A = R^{\mathrm{T}}R$, this may be done by solving the system $Ry = x$ and calculating $y^{\mathrm{T}}y$. It is possible to accomplish the reduction to trapezoidal form by bringing A into the high-speed storage of the computer in blocks; however, the procedure is more complicated than the analogous procedure for forming the normal equations (cf. Exercise 3.6).

The error analysis leading to Theorem 3.9 is given by Wilkinson (1965b).

The method of orthogonal triangularization is analogous to Gaussian elimination. An alternative is to proceed as in Section 3.3 and attempt to compute the QR factorization directly from the equation $A = QR$ (or from $Q = AR^{-1}$). This leads to the Gram–Schmidt orthogonalization. There are two ways of arranging the calculations; one gives the classical Gram–Schmidt algorithm (Exercise 1.10), the other gives what is called the modified Gram–Schmidt algorithm (Exercise 1.11). Since both methods form Q as a linear combination of the columns of A either one will produce a matrix Q whose columns deviate from orthogonality when A is ill conditioned.

Since $A^{\dagger} = R^{-1}Q^{\mathrm{H}}$, the Gram–Schmidt orthogonalization can be used to solve the linear least squares problem. In this connection the classical Gram–Schmidt algorithm is unstable; the modified Gram–Schmidt algorithm, however, is quite stable, even when A is ill conditioned. This is because the deviation from orthogonality of the columns of Q is in some sense proportional to $\varkappa(A)$, which appears in the final bounds anyway. In fact the *a priori* error bounds for least squares solutions computed by modified Gram–Schmidt orthogonalization are better than the corresponding bounds for the method of orthogonal triangularization.

The above considerations illustrate the dangers of using the term "stable" without specifying the problem involved. As a means of computing the QR decomposition of a matrix, the modified Gram–Schmidt algorithm is unstable; as a means of solving least squares problems, it is stable.

Rice (1966) was the first to point out the superior numerical properties of the modified Gram–Schmidt algorithm. Björk (1967a) gives detailed error analyses.

4. THE ITERATIVE REFINEMENT OF LEAST SQUARES SOLUTIONS

In this section we shall consider a method for improving an approximate solution to the linear least squares problem. The method amounts to applying Algorithm 4.5.1 to a system of linear equations associated with the linear least squares problem. As such the method has much in common with Algorithm 4.5.1. It requires that a residual (not the residual of the least squares problem) be computed in double precision. Applied iteratively it converges to nearly the true solution, provided that the least squares matrix is not too ill conditioned. Although the method requires that the least squares matrix be triangularized as described in the last section, the same triangularization may be used repeatedly in each iteration. However, the original least squares matrix must also be retained to compute the residual.

We shall again be concerned with solving the linear least squares problem of minimizing

$$\| b - Ax \|_2^2, \tag{4.1}$$

where A is an $m \times n$ matrix with linearly independent columns. The method is based on the following trivial observation. Let x be a solution of the linear least squares problem and let $r = b - Ax$ be the associated residual vector. Then the $(n + m)$-vector $(x^{\mathrm{T}}, r^{\mathrm{T}})^{\mathrm{T}}$ satisfies the equation

$$\begin{pmatrix} A & I_m \\ 0 & A^{\mathrm{T}} \end{pmatrix} \begin{pmatrix} x \\ r \end{pmatrix} = \begin{pmatrix} b \\ 0 \end{pmatrix}. \tag{4.2}$$

To see this, observe that the equation $Ax + I_m r = b$ follows from the definition of the residual, while the equation $A^{\mathrm{T}}r = 0$ follows from Corollary 2.3.

The system (4.2) is square and, as we shall see, nonsingular. This suggests that, given an approximate solution x and an approximate residual r, we attempt to improve them by the procedure described in Section 4.5. Specifically, we compute the "residual"

$$\begin{pmatrix} s \\ t \end{pmatrix} = \begin{pmatrix} b \\ 0 \end{pmatrix} - \begin{pmatrix} A & I_m \\ 0 & A^{\mathrm{T}} \end{pmatrix} \begin{pmatrix} x \\ r \end{pmatrix} = \begin{pmatrix} b - Ax - r \\ -A^{\mathrm{T}}r \end{pmatrix}$$

in double precision, then solve the system

$$\begin{pmatrix} A & I_m \\ 0 & A^{\mathrm{T}} \end{pmatrix} \begin{pmatrix} g \\ h \end{pmatrix} = \begin{pmatrix} s \\ t \end{pmatrix} \tag{4.3}$$

in single precision. The improved solution and residual will be given by $x + g$ and $r + h$.

The chief difficulty with this scheme is that the system (4.3) is of order $n + m$. Since m may be very large, the solution of the system by the techniques of Chapter 3 may be prohibitively expensive. Fortunately the special form of the system allows its efficient solution in terms of the output of Algorithm 3.8.

To this end, suppose that Algorithm 3.8 has been applied to yield an orthogonal matrix Q (in factored form) satisfying

$$QA = \begin{pmatrix} R \\ 0 \end{pmatrix}, \tag{4.4}$$

where R is a nonsingular upper triangular matrix of order n. Let

$$Qs = u = \begin{pmatrix} u_1 \\ u_2 \end{pmatrix}, \tag{4.5}$$

and

$$Qh = k = \begin{pmatrix} k_1 \\ k_2 \end{pmatrix},$$

where $u_1, k_1 \in \mathbb{R}^n$. Now the system (4.3) may be written in the form

$$Ag + h = s \tag{4.6}$$

and

$$A^{\mathrm{T}}h = t. \tag{4.7}$$

If Equation (4.6) is premultiplied by Q, there results

$$\begin{pmatrix} R \\ 0 \end{pmatrix} g + \begin{pmatrix} k_1 \\ k_2 \end{pmatrix} = \begin{pmatrix} u_1 \\ u_2 \end{pmatrix},$$

or equivalently

$$Rg + k_1 = u_1 \tag{4.8}$$

and

$$k_2 = u_2. \tag{4.9}$$

On the other hand, since $Q^{\mathrm{T}}Q = I$, we may write (4.7) in the form

$$t = A^{\mathrm{T}}Q^{\mathrm{T}}Qh = (R^{\mathrm{T}}, 0)\begin{pmatrix} k_1 \\ k_2 \end{pmatrix},$$

or

$$R^{\mathrm{T}}k_1 = t. \tag{4.10}$$

Thus if u is calculated from (4.5), we may find k_1 by solving the lower triangular system (4.10), find k_2 from (4.9), and find g by solving the upper triangular system (4.8). Once k is known, h may be found as $h = Q^{\mathrm{T}}k$.

The products $u = Qs$ and $h = Q^{\mathrm{T}}k$ may be calculated from the factored form of Q. The m-vector h may overwrite the m-vector s, and the n-vector g may overwrite the n-vector t. In fact, as shown below, the calculations may be arranged so that only a single additional n-vector is needed to hold k_1.

All these considerations may be summed up in the following algorithm.

ALGORITHM 4.1. Let $A \in \mathbb{R}^{m \times n}$ have linearly independent columns, and suppose Q and R satisfying (4.4) have been determined from Algorithm 3.8. Let x and r be approximate solutions and residuals to the least squares problem of minimizing (4.1). Provided A is not too ill conditioned, this algorithm overwrites x and r with improved approximations. The m-vector s is assumed to be partitioned in the form $s = (s_1^{\mathrm{T}}, s_2^{\mathrm{T}})^{\mathrm{T}}$, where $s_1 \in \mathbb{R}^n$.

1) Compute $s = b - Ax - r$ and $t = -A^{\mathrm{T}}r$ in double precision and round to single precision
2) $s \leftarrow u = Qs$
3) $k_1 = R^{-\mathrm{T}}t$
4) $t \leftarrow s_1 - k_1$
5) $t \leftarrow g = R^{-1}t$
6) $s_1 \leftarrow k_1$
7) $s \leftarrow h = Q^{\mathrm{T}}s$
8) $x \leftarrow x + g$
9) $r \leftarrow r + h$

Obviously the numerical properties of this algorithm will depend on the effects of rounding error on our method for solving the system (4.3). It can be shown that the computed g and h are near the exact solution of a slightly perturbed system whose matrix is given by

$$\begin{pmatrix} A + H_1 & I_m \\ 0 & A^{\mathrm{T}} + H_2 \end{pmatrix}. \tag{4.11}$$

Thus the analysis of Section 4.5 applies to show that if the computations are performed in t-digit arithmetic and

$$10^{-t}\varkappa\left[\begin{pmatrix} A & I_m \\ 0 & A^{\mathrm{T}} \end{pmatrix}\right] \tag{4.12}$$

is sufficiently less than unity, then the new x and r will be an improvement over the old.

However, this analysis does not adequately describe the properties of Algorithm 4.1. The problem is that the condition number appearing in (4.12) may be as large as $\varkappa^2(A)$. Now it is true that this condition number reflects the sensitivity of the system (4.3) to random perturbations. However, the perturbations in (4.11) are far from random; they leave the zero elements of the matrix undisturbed. An analysis that takes into account the special nature of the perturbations in (4.11) is beyond the scope of this book. However, the results of such an analysis are such as to show that if $\varkappa(A) \cdot 10^{-t}$ is sufficiently less than unity, then the new x and r will be an improvement over the old.

Algorithm 4.1 may be applied iteratively to yield a sequence of improved solutions. The iteration should be terminated when the corrections g and h

become negligible compared to the iterates x and r. Erratic behavior on the part of the iterates indicates that the matrix A is too ill conditioned for the method to work (cf. page 205).

NOTES AND REFERENCES

The natural generalization of the method of iterative refinement discussed in Section 4.5 is the following algorithm. Given the approximate solution x of the least squares problem of minimizing $\| b - Ax \|_2$,

1) Compute $r = b - Ax$ in double precision
2) Compute $d = A^{\dagger} r$ in single precision
3) $x \leftarrow x + d$

This algorithm was proposed by Golub (1965) and analyzed by Golub and Wilkinson (1966), who showed that the method is satisfactory only when the residual is small.

The observation that the least squares solution and residual satisfy (4.2) is due to Golub. Björk (1967b, 1968) describes and analyzes the method of iterative refinement given in this section and extends it to cover least squares problems with linear equality constraints. Björk and Golub (1967) have published an ALGOL program.

EIGENVALUES AND EIGENVECTORS

CHAPTER 6

The last two chapters of this book are devoted to the algebraic eigenvalue problem. In this chapter we shall develop some of the underlying theory, and in the next chapter we shall describe some of the more important computational techniques, notably the widely used QR algorithm. Because of the broad nature of this subject, the material in these two chapters represents only a sampling from the modern literature.

The eigenvalues of a matrix with real elements may be complex numbers. Hence a natural setting for the discussion of the eigenvalue problem is a space in which vectors have complex components. Fortunately this space, which is discussed in Section 1, differs little from the space $\mathbb{R}^n$ which has heretofore occupied our attention.

In Section 2 we shall introduce the notions of eigenvalue and eigenvector and discuss their elementary properties. In Section 3 we shall discuss how far a matrix may be simplified by transformations that preserve its eigenvalues. While these reductions are not themselves constructive, they have important theoretical and practical consequences. In Section 4 we shall investigate the sensitivity of eigenvalues and eigenvectors of a matrix to perturbations in the elements of the matrix. The principal object of this section is, of course, to gain insight into the nature of ill-conditioned eigenvalue problems; however, the techniques developed in Section 4 also suggest practical ways of refining approximate eigenvalues and eigenvectors.

Among matrices with complex elements, the natural generalization of the class of symmetric matrices is the class of Hermitian matrices. The eigenvalues and eigenvectors of Hermitian matrices have many special properties not shared by general matrices, and in Section 5 we shall discuss some of these properties.

Closely related to the notion of the eigenvalues of a square matrix is the notion of the singular values of a general matrix. In Section 6 we shall establish the existence of what is called the singular value decomposition of a matrix and examine its application to degenerate least squares problems.

1. THE SPACE $\mathbb{C}^n$

In this chapter and the next we shall be concerned with certain scalars and vectors associated with a square matrix called the eigenvalues and eigenvectors of the matrix. Although most matrices encountered in practice have elements that are real numbers, it is possible for such a real matrix to have an eigenvalue that is a complex number. Thus the most natural setting for discussing the eigenvalue problem is a space consisting of vectors whose components are complex numbers, which we shall denote by $\mathbb{C}^n$. This section is devoted to describing this space and its properties.

The first step is to describe the complex numbers. Historically the complex numbers were introduced to deal with polynomial equations, such as $x^2 + 1 = 0$, which have no real solutions. Informally, the complex numbers may be defined as follows. Let the symbol i denote a quantity satisfying the equation

$$i^2 = -1. \tag{1.1}$$

Then the set of complex numbers $\mathbb{C}$ is the set of all "numbers" of the form

$$\alpha + \beta i,$$

where α and β are real numbers. Addition, subtraction, and multiplication are performed among complex numbers by treating them as binomials in the unknown i and using (1.1) to simplify the resulting expressions. Thus if $\lambda = \alpha + \beta i$ and $\mu = \gamma + \delta i$ are complex numbers,

$$\lambda + \mu = (\alpha + \beta i) + (\gamma + \delta i) = \alpha + \gamma + \beta i + \delta i = (\alpha + \gamma) + (\beta + \delta)i,$$

and

$$\begin{aligned}\lambda\mu &= (\alpha + \beta i)(\gamma + \delta i) = \alpha\gamma + (\alpha\delta + \beta\gamma)i + \beta\delta i^2 \\ &= (\alpha\gamma - \beta\delta) + (\alpha\delta + \beta\gamma)i.\end{aligned}$$

Complex numbers are sometimes referred to as imaginary numbers, especially when they are of the form $0 + \beta i$. This is because no real number can satisfy Equation (1.1) and therefore the symbol i must be an "imaginary" number. From a more formal point of view, the complex numbers are simply mathematical objects that satisfy certain laws of composition. In fact, we could define the complex numbers as the set of all ordered pairs of real numbers with a sum defined by

$$(\alpha, \beta) + (\gamma, \delta) = (\alpha + \gamma, \beta + \delta)$$

and a product defined by

$$(\alpha, \beta)(\gamma, \delta) = (\alpha\gamma - \beta\delta, \alpha\delta + \beta\gamma).$$

The advantage of our informal approach is that it leads naturally to the "right" definitions of addition and multiplication.

The complex numbers share many of the properties of the real numbers. In the first place the complex numbers in some sense contain the real numbers, for the correspondence $\alpha \leftrightarrow \alpha + 0i$ is preserved under addition and multiplication. Accordingly we shall identify the real number α with the complex number $\alpha + 0i$. This identification justifies calling the number α the *real part* of the complex number $\alpha + \beta i$. Naturally enough, the number β is called the *imaginary part* of $\alpha + \beta i$.

Addition and multiplication of complex numbers satisfy the same commutative, associative, and distributive laws that the real numbers satisfy. The complex numbers have a zero element $0 + 0i$, which we shall denote simply by 0. If $\lambda = \alpha + \beta i$ is a complex number, the number $-\lambda = -\alpha + (-\beta)i = -\alpha - \beta i$, and no other, satisfies the equation

$$\lambda + (-\lambda) = 0.$$

Thus each complex number has a unique additive inverse. The nonzero complex numbers also have multiplicative inverses. For if $\lambda = \alpha + \beta i \neq 0$, then $\alpha^2 + \beta^2 \neq 0$, and the number

$$\lambda^{-1} = \frac{\alpha - \beta i}{\alpha^2 + \beta^2} \tag{1.2}$$

is well defined. It is easily verified that

$$\lambda\lambda^{-1} = 1.$$

We may summarize these considerations by saying that the complex numbers, like the real numbers, satisfy the axioms of a field.

Just as a real number may be pictured as a point lying on a line, so may a complex number be pictured as a point lying in a plane. In fact with each complex number $\alpha + \beta i$ we may associate a vector $(\alpha, \beta)^{\mathrm{T}} \in \mathbb{R}^2$. Since this association is preserved by addition, the addition of complex numbers may be interpreted as the addition of vectors in $\mathbb{R}^2$. The operation of multiplication may also be interpreted geometrically. For example, the operation of multiplying by i is equivalent to rotating the complex number $\alpha + \beta i$ ninety degrees counterclockwise (Fig. 1). When the plane $\mathbb{R}^2$ is regarded as consisting of complex numbers, it is usually referred to as the *complex plane.*

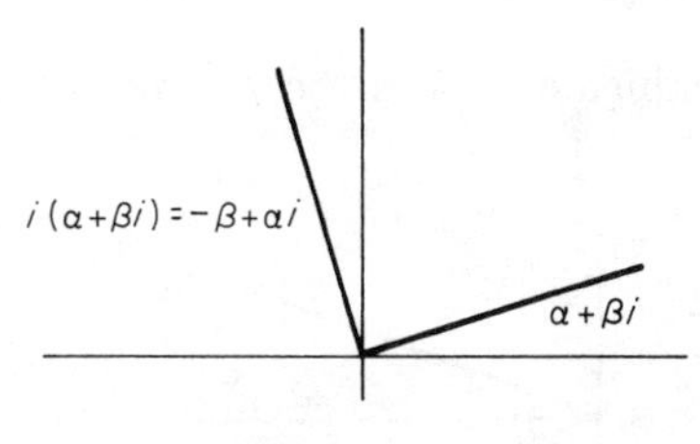

Fig. 1

The absolute value of a real number is its distance from the origin. The interpretation of complex numbers as points in the complex plane suggests that we define the absolute value of a complex number as its distance from the origin in the plane. This simply amounts to setting

$$|\alpha + \beta i| = \|(\alpha, \beta)^{\mathrm{T}}\|_2 = \sqrt{\alpha^2 + \beta^2}. \tag{1.3}$$

Because of the properties of the 2-norm, we have the following relations among any two complex numbers λ and μ:

$$|\lambda| \geq 0,$$

$$|\lambda| = 0 \iff \lambda = 0,$$

$$|\lambda + \mu| \leq |\lambda| + |\mu|.$$

Moreover, it is easy to show that

$$|\lambda\mu| = |\lambda||\mu|.$$

Thus the absolute value of a complex number, as defined by (1.3), shares most of the properties of the absolute value of a real number.

One property not shared by the complex absolute value is the useful formula

$$|\lambda|^2 = \lambda \cdot \lambda. \tag{1.4}$$

This formula can fail because, when λ is complex, λ^2 need not be positive, or even real. The proper generalization of (1.4) may be obtained by introducing the notion of the conjugate of a complex number. Specifically, if $\lambda = \alpha + \beta i$ is a complex number, we define the *conjugate* of λ to be the number

$$\bar{\lambda} = \alpha - \beta i.$$

Geometrically, the conjugate of λ is the number obtained by reflecting λ in the x-axis (Fig. 2).

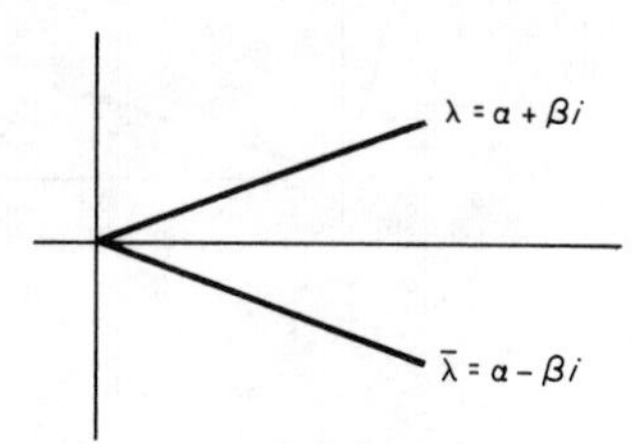

Fig. 2

The conjugation operation has a number of easily verified properties, among the more important of which are

$$\overline{\lambda + \mu} = \bar{\lambda} + \bar{\mu}$$

and

$$\overline{\lambda\mu} = \bar{\lambda}\bar{\mu}.$$

We also have the following generalization of (1.4):

$$|\lambda|^2 = \lambda\bar{\lambda}.$$

The use of the conjugate often results in considerable simplification of complicated expressions. For example, (1.2) may be derived as follows:

$$\frac{1}{\lambda} = \frac{1}{\lambda} \cdot 1 = \frac{1}{\lambda} \frac{\bar{\lambda}}{\bar{\lambda}} = \frac{\bar{\lambda}}{|\lambda|^2}.$$

We have already mentioned that the complex numbers were first introduced to aid in the solution of polynomial equations. The fundamental theorem of algebra asserts that any polynomial with complex coefficients has a zero. We shall need this result in the next section to establish the existence of eigenvalues. Here we combine the results we shall need into a single theorem, which we state without proof.

THEOREM 1.1. Let

$$\pi(\lambda) = \lambda^n + \alpha_{n-1}\lambda^{n-1} + \cdots + \alpha_0$$

be a polynomial in λ with complex coefficients $\alpha_0, \alpha_1, \ldots, \alpha_{n-1}$. Then there are unique numbers $\lambda_1, \lambda_2, \ldots, \lambda_n$ such that

$$\pi(\lambda) = (\lambda - \lambda_1)(\lambda - \lambda_2) \cdots (\lambda - \lambda_n).$$

The numbers λ_i are the only numbers for which $\pi(\lambda) = 0$. If the coefficients of π are real, then the nonreal λ_i may be grouped in conjugate pairs (e.g., $\lambda_1 = \bar{\lambda}_2, \lambda_3 = \bar{\lambda}_4, \ldots$).

The number n is called the degree of π and the numbers λ_i are called the zeros of π. Thus the theorem states that a polynomial of degree n has n zeros. It should be noted that the λ_i need not be distinct. If r of the λ_i are identical, that number is said to be a *zero of multiplicity r*. The last part of the theorem states that if the coefficients of π are real, then the zeros of π must be distributed in the complex plane symmetrically with respect to the x-axis.

We turn now to the definition of complex vectors and matrices.

DEFINITION 1.2. A *complex n-vector x* is a collection of complex numbers $\xi_1, \xi_2, \ldots, \xi_n$ arranged in order in a column:

$$x = \begin{pmatrix} \xi_1 \\ \xi_2 \\ \cdot \\ \cdot \\ \cdot \\ \xi_n \end{pmatrix}.$$

The set of all complex n-vectors is denoted by $\mathbb{C}^n$. A *complex* $m \times n$ *matrix* is a rectangular array of numbers having m rows and n columns. The set of all $m \times n$ matrices is denoted by $\mathbb{C}^{m \times n}$.

It is not accidental that the wording of Definition 1.2 is almost identical with the wording of Definitions 1.1.1 and 1.3.1; rather it indicates a deep similarity between the spaces $\mathbb{R}^n$ and $\mathbb{C}^n$. In the first place we may expropriate the nomenclature and notational conventions of $\mathbb{R}^n$ for $\mathbb{C}^n$. Thus we shall refer to complex numbers as scalars and denote them by lower case Greek letters. We shall also speak of the components of a complex vector and the elements of a complex matrix. In fact, except where it is important to make a distinction, we shall drop the qualifier "complex" when referring to complex vectors and matrices.

What is more important is that the entire content of Chapter 1 remains valid in $\mathbb{C}^n$. This is because in making definitions and in stating and proving theorems we have used only the field axioms for the real numbers. Since these axioms are also satisfied by the complex numbers, the development goes through in $\mathbb{C}^n$ without change.

Only when we come to the geometry of $\mathbb{C}^n$ does the similarity between $\mathbb{R}^n$ and $\mathbb{C}^n$ begin to break down. The most obvious discrepancy is in the graphical representation of vectors. We have seen that the space $\mathbb{C}^1$, which we identify with $\mathbb{C}$, requires two real dimensions for its representation as the complex plane, and analogously the space $\mathbb{C}^2$ requires four. Thus it is impossible to draw configurations of complex 2-vectors, even in three-dimensional perspective. Fortunately the identity between $\mathbb{R}^n$ and $\mathbb{C}^n$ is so deep that it is usually possible to think and draw real, while defining and proving complex.

A more subtle difficulty lies in the problem of making the correct definition of the inner product in $\mathbb{C}^n$. In $\mathbb{R}^n$ the natural generalization of (1.4) is the formula

$$\| x \|_2^2 = x^{\mathrm{T}} x. \tag{1.5}$$

Unfortunately, this formula does not extend directly to $\mathbb{C}^n$, for $x^{\mathrm{T}}x$ need no longer be a real number. Just as we salvaged the formula (1.4) by introducing the notion of the conjugate of a complex number, so we may salvage (1.5) by introducing the notion of conjugate transpose.

DEFINITION 1.3. Let $A \in \mathbb{C}^{m \times n}$. The *conjugate* of A is the $m \times n$ matrix $\bar{A}$ whose elements are given by $\bar{\alpha}_{ij}$. The *conjugate transpose* of A is the $n \times m$

matrix A^{H} defined by

$$A^{\mathrm{H}} = \bar{A}^{\mathrm{T}}.$$

Thus the conjugate transpose of a matrix A is the matrix obtained by transposing A and conjugating its elements. It is easily verified that the operation of taking the conjugate transpose has the usual properties of transposition [e.g., $(AB)^{\mathrm{H}} = B^{\mathrm{H}}A^{\mathrm{H}}$], with the exception of the identity

$$(\lambda A)^{\mathrm{H}} = \bar{\lambda} A^{\mathrm{H}}.$$

The conjugate transpose of A is often written A^*, and in more abstract settings it is usually called the adjoint of A.

EXAMPLE 1.4. If α is a scalar, then $\alpha^{\mathrm{H}} = \bar{\alpha}$. The scalar α is real if and only if $\alpha^{\mathrm{H}} = \alpha$.

Recall that a symmetric matrix is a square matrix A satisfying $A^{\mathrm{T}} = A$. A matrix having the analogous property with respect to the conjugate transpose is called a Hermitian matrix.

DEFINITION 1.5. The matrix $A \in \mathbb{C}^{n \times n}$ is *Hermitian* if $A^{\mathrm{H}} = A$.

In terms of scalars A is Hermitian if and only if $\alpha_{ij} = \bar{\alpha}_{ji}$. Since this implies that $\alpha_{ii} = \bar{\alpha}_{ii}$ is real (cf. Example 1.4), it follows that the diagonal elements of a Hermitian matrix are real. A real symmetric matrix is always Hermitian, but a Hermitian matrix is symmetric only if it is real.

One of the reasons for introducing Hermitian matrices is to extend the definition of positive definiteness. The property of Hermitian matrices that allows us to do this is contained in the following theorem.

THEOREM 1.6. Let $A \in \mathbb{C}^{n \times n}$ be Hermitian. Then for any $x \in \mathbb{C}^n$, $x^{\mathrm{H}}Ax$ is real.

PROOF. We have

$$(x^{\mathrm{H}}Ax)^{\mathrm{H}} = x^{\mathrm{H}}A^{\mathrm{H}}x = x^{\mathrm{H}}Ax.$$

Hence by Example 1.4, $x^{\mathrm{H}}Ax$ is real. ■

DEFINITION 1.7. A Hermitian matrix A is *positive definite* if

$$x \neq 0 \;\Rightarrow\; x^{\mathrm{H}}Ax > 0.$$

It is *positive semidefinite* if

$$x \neq 0 \;\Rightarrow\; x^{\mathrm{H}}Ax \geq 0.$$

Of course it is Theorem 1.6 that allows us to compare $x^{\mathrm{H}}Ax$ with zero in Definition 1.7. A trivial example of a positive definite matrix is the identity matrix. A less trivial example is any matrix of the form $B^{\mathrm{H}}B$, where B has linearly independent columns (cf. Example 3.3.6).

If x and y are vectors in $\mathbb{C}^n$, we define the *inner product* of x and y as the number

$$y^{\mathrm{H}}x.$$

Since the identity matrix is positive definite,

$$x^{\mathrm{H}}x = x^{\mathrm{H}}Ix \geq 0.$$

This justifies defining a function $\| \cdot \|_2 : \mathbb{C}^n \to \mathbb{R}$ by

$$\| x \|_2^2 = x^{\mathrm{H}}x. \tag{1.6}$$

The proofs in Chapter 4 extend immediately to show that the function $\| \cdot \|_2$ is a homogeneous, definite function that satisfies the triangle inequality. Hence $\| \cdot \|_2$ defined by (1.6) is a norm on $\mathbb{C}^n$. It also satisfies the Cauchy inequality

$$| y^{\mathrm{H}}x | \leq \| x \|_2 \, \| y \|_2 .$$

More generally all the definitions and theorems about norms on $\mathbb{R}^n$ extend to $\mathbb{C}^n$; however, in dealing with the 2-norm, we must take care to replace the transpose by the conjugate transpose. Since the concept of orthogonality in $\mathbb{R}^n$ was defined in terms of the inner product $x^{\mathrm{T}}y$, this definition must be recast for $\mathbb{C}^n$.

DEFINITION 1.8. Two complex n-vectors x and y are orthogonal if $x^{\mathrm{H}}y = 0$.

With this definition all the theory of Chapter 4, including the theory of least squares, extends to $\mathbb{C}^n$. However it is customary to restrict the term "orthogonal matrix" to mean a real square matrix with orthonormal columns. A complex square matrix with orthonormal columns is called a unitary matrix.

DEFINITION 1.9. Let $A \in \mathbb{C}^{n\times n}$. Then A is *unitary* if

$$A^{\mathrm{H}}A = I.$$

We conclude this section with a word about computations with complex matrices. In practice it is rather unusual to encounter problems involving complex matrices, and when one does, it is often possible to reduce the problem to the real case by various tricks. For example, if

$$C = A + Bi,$$

$$z = x + yi,$$

and

$$r = p + qi,$$

where A, B, x, y, p, and q are real and of order n, then

$$Cz = r \tag{1.7}$$

if and only if

$$\begin{pmatrix} A & -B \\ B & A \end{pmatrix} \begin{pmatrix} x \\ y \end{pmatrix} = \begin{pmatrix} p \\ q \end{pmatrix}. \tag{1.8}$$

Thus the problem of solving the complex system (1.7) of order n may be reduced to that of solving the real system (1.8) of order $2n$.

However, it may be impractical to recast a complex problem as a real problem. In the example above, for instance, the system (1.7) requires about $2n^2$ storage locations for its representation on a computer, while the real system (1.8) requires $4n^2$ locations, a distinct drawback if storage is at a premium. Fortunately all the algorithms presented so far extend with obvious and trivial modifications to the complex case. The one exception is Algorithm 5.3.5 for constructing elementary reflectors and those algorithms that depend on it. The modifications necessary to work with elementary reflectors over the complex numbers are given in the exercises.

Of course, in practice the algorithms must be executed on a computer. Most computers do not have the ability to manipulate complex numbers directly; rather a complex number must be represented as a pair of floating-point numbers, and arithmetic operations with complex numbers must be performed by precoded subroutines. In some languages (PL/I for example) one need never see this auxiliary arithmetic apparatus; rather one may describe his variables as complex and allow the compiler to generate the necessary code. Should the reader ever have to code his own routines for complex arithmetic, he should recognize that it is a rather tricky business. For example, the use of (1.2), as it stands, to compute λ^{-1} can lead to unnecessary overflows.

The effect of rounding error in the complex case is much the same as in the real case. With properly coded subroutines, floating-point complex arithmetic may be made to satisfy bounds of the form given for real arithmetic in Chapter 2. Since these bounds are the basis for the results on rounding error cited in this book, it follows that they continue to hold in the complex case.

EXERCISES

1. Regarded as a point in the complex plane, the complex number δ may be represented by its polar coordinates (ϱ, ϕ). Show that $\varrho = |\delta|$. The angle ϕ is called the argument of δ and is written $\arg(\delta)$.

2. Let $\delta_1, \delta_2 \in \mathbb{C}$. Prove that

$$|\delta_1 \delta_2| = |\delta_1||\delta_2|$$

and

$$\arg(\delta_1 \delta_2) = \arg(\delta_1) + \arg(\delta_2).$$

Use this result to interpret the multiplication of complex numbers geometrically.

3. Show that $\arg(\bar{\delta}) = -\arg(\delta)$.

4. Find the arguments of the following complex numbers:

(a) 1, (b) -1,
(c) i, (d) $1 + i$.

5. Describe geometrically the following sets of complex numbers:

(a) $\{\delta : |\delta| = 1\}$, (b) $\{\alpha + \beta i : |\alpha| + |\beta| < 1\}$,
(c) $\{\delta : \delta^2 \text{ is real}\}$, (d) $\{\delta : |\delta - \alpha| < \varrho\}$.

6. What is $\sqrt{i}$?

7. Conclude from Theorem 1.1 that every nonzero complex number has two distinct square roots. Give an explicit representation of them.

8. Factor the following polynomials as in Theorem 1.1.

   (a) $\lambda^2 - 1$, (b) $\lambda^2 + 1$,
   (c) $\lambda^4 - 1$, (d) $\lambda^4 + 1$.

9. Show that the quadratic polynomial $(\lambda - \lambda_1)(\lambda - \bar{\lambda}_1)$ has real coefficients. Conclude from Theorem 1.1 that if $\pi(\lambda) = \lambda^n + \alpha_{n-1}\lambda^{n-1} + \cdots + \alpha_0$ has real coefficients, then π may be written in the form $\pi(\lambda) = \pi_1(\lambda)\pi_2(\lambda) \cdots \pi_k(\lambda)$ where each π_k is a polynomial with real coefficients of degree at most 2.

10. Give INFL codes for performing the complex arithmetic operations using only real arithmetic. The codes should not be affected by underflows and overflows provided the absolute values of both the operands and the results can be represented in floating-point form.

11. In $\mathbb{C}^n$ the space $\mathbb{R}^n$ may be identified with the set of all complex n-vectors having only real components. With this identification, is $\mathbb{R}^n$ a subspace of $\mathbb{C}^n$?

12. Which of the following represents the vector z?

    (a) $(\zeta_1, \zeta_2, \ldots, \zeta_n)^{\mathrm{T}}$,
    (b) $(\zeta_1, \zeta_2, \ldots, \zeta_n)^{\mathrm{H}}$,
    (c) $(\zeta_1, \zeta_2, \ldots, \zeta_n)^{\mathrm{H}}$.

13. Prove that $A = \bar{A}$ if and only if the elements of A are real.

14. Let A be Hermitian. When is λA Hermitian?

15. Prove that A is singular if and only if A^{H} is singular.

16. Show by example that $\mathcal{R}(A^{\mathrm{H}})$ need not be equal to $\mathcal{R}(A^{\mathrm{T}})$.

17. Let $A \in \mathbb{C}^{l \times m}$ and $B \in \mathbb{C}^{m \times n}$. Prove that

    (a) $(\lambda A)^{\mathrm{H}} = \bar{\lambda} A^{\mathrm{H}}$,
    (b) $(A^{\mathrm{H}})^{\mathrm{H}} = A$,
    (c) $(AB)^{\mathrm{H}} = B^{\mathrm{H}} A^{\mathrm{H}}$.

18. Prove the Cauchy inequality in $\mathbb{C}^n$.

19. In Algorithms 5.3.5 and 5.3.6 let the transpose be replaced by the conjugate transpose, "sign" be replaced by "arg," and the statement $\pi = \sigma v_1$ be replaced by $\pi = \bar{\sigma} v_1$. Show that these modified algorithms produce "elementary reflectors" of the form $R = I - \pi^{-1} u u^{\mathrm{H}}$ such $Rx = -\sigma e_1$.

2. EIGENVALUES AND EIGENVECTORS

To motivate the definitions of eigenvalue and eigenvector and to give a simple example of one of their applications, we begin by considering the 2×2 matrix

$$A = \begin{pmatrix} 3 & -1 \\ -1 & 3 \end{pmatrix}.$$

If we set

$$x_1 = (1, 1)^{\mathrm{T}}$$

and

$$x_2 = (1, -1)^{\mathrm{T}},$$

then it is easy to verify that

$$Ax_1 = 2x_1 \tag{2.1}$$

and

$$Ax_2 = 4x_2. \tag{2.2}$$

In other words the linear transformation A simply multiplies the vector x_1 by a factor of two and the vector x_2 by a factor of four. We call the number two an eigenvalue of A corresponding to the eigenvector x_1 and the number four an eigenvalue corresponding to the eigenvector x_2.

A knowledge of the eigenvalues and eigenvectors of a matrix can be used to simplify computations with the matrix. For example, any vector $y \in C^2$ can be written in the form

$$y = \gamma_1 x_1 + \gamma_2 x_2,$$

where

$$\gamma_1 = \tfrac{1}{2}(\eta_1 + \eta_2),$$

and

$$\gamma_2 = \tfrac{1}{2}(\eta_1 - \eta_2).$$

Then it follows from (2.1) and (2.2) that

$$Ay = 2\gamma_1 x_1 + 4\gamma_2 x_2,$$

and more generally that

$$A^k y = 2^k \gamma_1 x_1 + 4^k \gamma_2 x_2. \tag{2.3}$$

Thus if we know the components of y with respect to the basis $\{x_1, x_2\}$ for $\mathbb{C}^2$, we can compute $A^k y$ without performing any matrix multiplications.

Moreover, we can draw useful theoretical consequences from (2.3). Suppose $\gamma_2 \neq 0$ (for a vector chosen at random this will almost certainly be the case). Then when k is sufficiently large the second term in the right-hand side of (2.3) dominates the first. In other words, *given almost any vector y, the vector $A^k y$ will ultimately tend to lie along x_2.* Embellishments of this simple observation form the theoretical basis for the computational methods to be discussed in the next chapter.

We now turn to the formal definition of eigenvalue and eigenvector.

DEFINITION 2.1. Let $A \in \mathbb{C}^{n\times n}$ and $x \in \mathbb{C}^n$. Then x is an *eigenvector* of A corresponding to the *eigenvalue* λ if

1. $x \neq 0$,
2. $Ax = \lambda x$.

The set of eigenvalues of A is denoted by $\lambda(A)$.

The requirement that $x \neq 0$ is necessary, for if $x = 0$, then any number λ satisfies the equation $Ax = \lambda x$. An eigenvector can correspond to only one eigenvalue, but an eigenvalue can have many eigenvectors. For example, if x is an eigenvector with eigenvalue λ, then so is αx for any $\alpha \neq 0$. More generally if $x_1, x_2, \ldots, x_k$ are eigenvectors with the common eigenvalue λ, then

$$\begin{aligned} A(\alpha_1 x_1 + \alpha_2 x_2 + \cdots + \alpha_k x_k) &= \alpha_1 A x_1 + \alpha_2 A x_2 + \cdots + \alpha_k A x_k \\ &= \alpha_1 \lambda x_1 + \alpha_2 \lambda x_2 + \cdots + \alpha_k \lambda x_k \\ &= \lambda(\alpha_1 x_1 + \alpha_2 x_2 + \cdots + \alpha_k x_k). \end{aligned}$$

Hence if $\alpha_1 x_1 + \alpha_2 x_2 + \cdots + \alpha_k x_k \neq 0$, it is also an eigenvector with eigenvalue λ.

As was indicated in the example at the beginning of this chapter, eigenvalues and eigenvectors may be used to simplify computations with powers of a matrix.

EXAMPLE 2.2. Suppose that the matrix $A \in \mathbb{C}^{n \times n}$ has n linearly independent eigenvectors $x_1, x_2, \ldots, x_n$ corresponding to eigenvalues $\lambda_1, \lambda_2, \ldots, \lambda_n$. Then any vector y may be written in the form

$$y = \gamma_1 x_1 + \gamma_2 x_2 + \cdots + \gamma_n x_n,$$

where the γ_i are the components of y with respect to the basis $\{x_1, x_2, \ldots, x_n\}$. From the fact that $Ax_i = \lambda_i x_i$ it follows that

$$A^k y = \lambda_1{}^k \gamma_1 x_1 + \lambda_2{}^k \gamma_2 x_2 + \cdots + \lambda_n{}^k \gamma_n x_n.$$

In particular, if $|\lambda_1| > |\lambda_i|$ $(i = 2, 3, \ldots, n)$ and $\gamma_1 \neq 0$, then $A^k y$ will tend to lie along the vector x_1 when k is large.

While the above example indicates that eigenvalues and eigenvectors have fruitful applications within matrix theory, the reader may legitimately ask if they are useful in other fields. The following example, taken from the subject of differential equations, is just one of many applications.

EXAMPLE 2.3. Let the notation $y(\tau)$ mean a vector y whose components $\eta_1(\tau), \ldots, \eta_n(\tau)$ are functions of the scalar τ. The derivative of y is the vector

$$\frac{dy}{d\tau} = \left(\frac{d\eta_1}{d\tau}, \frac{d\eta_2}{d\tau}, \ldots, \frac{d\eta_n}{d\tau}\right)^{\mathrm{T}}.$$

With this notation a system of homogeneous differential equations with constant coefficients can be written in the form

$$\frac{dy}{d\tau} = Ay. \tag{2.4}$$

Particular solutions of this system may be obtained from a knowledge of the eigenvalues and eigenvectors of A. Specifically, suppose that $Ax = \lambda x$, and set

$$y(\tau) = e^{\lambda\tau} x.$$

Then

$$\frac{dy(\tau)}{d\tau} = \frac{de^{\lambda\tau}}{d\tau}x = \lambda e^{\lambda\tau}x = \lambda y(\tau).$$

On the other hand,

$$Ay(\tau) = e^{\lambda\tau}Ax = \lambda e^{\lambda\tau}x = \lambda y(\tau).$$

Thus the vector $y(\tau)$ satisfied the differential equation (2.4).

Some of the elementary consequences of Definition 2.1 have important practical implications. We begin by exhibiting a class of transformations that do not change the eigenvalues of a matrix.

THEOREM 2.4. Let $A, P \in \mathbb{C}^{n\times n}$ with P nonsingular. Then λ is an eigenvalue of A with eigenvector x if and only if λ is an eigenvalue of $P^{-1}AP$ with eigenvector $P^{-1}x$.

PROOF. Since P is nonsingular, $P^{-1}x$ is nonzero along with x. Moreover, the equations

$$Ax = \lambda x$$

and

$$(P^{-1}AP)(P^{-1}x) = \lambda(P^{-1}x)$$

are obviously equivalent. ∎

We shall say that the square matrices A and B are *similar* if there is a nonsingular matrix P such that $B = P^{-1}AP$. In this terminology Theorem 2.4 says that similar matrices have the same eigenvalues. This suggests the possibility of trying to compute the eigenvalues of a matrix A by applying a sequence of similarity transformations which reduce it to some simple form from which the eigenvalues may be read off. In the next section we shall analyze theoretically how far we may go with such a procedure. In Chapter 7 we shall give algorithms for reducing matrices to simple forms by similarity transformations.

It should be noted that the eigenvalues and eigenvectors of a matrix are *not* preserved when the matrix is premultiplied or postmultiplied by another matrix. In particular the elementary row operations, such as interchanging two rows of a matrix, may change the eigenvalues.

The next theorem shows how the eigenvalues of a matrix behave under some other transformations.

THEOREM 2.5. Let $A \in \mathbb{C}^{n \times n}$ and let λ be an eigenvalue of A with eigenvector x. Then

1. $\alpha\lambda$ is an eigenvalue of αA with eigenvector x,
2. $\lambda - \mu$ is an eigenvalue of $A - \mu I$ with eigenvector x,
3. if A is nonsingular, then $\lambda \neq 0$ and λ^{-1} is an eigenvalue of A^{-1} with eigenvector x.

PROOF. For parts 1 and 2 note that the equation $Ax = \lambda x$ implies the equations $\alpha Ax = \alpha\lambda x$ and $(A - \mu I)x = (\lambda - \mu)x$. For part 3 note that $\lambda = 0$ implies $Ax = 0 \cdot x = 0$. Hence the homogeneous equation $Ax = 0$ has a nontrivial solution and A is singular. Since A is assumed nonsingular, we must have $\lambda \neq 0$. Then the equation $Ax = \lambda x$ implies that $A^{-1}x = \lambda^{-1}x$. ∎

Part 1 of the theorem says that multiplying a matrix by a constant α multiplies each eigenvalue by α. Part 2 says that the effect of subtracting a constant μ from the diagonal elements of a matrix is to subtract μ from the eigenvalues. Part 3 says that if a matrix is inverted, its eigenvalues are inverted. In all these transformations the eigenvectors are left unchanged.

The results of the last two theorems have been proved by exhibiting an eigenvector of the transformed matrix. This may not always be possible, and it is convenient to have an alternate characterization for the eigenvalues of a matrix.

THEOREM 2.6. The number λ is an eigenvalue of A if and only if $A - \lambda I$ is singular.

PROOF. If λ is an eigenvalue of A, then $Ax = \lambda x$ for some nonzero x. But this is equivalent to the equation $(A - \lambda I)x = 0$, which implies that $A - \lambda I$ is singular. Conversely if $A - \lambda I$ is singular, then the equation $(A - \lambda I)x = 0$ has a nontrivial solution, and this solution satisfies the equation $Ax = \lambda x$. ∎

A corollary of this theorem is the following result concerning the eigenvalues of the transpose and the conjugate transpose of a matrix.

THEOREM 2.7. Let $A \in \mathbb{C}^{n \times n}$.

1. $\lambda(A^{\mathrm{T}}) = \lambda(A)$.
2. $\lambda(A^{\mathrm{H}}) = \{\bar{\lambda} : \lambda \in \lambda(A)\}$.

PROOF. For part 1, note that the matrix $A - \lambda I$ is singular if and only if the matrix $(A - \lambda I)^{\mathrm{T}} = A^{\mathrm{T}} - \lambda I$ is singular. Hence by Theorem 2.6, λ is an eigenvalue of A if and only if λ is an eigenvalue of A^{T}. For part 2, note that $A - \lambda I$ is singular if and only if $(A - \lambda I)^{\mathrm{H}} = A^{\mathrm{H}} - \bar{\lambda} I$ is singular. ■

By the above corollary, if λ is an eigenvalue of A, there is a nonzero vector y such that

$$A^{\mathrm{H}} y = \bar{\lambda} y.$$

We shall call y a *left eigenvector* of A corresponding to the eigenvalue λ. This terminology is justified by the fact that the row vector y^{H} satisfies the equation

$$y^{\mathrm{H}} A = \lambda y^{\mathrm{H}}.$$

We shall also call the row vector y^{H} a left eigenvector. The reader is cautioned that some authors use the term left eigenvector for an eigenvector of A^{T}.

As we mentioned above, some of our computational methods will involve reducing a matrix to a simpler form by similarity transformations. One particularly important form is the quasi-triangular form, which is a special case of the block triangular form.

DEFINITION 2.8. The square matrix T is *block upper triangular* if it can be partitioned in the form

$$T = \begin{pmatrix} T_{11} & T_{12} & \cdots & T_{1n} \\ 0 & T_{22} & \cdots & T_{2n} \\ \vdots & \vdots & & \vdots \\ 0 & 0 & \cdots & T_{nn} \end{pmatrix}, \tag{2.5}$$

where each *diagonal block* T_{ii} is square. If each diagonal block is of order at most two, then T is said to be in *quasi-triangular* form.

When all the diagonal blocks of a block triangular matrix are of order unity, the matrix is triangular. A block triangular matrix is singular if and only if one of its diagonal blocks is singular. This fact is a generalization of Theorem 3.1.2 and is proved in the same way. Using it, we may establish the following relation between the eigenvalues of a block triangular matrix and the eigenvalues of its diagonal blocks.

THEOREM 2.9. Let the block triangular matrix T be partitioned as in (2.5). Then

$$\lambda(T) = \bigcup_{i=1}^{n} \lambda(T_{ii}).$$

PROOF. First note that the matrix $T - \lambda I$ is also block triangular with diagonal blocks $T_{ii} - \lambda I$. Thus $T - \lambda I$ is singular if and only if at least one of the matrices $T_{ii} - \lambda I$ is singular, which, in view of Theorem 2.6, establishes the result. ■

Since the "diagonal blocks" of a triangular matrix are its diagonal elements, we have the following corollary.

COROLLARY 2.10. The eigenvalues of a triangular matrix are its diagonal elements.

In Corollary 2.10 we have for the first time established that there are general classes of matrices that have eigenvalues. Moreover, we have shown that a triangular matrix of order n has exactly n eigenvalues, provided that we count repeated diagonal elements once for each occurrence. This result is not particular to triangular matrices but is true of all matrices. However its proof involves the use of determinants, which are discussed in Appendix 2.

Let A be of order n. Then λ is an eigenvalue of A if and only if $\lambda I - A$ is singular. However, $\lambda I - A$ is singular if and only if

$$\det(\lambda I - A) = 0. \tag{2.6}$$

Thus the eigenvalues of A are precisely those values of λ that satisfy Equation (2.6), which is called the *characteristic equation* of A. Now the function

$$\pi(\lambda) = \det(\lambda I - A) \tag{2.7}$$

is a polynomial in λ whose leading term is λ^n. It follows from Theorem 1.1 that the characteristic equation has n roots, some of which may be repeated. These roots are the eigenvalues of A. If a root λ is repeated k times, λ is said to be a *multiple eigenvalue of A of multiplicity k*. If $k = 1$, the eigenvalue λ is said to be *simple*. We may summarize these considerations in the following theorem.

THEOREM 2.11. Let $A \in \mathbb{C}^{n \times n}$. Then A has exactly n eigenvalues, which are the zeros of the *characteristic polynomial* (2.7) with multiplicities counted.

Since the characteristic polynomial of a matrix with real elements has real coefficients, it follows from Theorems 1.1 and 2.11 that the complex eigenvalues of a real matrix occur in conjugate pairs.

The phenomenon of multiple eigenvalues is a complicating factor. For example, we know from Theorem 2.4 that similar matrices have the same eigenvalues. But it is conceivable that a similarity transformation could change the multiplicity of an eigenvalue. That this cannot happen is a consequence of the following theorem.

THEOREM 2.12. If A and B are similar, then A and B have the same characteristic polynomial.

PROOF. Let $B = P^{-1}AP$. Then

$$\begin{aligned}\det(\lambda I - B) = \det(\lambda I - P^{-1}AP) &= \det[P^{-1}(\lambda I - A)P] \\ &= \det(P)\det(P^{-1})\det(\lambda I - A) \\ &= \det(\lambda I - A). \quad \blacksquare\end{aligned}$$

Another problem involving multiple eigenvalues is the following. Let λ be an eigenvalue of A of multiplicity k. From the discussion immediately following Definition 2.1, we know that the set of all eigenvectors corresponding to λ, along with the zero vector, form a subspace, call it $\mathfrak{X}_\lambda$. We shall show in the next section that the dimension of $\mathfrak{X}_\lambda$ cannot exceed the multiplicity k of the eigenvalue λ. However, it may fall short of it, as the following example shows.

EXAMPLE 2.13. Let $J_\lambda^{(k)} \in \mathbb{C}^{k\times k}$ have the form

$$J_\lambda^{(k)} = \begin{pmatrix} \lambda & 1 & 0 & \cdots & 0 & 0 \\ 0 & \lambda & 1 & \cdots & 0 & 0 \\ \vdots & \vdots & \vdots & & \vdots & \vdots \\ 0 & 0 & 0 & \cdots & \lambda & 1 \\ 0 & 0 & 0 & \cdots & 0 & \lambda \end{pmatrix}. \tag{2.8}$$

Then $J_\lambda^{(k)}$ has λ as an eigenvalue of multiplicity k. But every eigenvector of $J_\lambda^{(k)}$ is a multiple of e_1; for if we write componentwise the equation $J_\lambda^{(k)}x = \lambda x$, we obtain the equations

$$\begin{aligned} \lambda\xi_1 + \xi_2 &= \lambda\xi_1, \\ \lambda\xi_2 + \xi_3 &= \lambda\xi_2, \\ &\vdots \\ \lambda\xi_{k-1} + \xi_k &= \lambda\xi_{k-1}. \end{aligned}$$

These equations imply that $\xi_2 = \xi_3 = \cdots = \xi_k = 0$. A matrix of the form (2.8) is called a *Jordan block*.

The state of affairs represented by Example 2.13 is important enough to have a name.

DEFINITION 2.14. A square matrix A is *defective* if it has an eigenvalue of multiplicity k having fewer than k linearly independent eigenvectors.

Thus a defective matrix is one without enough eigenvectors. We shall see later that there are theoretical and computational difficulties associated with defective matrices.

It is often important to have bounds on the eigenvalues of a matrix. In later sections we shall derive some rather sophisticated theorems for localizing eigenvalues in the complex plane. However, one useful result may be had immediately from the theory of norms.

THEOREM 2.15. Let $\|\cdot\|$ denote a consistent matrix norm on $\mathbb{C}^{n\times n}$. Then for any $A \in \mathbb{C}^{n\times n}$

$$\lambda \in \lambda(A) \quad\Rightarrow\quad |\lambda| \le \|A\|.$$

PROOF. By Theorem 4.2.6 there is a vector norm $\nu : C^n \to R$ that is consistent with $\| \cdot \|$. Let $Ax = \lambda x$, where $x \neq 0$. Then

$$|\lambda| \nu(x) = \nu(\lambda x) = \nu(Ax) \leq \| A \| \nu(x),$$

and since $\nu(x) \neq 0$, the result follows. ■

Geometrically, Theorem 2.15 may be interpreted as saying that all the eigenvalues of A lie in a disk in the complex plane that is centered about the origin and has radius $\| A \|$. Another, more succinct, expression of Theorem 2.15 may be obtained by introducing the notion of the spectral radius of a matrix.

DEFINITION 2.16. Let $A \in \mathbb{C}^{n \times n}$. Then the *spectral radius* of A is the number

$$\varrho(A) = \max\{|\lambda| : \lambda \in \lambda(A)\}.$$

Thus the spectral radius of A is the size of the largest eigenvalue of A. Theorem 2.15 says that if $\| \cdot \|$ is any consistent matrix norm, then

$$\varrho(A) \leq \| A \|.$$

In the next section we will prove a sort of converse to this result.

EXERCISES

1. Let

$$A = \begin{pmatrix} 3 & -1 \\ -1 & 3 \end{pmatrix}.$$

Find the most general form of a solution of the differential equation

$$\frac{dy}{dt} = Ay.$$

2. Let $A \in \mathbb{C}^{n \times n}$. Show that, for $k > 0$, if λ is an eigenvalue of A with eigenvector x, then λ^k is an eigenvalue of A^k with eigenvector x.

3. A matrix is *idempotent* if $A^2 = A$. Show that if A is idempotent and $\lambda \in \lambda(A)$, then $\lambda = 0$ or $\lambda = 1$.

4. A matrix $A \in \mathbb{C}^{n \times n}$ is *nilpotent* if $A^k = 0$ for some $k > 0$. Prove that zero is the only eigenvalue of a nilpotent matrix.

5. Show that if U is unitary and $\lambda \in \lambda(U)$, then $|\lambda| = 1$.

6. Prove that if $A \in \mathbb{R}^{n \times n}$, then the eigenvectors corresponding to complex eigenvalues occur in conjugate pairs.

7. Prove that $Ax = \lambda x$ if and only if $x \in \mathfrak{N}(A - \lambda I)$.

8. Let $\lambda \in \lambda(A)$, where $A \in \mathbb{C}^{n \times n}$. Prove that the maximum number of linear independent eigenvectors corresponding to λ is $n -$ rank$(A - \lambda I)$.

9. Prove that $\lambda \in \lambda(A)$ if and only if $\bar{\lambda} \in \lambda(\bar{A})$.

10. Let λ be an eigenvalue of A with eigenvector x. Show that

$$\lambda = \frac{x^{\mathrm{H}} A x}{x^{\mathrm{H}} x}. \qquad (*)$$

The right-hand side of (*) is called a *Rayleigh quotient*.

11. Prove that the eigenvalues of a Hermitian matrix are real. [*Hint*: Use Theorem 1.6 and the previous exercise.]

12. Prove that the eigenvalues of a positive definite matrix are positive and the eigenvalues of a positive semidefinite matrix are nonnegative.

13. Prove that if x is a right eigenvector of A corresponding to λ_1 and y is a left eigenvector corresponding to $\lambda_2 \neq \lambda_1$, then $y^{\mathrm{H}}x = 0$ (i.e., x and y are orthogonal). [*Hint*: Show that both $y^{\mathrm{H}}Ax = \lambda_1 y^{\mathrm{H}}x$ and $y^{\mathrm{H}}Ax = \lambda_2 y^{\mathrm{H}}x$.]

14. Let $A \in \mathbb{C}^{m \times n}$ and $B \in \mathbb{C}^{n \times m}$. Show that the matrices

$$\begin{pmatrix} AB & 0 \\ B & 0 \end{pmatrix}, \qquad \begin{pmatrix} 0 & 0 \\ B & BA \end{pmatrix}$$

are similar. Conclude that the nonzeroeigen values of AB are the same as those of BA.

15. What is the characteristic polynomial of

$$A = \begin{pmatrix} \alpha & \beta \\ \gamma & \delta \end{pmatrix}.$$

16. In four-digit arithmetic compute the characteristic polynomial of the matrix

$$A = \begin{pmatrix} 1 & .9\times 10^{-2} \\ .9\times 10^{-2} & 1 \end{pmatrix}.$$

Compare the eigenvalues of A with the zeros of the computed characteristic polynomial.

17. Consider the following scheme for finding the eigenvalues of the 2×2 matrix

$$A = \begin{pmatrix} \alpha & \beta \\ \gamma & \delta \end{pmatrix}.$$

1. Compute the coefficients of the characteristic polynomial π of the matrix

$$A - \alpha I = \begin{pmatrix} 0 & \beta \\ \gamma & \delta - \alpha \end{pmatrix}.$$

2. Compute the zeros ϱ_1 and ϱ_2 of π.
3. Set $\lambda_i = \varrho_i + \alpha$ $(i = 1, 2)$.

Implement this scheme in INFL and apply it in four-digit arithmetic to the matrix of Exercise 16.

18. Let $\pi(\lambda) = \lambda^n + \alpha_{n-1}\lambda^{n-1} + \cdots + \alpha_0$ be a polynomial with zeros $\varrho_1, \varrho_2, \ldots, \varrho_n$. Let

$$C_\pi = \begin{pmatrix} -\alpha_{n-1} & -\alpha_{n-2} & \cdots & -\alpha_1 & -\alpha_0 \\ 1 & 0 & \cdots & 0 & 0 \\ 0 & 1 & \cdots & 0 & 0 \\ \vdots & \vdots & & \vdots & \vdots \\ 0 & 0 & \cdots & 1 & 0 \end{pmatrix}.$$

Show that the characteristic polynomial of C_π is $\pi(\lambda)$. Conclude that the matrix C_π, which is called the *companion matrix* of π, has eigenvalues $\varrho_1, \varrho_2, \ldots, \varrho_n$. What are the eigenvectors of C_π? [*Hint*: Normalize x_i so that $\xi_n^{(i)} = 1$ and solve for the other components using the equation $Ax_i = \varrho_i x_i$.]

19. Show that the matrix

$$C_\pi' = \begin{pmatrix} 0 & 0 & \cdots & 0 & -\alpha_0 \\ 1 & 0 & \cdots & 0 & -\alpha_1 \\ 0 & 1 & \cdots & 0 & -\alpha_2 \\ \vdots & \vdots & & \vdots & \vdots \\ 0 & 0 & \cdots & 1 & -\alpha_{n-1} \end{pmatrix}.$$

can also be regarded as a companion matrix. Discuss other variants.

20. Find the inverses of C_π and C_π' in Exercises 18 and 19.

21. What are the left eigenvectors of $J_\lambda^{(k)}$? Show that if $k > 1$, x is a right eigenvector of $J_\lambda^{(k)}$, and y is a left eigenvector, then $x^{\mathrm{H}}y = 0$.

22. An upper Hessenberg matrix is unreduced if its subdiagonal elements are nonzero. Show that if $A \in \mathbb{C}^{n\times n}$ is an unreduced Hessenberg matrix, then $\operatorname{rank}(A) \geq n - 1$. Conclude that an unreduced Hessenberg matrix with multiple eigenvalues is defective.

23. Find a matrix A such that for any matrix norm $\|\cdot\|$, we have $\varrho(A) < \|A\|$. [*Hint*: Take $\varrho(A) = 0$.]

24. Let $A \in \mathbb{C}^{n\times n}$ have nonzero eigenvalues $\lambda_1, \lambda_2, \ldots, \lambda_n$ ordered so that $|\lambda_1| \geq |\lambda_2| \geq \cdots \geq |\lambda_n| > 0$. Let $\|\cdot\|$ denote a consistent matrix norm on $\mathbb{C}^{n\times n}$ and let $\varkappa(A) = \|A\|\,\|A^{-1}\|$ be the condition number of A with respect to $\|\cdot\|$. Show that $\varkappa(A) \geq |\lambda_1/\lambda_n|$.

25. Let

$$A = \begin{pmatrix} 3 & -1 \\ -1 & 3 \end{pmatrix}.$$

For $\mu = 1.9$, 1.99, 1.999, and 1.9999, calculate $x^{(\mu)} = (A - \mu I)^{-1}e$, and $y^{(\mu)} = x^{(\mu)}/\xi_1^{(\mu)}$. Explain your results. [*Hint*: What are the eigenvalues and eigenvectors of $(A - \mu I)^{-1}$.]

26. Let $A \in \mathbb{C}^{n\times n}$ and let $p(\lambda) = \pi_0 + \pi_1\lambda + \cdots + \pi_k\lambda^k$ be a polynomial. Define $p(A)$ by $p(A) = \pi_0 I + \pi_1 A + \cdots + \pi_k A^k$. Show that if λ is an eigenvalue of A with eigenvector x, then $p(\lambda)$ is an eigenvalue of $p(A)$ with eigenvector x.

27. Let $A \in \mathbb{C}^{n\times n}$ and let $r(\lambda) = p(\lambda)/q(\lambda)$, where p and q are polynomials. If $q(A)$ is nonsingular, define $r(A) = [q(A)]^{-1}p(A)$. Show that if

$q(A)$ is nonsingular and λ is an eigenvalue of A with eigenvector x, then $r(\lambda)$ is an eigenvalue of $r(A)$ with eigenvector x.

28. Let A, p, q, and r be as in Exercise 27. Show that if A is triangular and if for any $\lambda \in \lambda(A)$, $q(\lambda) \neq 0$, then $r(A)$ is well defined. Moreover any eigenvalue of $r(A)$ may be written in the form $r(\lambda)$, where $\lambda \in \lambda(A)$. [*Hint*: $p(A)$, $q(A)$, and $r(A)$ are also triangular. What are their diagonal elements?]

29. Let $T \in \mathbb{C}^{n\times n}$ be upper triangular with distinct diagonal elements. Let

$$t_i = (\tau_{1i}, \tau_{2i}, \ldots, \tau_{i-1,i})^{\mathrm{T}}.$$

Show that the vector

$$x_i = \begin{pmatrix} (T^{[i-1]} - \tau_{ii}I)^{-1}t_i \\ -1 \\ 0 \end{pmatrix}$$

is an eigenvector of T corresponding to the eigenvalues τ_{ii}. Conclude that the matrix $X = (x_1, x_2, \ldots, x_n)$ of eigenvectors of T is upper triangular.

3. REDUCTION OF MATRICES BY SIMILARITY TRANSFORMATIONS

In this section we shall investigate the problem of reducing a matrix to simple forms by similarity transformations. We have already suggested that such reductions may have computational applications. If, for example, a matrix can be reduced to triangular form, then the diagonal elements of the triangular matrix are the eigenvalues of the original matrix. The results of this section are primarily theoretical, saying how far one can reduce various classes of matrices using various classes of similarity transformations, but not giving constructive procedures for accomplishing the reductions in the absence of a knowledge of the eigenvalues and eigenvectors of the matrix. Nonetheless, the deflation technique to be described in this section is useful in reducing the size of an eigenvalue problem when an eigenvalue and eigenvector are known.

The first result that we shall establish is that if a matrix is not defective, it may be reduced to diagonal form by a similarity transformation. Before proving this result, we first point out a notational device that often simplifies manipulations with eigenvalues and eigenvectors. Let $A \in \mathbb{C}^{n\times n}$ and let the vectors $x_1, x_2, \ldots, x_k$ satisfy the equations

$$Ax_i = \lambda_i x_i \qquad (i = 1, 2, \ldots, k). \tag{3.1}$$

Then if we set

$$X = (x_1, x_2, \ldots, x_k)$$

and

$$\Lambda = \operatorname{diag}(\lambda_1, \lambda_2, \ldots, \lambda_k),$$

it is easily verified that the set of equations (3.1) is equivalent to the single matrix equation

$$AX = X\Lambda.$$

We begin our formal development with a lemma, which is interesting in its own right.

LEMMA 3.1. Let $x_1, x_2, \ldots, x_k$ be eigenvectors of A corresponding to the distinct eigenvalues $\lambda_1, \lambda_2, \ldots, \lambda_k$. Then the vectors x_i are linearly independent.

PROOF. The proof is by induction on k. For $k = 1$, the theorem is trivially true, since $x_1 \neq 0$. Assume that the theorem holds for $k - 1$ eigenvectors and let $x_1, x_2, \ldots, x_k$ be eigenvectors corresponding to the distinct eigenvalues $\lambda_1, \lambda_2, \ldots, \lambda_k$. Then $x_1, x_2, \ldots, x_{k-1}$ are linearly independent. Let $X = (x_1, x_2, \ldots, x_{k-1})$ and suppose that $x_k = Xc$ for some nonzero vector $c \in \mathbb{C}^{k-1}$. Then

$$\lambda_k x_k = Ax_k = AXc = X\Lambda c,$$

where $\Lambda = \operatorname{diag}(\lambda_1, \lambda_2, \ldots, \lambda_{k-1})$. It follows that

$$0 = \lambda_k x_k - X\Lambda c = \lambda_k Xc - X\Lambda c = X(\lambda_k I - \Lambda)c.$$

Since λ_k is distinct from $\lambda_1, \lambda_2, \ldots, \lambda_{k-1}$, the matrix $\lambda_k I - \Lambda$ is nonsingular and $d = (\lambda_\lambda I - \Lambda)c \neq 0$. However, $Xd = 0$, which contradicts the linear independence of $x_1, x_2, \ldots, x_{k-1}$. ■

We are now in a position to show that a matrix can be reduced to diagonal form by a similarity transformation if and only if it is nondefective.

THEOREM 3.2. Let $A \in \mathbb{C}^{n\times n}$. Then A is nondefective if and only if there is a nonsingular matrix X such that

$$X^{-1}AX = \Lambda \equiv \mathrm{diag}(\lambda_1, \lambda_2, \ldots, \lambda_n), \tag{3.2}$$

where the numbers $\lambda_1, \lambda_2, \ldots, \lambda_n$ are the eigenvalues of A. The *ith* column of X is an eigenvector corresponding to λ_i, and the ith row of X^{-1} is a left eigenvector corresponding to λ_i.

PROOF. Suppose that A is nondefective. Let $\mu_1, \mu_2, \ldots, \mu_r$ denote the distinct eigenvalues of A, and let μ_i have multiplicity m_i. Because A is nondefective, μ_i has a set of m_i linearly independent eigenvectors, which we take to be the columns of the matrix $X_i \in \mathbb{C}^{n\times m_i}$. Then

$$AX_i = X_i\Lambda_i,$$

where $\Lambda_i = \mathrm{diag}(\mu_i, \mu_i, \ldots, \mu_i)$. Moreover if we set $X = (X_1, X_2, \ldots, X_r)$, then

$$AX = X\Lambda, \tag{3.3}$$

where $\Lambda = \mathrm{diag}(\Lambda_1, \Lambda_2, \ldots, \Lambda_r)$ consists of the eigenvalues of A.

Now the columns of X are linearly independent. For if not, there exist a a vector $c \neq 0$ such that $Xc = 0$. If c is partitioned conformally with X, we have

$$X_1c_1 + X_2c_2 + \cdots + X_rc_r = 0, \tag{3.4}$$

but since $c \neq 0$, not all the c_i are zero. If c_i is nonzero, then X_ic_i is an eigenvector of A corresponding to μ_i. Thus (3.4) exhibits a nontrivial linear combination of eigenvectors corresponding to distinct eigenvalues which is zero, contradicting Lemma 3.1.

Since the columns of X are linearly independent, X is nonsingular. Hence upon premultipling (3.3) by X^{-1}, we obtain (3.2).

Conversely suppose that (3.2) holds. Then upon premultiplying by X we obtain (3.3), which is equivalent to the equations $Ax_i = \lambda_i x_i$. Thus the columns of X represent a set of n linearly independent eigenvectors, one for each eigenvalue counting multiplicities. Hence A is nondefective.

The conclusion that the rows of X^{-1} are left eigenvectors of A follows upon postmultiplying (3.2) by X^{-1}. ■

The above theorem implies that the eigenvectors of a nondefective matrix of order n span $\mathbb{C}^n$. For this reason a nondefective matrix is often said to have a complete set of eigenvectors. Nondefective matrices are also called *diagonalizable* or *normalizable*.

The decomposition (3.2) is essentially unique. If λ_{i_1} is an eigenvalue of multiplicity k and $\lambda_{i_1} = \lambda_{i_2} = \cdots = \lambda_{i_k}$, then the columns $x_{i_1}, x_{i_2}, \ldots, x_{i_k}$ must form a basis for the null space of $A - \lambda_{i_1} I$, and any basis will do. Beyond this freedom to select a basis, the columns of X are uniquely determined. In particular, if λ_i is a simple eigenvalue, x_i is determined up to a constant multiple. Of course one may interchange the columns of X, which amounts to changing the ordering of the eigenvalues in (3.2).

Theorem 3.2 would lose a good deal of its impact if nondefective matrices were scarce. Fortunately any matrix with distinct eigenvalues has one eigenvector for each eigenvalue and hence is nondefective. Since a matrix taken at random is more likely than not to have distinct eigenvalues, in some sense most matrices are nondefective.

In practice, general similarity transformations have one drawback. Although the transforming matrix P must be nonsingular, it may be ill conditioned with respect to inversion, thus making it difficult to form $P^{-1}AP$. One cure that suggests itself immediately is to restrict the transformations to some class of matrices that is automatically well conditioned. Of course we should not expect to be able to obtain as complete a reduction with such a restricted class of transformations; however, we may still be able to obtain a significant simplification of the problem.

A very natural class of transformations is the class of unitary transformations. This class has three advantages. First, unitary matrices are easy to invert; for if U is unitary, then $U^{-1} = U^{\mathrm{H}}$ so that $U^{-1}AU = U^{\mathrm{H}}AU$ may be easily computed. Secondly, unitary matrices are perfectly conditioned with respect to the 2-norm. In fact, from Theorem 5.1.5,

$$\varkappa(U) = \| U \|_2 \, \| U^{\mathrm{H}} \|_2 = 1 \cdot 1 = 1.$$

A third, less obvious advantage of unitary transformations is that they lend themselves naturally to backward error analyses. For example, suppose that having computed $U^{\mathrm{H}}AU$, we introduce an error F into the result. If

we set $E = UFU^{\mathrm{H}}$, then $\| E \|_2 = \| F \|_2$ and

$$U^{\mathrm{H}}(A + E)U = U^{\mathrm{H}}AU + F.$$

In other words a perturbation in the result can be accounted for by a perturbation *of the same size* in the original problem.

We begin our discussion of unitary similarity transformations by showing how they may be used to eliminate a known eigenvalue from a problem. This process, which is one of many *deflation* techniques, produces a matrix, of order one less than the original matrix, that contains all the eigenvalues of the original matrix except the one that was eliminated. We shall first give a general description of the process, and then sketch the practical details.

Let x be an eigenvector of $A \in \mathbb{C}^{n\times n}$ corresponding to the eigenvalue λ, and suppose that x has been normalized so that $\| x \|_2 = 1$. Then we can find a matrix $U \in \mathbb{C}^{n\times(n-1)}$ such that (x, U) is unitary. Since $Ax = \lambda x$,

$$A(x, U) = (\lambda x, AU)$$

and

$$(x, U)^{\mathrm{H}}A(x, U) = \begin{pmatrix} x^{\mathrm{H}} \\ U^{\mathrm{H}} \end{pmatrix}(\lambda x, AU) = \begin{pmatrix} \lambda x^{\mathrm{H}}x & x^{\mathrm{H}}AU \\ \lambda U^{\mathrm{H}}x & U^{\mathrm{H}}AU \end{pmatrix}.$$

Now $x^{\mathrm{H}}x = 1$, and since the columns of U are orthogonal to x, we have $U^{\mathrm{H}}x = 0$. Hence if we set

$$C = U^{\mathrm{H}}AU$$

and

$$h^{\mathrm{H}} = x^{\mathrm{H}}AU,$$

then

$$(x, U)^{\mathrm{H}}A(x, U) = \begin{pmatrix} \lambda & h^{\mathrm{H}} \\ 0 & C \end{pmatrix}. \tag{3.5}$$

The matrix on the right-hand side of (3.5) is block triangular and has for its eigenvalues the number λ and the eigenvalues of C. Hence the eigenvalues of C are the same as the eigenvalues of A, with the exception of λ.

We shall leave the details of the implementation of this method as an exercise. However, some comments are in order. In the first place we may determine the matrix (x, U) as an elementary reflector. For suppose that y is any eigenvector corresponding to the eigenvalue λ, and that an elementary reflector R has been determined so that $Ry = \pm \| y \|_2 e_1$. Then $Re_1 = \pm y/\| y \|_2$, and the first column of R is an eigenvector of length unity. Of course R will be determined in a factored form, and this form can be used to simplify the computation of RAR.

Rounding errors as such do not affect the practical algorithm very much. Much more important is the fact that we must use an approximate eigenvector in the calculations. If this is done, we obtain a matrix of the form

$$(x, U)^{\mathrm{H}} A(x, U) = \begin{pmatrix} x^{\mathrm{H}}Ax & h^{\mathrm{H}} \\ g & C \end{pmatrix}, \tag{3.6}$$

where $g = U^{\mathrm{H}}Ax$ is no longer zero. To complete the deflation we must set the components of g to zero, and, as we have seen above, this corresponds to introducing an error of size $\| g \|_2$ in the original matrix. Thus it is important to find conditions under which $\| g \|_2$ is small.

One obvious condition is that x be a very accurate eigenvector. However, if the eigenvector is ill conditioned, it will be impossible to compute an accurate approximation to it. Fortunately a weaker condition is sufficient. To formulate it suppose that x is an approximate eigenvector corresponding to the approximate eigenvalue μ. Define the residual vector

$$r = Ax - \mu x. \tag{3.7}$$

If μ and x are an exact eigenvalue–eigenvector pair, then $r = 0$. Thus the size of r in some sense measures the accuracy of x and μ. In terms of this measure there is, for fixed x, an optimum μ.

THEOREM 3.3. Let r be defined by (3.7). Then $\| r \|_2$ is a minimum when

$$\mu = \frac{x^{\mathrm{H}}Ax}{x^{\mathrm{H}}x}. \tag{3.8}$$

PROOF. The problem of minimizing $\| r \|_2^2 = \| Ax - \mu x \|_2^2$ is a linear least squares problem in the unknown μ whose normal equation is

$$x^{\mathrm{H}}x\mu = x^{\mathrm{H}}Ax. \quad \blacksquare$$

The number μ defined by (3.8) is precisely the estimate for λ that is obtained by setting g to zero in (3.6) [recall that in (3.6) $x^{\mathrm{H}}x = 1$]. It is called the *Rayleigh quotient* of the vector x, and it has many desirable properties as an estimate of an eigenvalue. Its importance for our deflation procedure is contained in the following theorem.

THEOREM 3.4. Let (x, U) be unitary and let g be defined as in (3.6). Then

$$\| g \|_2 = \| Ax - (x^{\mathrm{H}}Ax)x \|_2.$$

PROOF. Since (x, U) is unitary

$$\begin{aligned}\| Ax - (x^{\mathrm{H}}Ax)x \|_2 &= \| (x, U)^{\mathrm{H}}[Ax - (x^{\mathrm{H}}Ax)x] \|_2 \\ &= \left\| \begin{pmatrix} x^{\mathrm{H}}[Ax - (x^{\mathrm{H}}Ax)x] \\ U^{\mathrm{H}}[Ax - (x^{\mathrm{H}}Ax)x] \end{pmatrix} \right\|_2 \\ &= \left\| \begin{pmatrix} 0 \\ g - (x^{\mathrm{H}}Ax)U^{\mathrm{H}}x \end{pmatrix} \right\|_2 \\ &= \left\| \begin{pmatrix} 0 \\ g \end{pmatrix} \right\|_2.\end{aligned}$$

The last equality follows from the orthogonality of the columns of U to the vector x. ■

Thus the error induced by setting g to zero in the deflation procedure is the same size as the smallest residual that can be obtained from the approximate eigenvector x. This residual can be very small, even when x is a poor approximation.

EXAMPLE 3.5. Consider the matrix

$$A = \begin{pmatrix} 3 & -1 \\ -1 & 3 \end{pmatrix}$$

whose eigenvectors $(1, 1)^{\mathrm{T}}$ and $(1, -1)^{\mathrm{T}}$ correspond to eigenvalues 2 and 4, respectively. If we take as an approximate eigenvector the vector $(1, 1 + \varepsilon)$, then the deflating transformation is

$$R = \frac{1}{\sqrt{1 + (1 + \varepsilon)^2}} \begin{pmatrix} 1 & 1 + \varepsilon \\ 1 + \varepsilon & -1 \end{pmatrix},$$

and

$$R^{\mathrm{H}}AR = \frac{1}{2 + 2\varepsilon + \varepsilon^2} \begin{pmatrix} 4 + 4\varepsilon + 3\varepsilon^2 & -2\varepsilon - \varepsilon^2 \\ -2\varepsilon - \varepsilon^2 & 8 + 8\varepsilon + 3\varepsilon^2 \end{pmatrix}.$$

To see how the elements of $R^{\mathrm{H}}AR$ behave as ε approaches zero, note that when ε is small, ε^2 is negligible compared with ε. Hence we may drop all terms containing ε^2 to obtain

$$R^{\mathrm{H}}AR \cong \begin{pmatrix} 2 & -\varepsilon(1 + \varepsilon)^{-1} \\ -\varepsilon(1 + \varepsilon)^{-1} & 4 \end{pmatrix}.$$

Thus, as our theory above predicts, and error of ε in the approximate eigenvector causes the g vector [in this case the scalar $-\varepsilon(1+\varepsilon)^{-1}$] to be of order ε.

Two peculiarities of Example 3.5 are due to the fact that A is Hermitian. First the vectors g and h in (3.6) are equal. This is to be expected, since $R^{\mathrm{H}}AR$ is also Hermitian, and g and h occupy corresponding positions in the lower and upper halves of $R^{\mathrm{H}}AR$. The second peculiarity is the fact that the Rayleigh quotients which form the (1, 1)- and (2, 2)-elements of $R^{\mathrm{H}}AR$ are far more accurate approximate eigenvalues than the approximate eigenvectors we started with. The cause of this phenomenon will be explained in Section 5.

We turn now to the theoretical consequences of our deflation procedure. The first result concerns the relation between the left and right eigenvectors of a simple eigenvalue. We saw in Example 2.13 that the only eigenvectors of the matrix $J_\lambda^{(k)}$ were multiples of the vector e_1. Similarly we can show that the only left eigenvectors of $J_\lambda^{(k)}$ are multiples of e_k. Thus if $k > 1$, the left and right eigenvectors corresponding to the eigenvalue λ are orthogonal. The following theorem shows that this cannot happen to a simple eigenvalue.

THEOREM 3.6. Let λ be a simple eigenvalue of the matrix A with right and left eigenvectors x and y. Then $y^{\mathrm{H}}x \neq 0$.

PROOF. Without loss of generality we may take $\| x \|_2 = 1$. Then if $R = (x, U)$ is unitary we have as above

$$R^{\mathrm{H}}AR = \begin{pmatrix} \lambda & h^{\mathrm{H}} \\ 0 & C \end{pmatrix}.$$

We shall calculate a left eigenvector for $R^{\mathrm{H}}AR$ in the form $(1, z^{\mathrm{H}})$. From the equation

$$(1, z^{\mathrm{H}})R^{\mathrm{H}}AR = \lambda(1, z^{\mathrm{H}}), \tag{3.9}$$

we obtain the equation

$$h^{\mathrm{H}} + z^{\mathrm{H}}C = \lambda z^{\mathrm{H}},$$

or

$$z^{\mathrm{H}}(\lambda I - C) = h^{\mathrm{H}}. \tag{3.10}$$

Now since λ is simple it is distinct from the other eigenvalues of A, which are just the eigenvalues of C. Thus $\lambda I - C$ is nonsingular, and we may solve (3.10) for z, which then satisfies (3.9).

Since $(1, z^H)$ is a left eigenvector for R^HAR, $y^H = (1, z^H)R^H$ is a left eigenvector for A, and any other left eigenvector must be a multiple of y. But

$$y^Hx = (1, z^H)R^H(x, U)e_1 = (1, z^H)R^HRe_1 = (1, z^H)e_1 = 1 \neq 0,$$

which proves the theorem. ■

Note that the proof of Theorem 3.6 gives us an explicit representation for y in the form

$$y^H = (1, z^H)R^H = [1, h^H(\lambda I - C)^{-1}]\begin{pmatrix} x^H \\ U^H \end{pmatrix}. \tag{3.11}$$

We shall use this representation in the next section.

The next theorem answers the question of how far we may reduce a general matrix by unitary similarity transformations.

THEOREM 3.7. Let $A \in \mathbb{C}^{n \times n}$ have eigenvalues $\lambda_1, \lambda_2, \ldots, \lambda_n$. Then there is a unitary transformation U such that U^HAU is upper triangular with diagonal elements $\lambda_1, \lambda_2, \ldots, \lambda_n$, in that order.

PROOF. The proof is by induction on n. The theorem is trivially true for $n = 1$. Assume its truth for all matrices of order $n - 1$, and let $A \in \mathbb{C}^{n \times n}$. By the deflation procedure described above, we can find a unitary matrix R such that

$$R^HAR = \begin{pmatrix} \lambda_1 & h^H \\ 0 & C \end{pmatrix}.$$

The eigenvalues of C are $\lambda_2, \lambda_3, \ldots, \lambda_n$. By the induction hypotheses, we may find a unitary matrix S such that $T = S^HCS$ is upper triangular with diagonal elements $\lambda_2, \lambda_3, \ldots, \lambda_n$. Let

$$U = R\begin{pmatrix} 1 & 0 \\ 0 & S \end{pmatrix}.$$

Then

$$\begin{aligned} U^HAU &= \begin{pmatrix} 1 & 0 \\ 0 & S^H \end{pmatrix}\begin{pmatrix} \lambda_1 & h^H \\ 0 & C \end{pmatrix}\begin{pmatrix} 1 & 0 \\ 0 & S \end{pmatrix} \\ &= \begin{pmatrix} \lambda_1 & h^HS \\ 0 & S^HCS \end{pmatrix} = \begin{pmatrix} \lambda_1 & h^HS \\ 0 & T \end{pmatrix} \end{aligned}$$

which is the required reduction. ■

Theorem 3.7 lays the basis for the program, which we shall follow in the next chapter, of attempting to find the eigenvalue of a matrix by reducing it to triangular form using unitary similarity transformations. However, the result is useful in other ways. As an example, we use it to prove an important result in the theory of norms.

THEOREM 3.8. Let $A \in \mathbb{C}^{n \times n}$, and let $\varepsilon > 0$. Then there is a consistent matrix norm $\| \cdot \|$ (depending on A and ε) such that

$$\| A \| \leq \varrho(A) + \varepsilon. \tag{3.12}$$

PROOF. Let U be a unitary matrix such that $T = U^{\mathrm{H}}AU$ is upper triangular with diagonal elements $\lambda_1, \lambda_2, \ldots, \lambda_n$, the eigenvalues of A. Choose $\eta < 1$ so that

$$\eta \leq \frac{\varepsilon}{n-1} \min\{| \tau_{ij}^{-1} | : j = 2, 3, \ldots, n; i = 1, 2, \ldots, j-1\}.$$

Let $D = \operatorname{diag}(1, \eta, \eta^2, \ldots, \eta^{n-1})$. Then $D^{-1}TD$ has the form

$$D^{-1}TD = \begin{pmatrix} \lambda_1 & \tau_{12}\eta & \tau_{13}\eta^2 & \cdots & \tau_{1n}\eta^{n-1} \\ & \lambda_2 & \tau_{23}\eta & \cdots & \tau_{2n}\eta^{n-2} \\ & & \lambda_3 & \cdots & \tau_{3n}\eta^{n-3} \\ & \bigcirc & & \ddots & \vdots \\ & & & & \lambda_n \end{pmatrix}.$$

Hence by the definition of η

$$\begin{aligned} \| D^{-1}U^{\mathrm{H}}AUD \|_\infty = \| D^{-1}TD \|_\infty &\leq \max\{| \lambda_i |\} + (n-1) \max\{| \tau_{ij} | \eta\} \\ &\leq \varrho(A) + \varepsilon. \end{aligned} \tag{3.13}$$

Now the function $\| \cdot \| : \mathbb{C}^{n \times n} \to \mathbb{R}$ defined by

$$\| B \| = \| D^{-1}U^{\mathrm{H}}BUD \|_\infty$$

is a consistent matrix norm. Moreover, by (3.13) it satisfies (3.12). ■

A corollary of Theorem 3.8 is the important result that

$$\lim_{n \to \infty} A^n = 0 \quad \Leftrightarrow \quad \varrho(A) < 1.$$

The proof is indicated in the exercises.

In our discussion of unitary similarity transformations we have left one important point untreated. If the original matrix A is real, we do not wish to use a general unitary transformation R, since such a transformation would introduce complex elements into $R^{\mathrm{H}}AR$. Thus we must confine ourselves to real unitary transformations, that is to orthogonal transformations. If A has complex eigenvalues, then A obviously cannot be reduced to triangular form by orthogonal transformations, and it is important to establish just how far we can reduce A. Since the eigenvalues of A occur in conjugate pairs, it is natural to expect that we could reduce A to quasi-triangular form, in which the 2×2 blocks have conjugate eigenvalues. This is indeed the case.

THEOREM 3.9. Let $A \in \mathbb{R}^{n\times n}$. Then there is an orthogonal matrix U such that $U^{\mathrm{T}}AU$ is quasi-triangular. Moreover, U may be chosen so that any 2×2 diagonal block of $U^{\mathrm{T}}AU$ has only complex eigenvalues (which must therefore be conjugates).

The proof of Theorem 3.9 is based on a generalization of our deflation procedure in which a complex eigenvalue and its conjugate are removed at the same time. A sketch of this technique is given in the exercises.

Finally, for completeness, we give a statement of the reduction of a matrix to Jordan canonical form by similarity transformations. The notation $J_\lambda^{(k)}$ refers to the Jordan blocks introduced in Example 2.13.

THEOREM 3.10. Let $A \in \mathbb{C}^{n\times n}$. Then there are unique numbers $\lambda_1, \lambda_2, \ldots, \lambda_r \in \lambda(A)$ and unique positive integers $m_1, m_2, \ldots, m_r$ such that A is similar to the matrix

$$\operatorname{diag}(J_{\lambda_1}^{(m_1)}, J_{\lambda_2}^{(m_2)}, \ldots, J_{\lambda_r}^{(m_r)}). \tag{3.14}$$

The form (3.14) is called the *Jordan canonical form* of A. It is unique up to the ordering of the Jordan blocks. A typical Jordan form might be

$$\begin{pmatrix} 2 & 1 & & & & & \\ 0 & 2 & & & & \bigcirc & \\ & & 2 & 1 & 0 & & \\ & & 0 & 2 & 1 & & \\ & & 0 & 0 & 2 & & \\ & \bigcirc & & & & 4 & 1 \\ & & & & & 0 & 4 \end{pmatrix}. \tag{3.15}$$

Note that the same eigenvalue can appear in different blocks.

There is a good deal of terminology associated with the Jordan canonical form. First if X is the matrix that reduces A to its Jordan form (3.14), then $x_1, x_2, \ldots, x_{m_1}$ satisfy

$$Ax_1 = \lambda x_1$$

and

$$Ax_{i+1} = \lambda_1 x_{i+1} + x_i \qquad (i = 1, 2, \ldots, m_1 - 1),$$

with similar relations holding for the other vectors. The vectors x_i are called *generalized eigenvectors* or *principal vectors* of the matrix A.

Second, the polynomials

$$\pi_i(\lambda) = \det(\lambda I - J_{\lambda_i}^{(m_i)}) = (\lambda - \lambda_i)^{m_i}$$

are called the *elementary divisors* of A. They divide the characteristic polynomial of A which can be written in the form

$$\pi(\lambda) = \pi_1(\lambda)\pi_2(\lambda)\cdots\pi_r(\lambda).$$

The Jordan form is diagonal only when $m_1 = m_2 = \cdots = m_r = 1$, in which case each $\pi_i(\lambda)$ is linear. Thus another way of saying that A is defective is to say that it has nonlinear elementary divisors.

Finally a matrix of the form (3.15), in which some eigenvalue appears in more than one block is called a *derogatory matrix*. A general matrix is derogatory if its Jordan canonical form is derogatory.

EXERCISES

1. Show that, for $k > 1$, $J_\lambda^{(k)}$ cannot be diagonalized by a similarity transformation.

2. Let A be a nondefective matrix with eigenvalues $\lambda_1, \lambda_2, \ldots, \lambda_n$. Show that there are right eigenvectors $x_1, x_2, \ldots, x_n$ and left eigenvectors $y_1, y_2, \ldots, y_n$ such that $A = \sum_{i=1}^{n} \lambda_i x_i y_i^{\mathrm{H}}$.

3. Find the similarity transformation that diagonalizes the matrix

$$\begin{pmatrix} 1 & 1 \\ 0 & 1+\varepsilon \end{pmatrix}.$$

Discuss the dependence of this transformation on ε as ε tends to zero.

4. Write an INFL program which, given an eigenvector x of a matrix A, accomplishes the deflation described in the text.

5. Assume that a routine is at hand that, given a matrix A, produces an eigenvector of A. Describe how a matrix may be reduced to triangular form by applying successively the deflation technique described in the text. Write an INFL program to accomplish this reduction.

6. Show how, given a left eigenvector of the matrix A, one can construct a unitary matrix R such that

$$R^{\mathrm{H}}AR = \begin{pmatrix} \lambda & 0 \\ g & C \end{pmatrix}.$$

Give an INFL program implementing this technique.

7. Let x be an eigenvector of A corresponding to the eigenvalue λ. Let Q be any nonsingular matrix such that $Qx = \mu e_1$. Show that QAQ^{-1} has the form

$$QAQ^{-1} = \begin{pmatrix} \lambda & h^{\mathrm{H}} \\ 0 & C \end{pmatrix}.$$

8. In Exercise 7, take $Q = MP$ where M is an elementary lower triangular matrix of index unity and P is an elementary permutation. Show how to choose M and P so that MPx is a multiple of e_1. What is the best choice of P? Discuss the practical computation of QAQ^{-1}. Write an INFL program implementing this deflation technique. Compare operation counts with the method of Exercise 4.

9. Let $X = (X_1, X_2) \in \mathbb{C}^{n\times n}$ be nonsingular and let $A \in \mathbb{C}^{n\times n}$. Show that if $AX_1 = X_1M$, then

$$X^{-1}AX = \begin{pmatrix} M & B \\ 0 & C \end{pmatrix}.$$

Hence conclude that $\lambda(A) = \lambda(M) \cup \lambda(C)$.

10. In Exercise 3.9 let U be a unitary matrix such that $U^{\mathrm{H}}X_1$ is upper triangular. Show that

$$U^{\mathrm{H}}AU = \begin{pmatrix} P & Q \\ 0 & R \end{pmatrix}, \qquad (*)$$

and that $\lambda(P) = \lambda(M)$ and $\lambda(R) = \lambda(C)$.

11. Let $A \in \mathbb{R}^{n \times n}$ have the complex eigenvalue λ corresponding to the eigenvector x. Let $x_1 = x + \bar{x}$, $x_2 = i(x - \bar{x})$, and $X = (x_1, x_2)$. Then $X \in \mathbb{R}^{n \times 2}$ satisfies $AX = XM$ for some $M \in \mathbb{R}^{2 \times 2}$. Conclude from Exercise 3.10 there is an orthogonal matrix U such that $U^{\mathrm{T}}AU$ has the form (3.10.*), where $\lambda(P) = \{\lambda, \bar{\lambda}\}$.

12. Prove Theorem 3.9.

13. Let $A \in \mathbb{C}^{n \times n}$, and let $X \in \mathbb{C}^{n \times r}$ have orthonormal columns. Show that the matrix $B \in \mathbb{C}^{r \times r}$ that minimizes $\| AX - XB \|_{\mathrm{F}}$ is given by $B = X^{\mathrm{H}}AX$.

14. Let $A \in \mathbb{C}^{n \times n}$ and let $x \in \mathbb{C}^n$ with $\| x \|_2 = 1$. Let $r = Ax - \mu x$. Show that there is a matrix E with $\| E \|_{\mathrm{F}} = \| r \|_2$ such that x is an eigenvector of $A + E$ with eigenvalue μ.

15. A matrix is *normal* if $A^{\mathrm{H}}A = AA^{\mathrm{H}}$. Prove that a triangular, normal matrix is diagonal.

16. Prove that a matrix A is normal if and only if there is a unitary matrix U such that $U^{\mathrm{H}}AU$ is diagonal.

17. Let $\lambda(A) = \{\lambda_1, \lambda_2, \ldots, \lambda_n\}$ and $Ax_1 = \lambda_1 x_1$. Show that if $\lambda_1 \neq 0$ and $u_1^{\mathrm{H}}x_1 = \lambda_1$, then

$$\lambda(A - u_1 u_1^{\mathrm{T}}) = \{0, \lambda_2, \ldots, \lambda_n\}.$$

[*Hint*: Let $Ax_i = \lambda_i x_i$. To find the eigenvectors of $A - x_1 u_1^{\mathrm{T}}$, consider the vectors $x_i - \alpha_i x_1$.]

18. Let A have eigenvalues $\lambda_1, \lambda_2, \ldots, \lambda_n$ corresponding to eigenvectors $x_1, x_2, \ldots, x_n$. Show that if $v_1^{\mathrm{H}}x_1 = 1$, then $(I - x_1 v_1^{\mathrm{H}})A$ has eigenvalues $0, \lambda_2, \ldots, \lambda_n$ corresponding to eigenvectors x_1, and $x_1 - (v_1^{\mathrm{H}}x_i)x_1$ $(i = 2, 3, \ldots, n)$.

19. Write an INFL program that, given the eigenvector x of $A \in \mathbb{R}^{n \times n}$, effects the reduction of Exercise 18.

20. Let $x, y \in \mathbb{C}^n$ and $y^{\mathrm{H}}x = 1$. Show that there are matrices $X, Y \in \mathbb{C}^{n \times n}$ such that $Xe_1 = x$, $Ye_1 = y$, and $Y^{\mathrm{H}} = X^{-1}$.

21. Let λ be a simple eigenvalue of A. Construct a matrix X such that

$$X^{-1}AX = \begin{pmatrix} \lambda & 0 \\ 0 & C \end{pmatrix}.$$

[*Hint*: Let x and y be right and left eigenvectors corresponding to λ, and apply Exercise 20.]

22. Let $A \in \mathbb{C}^{n\times n}$. Show that for all sufficiently small ε, the matrix $A + \varepsilon I$ is nonsingular. [*Hint*: Without loss of generality A may be taken to be upper triangular.]

23. Give an explicit form for $(J_\lambda^{(n)})^k$.

24. Let $A \in \mathbb{C}^{n\times n}$ and let $\| \cdot \| : \mathbb{C}^{n\times n} \to \mathbb{R}$ be a consistent matrix norm. Show that

$$\lim_{k\to\infty} \| A^k \|^{1/k} = \varrho(A).$$

[*Hint*: Assume, without loss of generality, that A is in Jordan canonical form.]

25. Give an explicit form for $(J_\lambda^{(n)})^{-1}$.

26. Prove that

$$\lim_{n\to\infty} A^n = 0 \quad \Leftrightarrow \quad p(A) < 1.$$

[Hint: If $p(A) < 1$, choose a consistent norm $\| \cdot \|$ for which $\| A \| < 1$.]

NOTES AND REFERENCES

Most of the material in Sections 1–3 is classical and may be found in standard references. The decomposition of Theorem 3.9 is associated with the name of Schur.

The deflation technique described in Section 3 is only one of many for removing a known eigenvalue–eigenvector pair from an eigenvalue problem. Most of them are discussed by Wilkinson (AEP, Chapter IX).

Theorem 3.8 is due to Householder.

4. THE SENSITIVITY OF EIGENVALUES AND EIGENVECTORS

The methods for computing eigenvalues that we shall discuss in Chapter 7 yield a set of approximate eigenvalues of a matrix A that are the exact eigenvalues of a slightly perturbed matrix $A + E$. This result is in perfect analogy with the statement in Section 3.5 that the computed solution of a linear system is the exact solution of a slightly perturbed system. As in the case of linear systems, this backward rounding error analysis does not complete the job, for we have still the problem of assessing the effects of the perturbation E on the eigenvalues of A.

In Section 4.4 we used the theory of norms to obtain bounds on the error induced in the solution of a linear system by a perturbation in the matrix of the system. The theory of norms may also be applied to the eigenvalue problem. However, because of the greater complexity of the eigenvalue problem this approach leads to a number of different theorems of varying degree of applicability. Moreover, such an approach does not readily yield information about the sensitivity of the eigenvectors to perturbations. For these reasons we shall investigate the effects of perturbations on the eigenvalue by means of a technique commonly known as *first-order perturbation theory*. The technique has the drawback that it proves no theorems; its results are only estimates of the error. Moreover, our theory will not treat the important case of multiple eigenvalues. However, the estimates obtained from first-order perturbation theory are often more realistic than rigorous bounds obtained by other means. And, because the simplicity of the technique makes it applicable to a wide variety of problems, it is well worth knowing for its own sake.

The application of first-order perturbation theory to the eigenvalue problem involves a number of technical details that obscure the basic idea of the method itself. Therefore we begin by illustrating the method with two simple examples. The basis of the technique is the observation that if ε and η are small numbers, then the product $\varepsilon\eta$ is insignificant when compared to ε or η and may often be ignored. We have already used this observation in Example 3.5, where we simplified some rather complicated formulas by dropping terms containing ε^2.

We shall first use this observation to compute an approximation to $(1-\varepsilon)^{-1}$, where ε is small. When ε is zero, $(1-\varepsilon)^{-1} = 1^{-1} = 1$. By the continuity of the inversion of numbers, we should expect to find $(1-\varepsilon)^{-1}$ very near unity when ε is small. Thus we shall try to approximate $(1-\varepsilon)^{-1}$ by

$$(1-\varepsilon)^{-1} \cong 1 + \eta,$$

where η is a small number to be determined. Since $(1-\varepsilon)(1-\varepsilon)^{-1} = 1$, we must have

$$(1-\varepsilon)(1+\eta) = 1 - \varepsilon + \eta - \varepsilon\eta \cong 1. \tag{4.1}$$

Since $\varepsilon\eta$ is small compared with ε and η, we drop it from (4.1) and solve the resulting equation

$$1 - \varepsilon + \eta = 1$$

for η to obtain $\varepsilon = \eta$. Thus our approximation for $(1 - \varepsilon)^{-1}$ is

$$(1 - \varepsilon)^{-1} \cong 1 + \varepsilon. \tag{4.2}$$

Note that there are three steps in the above procedure. First a form is chosen for the approximate answer, in this case $1 + \eta$. Then the approximation is substituted into an equation, in this case $(1 + \varepsilon)(1 + \varepsilon)^{-1} = 1$, and all terms not linear in the small quantities are deleted (this is sometimes called dropping higher-order terms). Finally, the resulting linear equation is solved for the unknowns in the approximation.

How accurate is the approximation (4.2)? Since we dropped terms of degree 2 in deriving (4.1), it is reasonable to expect that the difference between $(1 - \varepsilon)^{-1}$ and $1 + \varepsilon$ will depend on ε^2. However, our derivation is in no sense a proof of this. In this case a proof is easy to construct, for

$$(1 - \varepsilon)^{-1} - (1 + \varepsilon) = \frac{\varepsilon^2}{1 - \varepsilon}.$$

Hence if $\varepsilon < \frac{1}{2}$, we have

$$|(1 - \varepsilon)^{-1} - (1 + \varepsilon)| < 2|\varepsilon^2|,$$

so that for small ε the approximation (4.2) is very good indeed.

As a second example of the technique, we consider the problem of approximating $(A - E)^{-1}$, where A is nonsingular. Again we first choose a form for our approximation. In this case it is convenient to seek $(A - E)^{-1}$ in the form

$$(A - E)^{-1} \cong A^{-1}(I + H).$$

Then from the equation $(A - E)(A - E)^{-1} = I$, we obtain

$$(A + E)A^{-1}(I + H) = I + H - EA^{-1} - EA^{-1}H \cong I.$$

Dropping the term $EA^{-1}H$, which is assumed small in comparison to E and H, and solving the resulting equation gives us

$$H = EA^{-1},$$

or

$$(A - E)^{-1} \cong A^{-1}(I + EA^{-1}). \tag{4.3}$$

Not only does (4.3) give us an approximation to $(A - E)^{-1}$, but it also gives us a means of deriving, in a loose manner, the condition number of a matrix with respect to inversion. From (4.3) we have

$$\| (A - E)^{-1} - A^{-1} \| \cong \| A^{-1}EA^{-1} \| \leq \| A^{-1} \|^2 \| E \|.$$

Hence we have the approximate inequality

$$\frac{\| (A - E)^{-1} - A^{-1} \|}{\| A^{-1} \|} \lesssim \| A^{-1} \| \| E \| = \varkappa(A) \frac{\| E \|}{\| A \|}.$$

This differs from the bound (4.3.11) only by the factor

$$\frac{1}{1 - \varkappa(A) \dfrac{\| E \|}{\| A \|}}$$

which is negligible when $\| E \|$ is small.

To establish rigorously the accuracy of the approximation (4.3) is more difficult than establishing the accuracy of (4.2) was. However, in much the same way as the bounds of Section 4.3 were established, we can show that if $\| A^{-1} \| \| E \| < \frac{1}{2}$, then

$$\| (A - E)^{-1} - A^{-1}(I + EA^{-1}) \| < 2 \| A^{-1} \|^3 \| E \|^2. \tag{4.4}$$

Thus as $\| E \|$ approach zero, the error in (4.3) approaches zero quadratically.

It is convenient to have a notation to express inequalities such as (4.4) in a compact form. The usual way to do this is by means of the O symbol.

DEFINITION 4.1. Let $P(E)$ and $Q(E)$ be matrices depending on E that are well defined for all sufficiently small E. For $\alpha > 0$ we write

$$P(E) = Q(E) + O(\| E \|^\alpha)$$

if there is a constant γ such that

$$\| P(E) - Q(E) \| \leq \gamma \| E \|^\alpha \tag{4.5}$$

for all sufficiently small E.

Thus we may rewrite (4.4) as

$$(A - E)^{-1} = A^{-1}(I + EA^{-1}) + O(\| E \|^2).$$

In this case the constant γ in the definition is $2 \| A^{-1} \|^3$, and by "sufficiently small E" is meant all E for which $\| A^{-1} \| \| E \| < \frac{1}{2}$.

We note that because of the equivalence of norms, Definition 4.1 is independent of the choice of norm. A change of norm may, of course, change the constant γ, but by choosing γ large enough we can ensure that the inequality (4.5) continues to be satisfied for the new norm.

With the above examples in mind, we shall now turn to the problem of assessing the effects of a perturbation on a simple eigenvalue of a matrix and its associated eigenvector. Let $A \in \mathbb{C}^{n \times n}$, and let λ be a simple eigenvalue of A and x be its associated eigenvector. Since $x \neq 0$, we may assume without loss of generality that $\| x \|_2 = 1$. Let $E \in \mathbb{C}^{n \times n}$ be a fixed perturbation. We wish to find approximations to an eigenvalue λ' and eigenvector x' of $A + E$ that are near λ and x.

As in the above examples, the first step is to determine a form for λ' and x'. The form for λ' is easy to choose, since λ and λ' can differ only by a scalar. Thus we take

$$\lambda' = \lambda + \mu, \tag{4.6}$$

where μ is presumed small.

It is important to choose a form for x' in such a way that $x - x'$ is as small as possible. Since a nonzero multiple of an eigenvector is still an eigenvector, it is natural to seek x' in the form

$$x' = \alpha y,$$

where y is a fixed vector to be determined, and choose α to make $x - x'$ as small as possible. If we use the 2-norm as a measure of size, we must solve the linear least squares problem of minimizing

$$\| x - x' \|_2 = \| x - \alpha y \|_2 .$$

Whatever the solution α is, the residual vector

$$r \equiv -q = x - x'$$

must be orthogonal to x. Thus we seek x' in the form

$$x' = x + q, \tag{4.7}$$

where q is orthogonal to x. If we let $U \in \mathbb{C}^{n \times (n-1)}$ be a matrix such that (x, U) is unitary, then q can be written in the form $q = Up$ for some $p \in \mathbb{C}^{n-1}$. Thus we finally seek our approximate eigenvector in the form

$$x' = x + Up, \tag{4.8}$$

where (x, U) is unitary and $p \in \mathbb{C}^{n-1}$.

We point out that the matrix (x, U) is precisely the matrix used in the deflation technique described in the last section. Hence we have

$$(x, U)^{\mathrm{H}} A(x, U) = \begin{pmatrix} \lambda & h^{\mathrm{H}} \\ 0 & C \end{pmatrix},$$

where

$$h^{\mathrm{H}} = x^{\mathrm{H}} AU$$

and

$$C = U^{\mathrm{H}} AU. \tag{4.9}$$

Moreover, the eigenvalues of C are the eigenvalues of A, excepting λ. Since λ is a simple eigenvalue of A, the eigenvalues of C are all different from λ.

The second step in our analysis is to substitute the forms (4.6) and (4.8) into the equation $(A + E)x' \cong \lambda' x'$ and simplify by dropping higher-order terms. In this case this means dropping all products of E, p, and μ in the relation

$$(A + E)(x + Up) \cong (\lambda + \mu)(x + Up) \tag{4.10}$$

to obtain

$$Ax + Ex + AUp = \lambda x + \mu x + \lambda Up.$$

Since $Ax = \lambda x$, this is equivalent to

$$Ex + AUp = \mu x + \lambda Up. \tag{4.11}$$

We first solve Equation (4.11) for p. Premultiplying by U^{H} and remembering that $U^{\mathrm{H}} U = I$ and $U^{\mathrm{H}} x = 0$, we obtain

$$g + Cp = \lambda p,$$

where C is defined by (4.9) and

$$g = U^{\mathrm{H}} Ex.$$

Since λ is not an eigenvalue of C, the matrix $\lambda I - C$ is nonsingular, and

$$p = (\lambda I - C)^{-1} g. \tag{4.12}$$

This completely determines x'.

It is of course possible to determine μ from (4.11). However, for reasons that will become apparent later, it is important to retain some of the higher-order terms. Therefore, we work directly with the relation (4.10). Multiplying (4.10) by x^{H} and remembering that $x^{\mathrm{H}}Ax = \lambda$ and $x^{\mathrm{H}}U = 0$, we get

$$\begin{aligned}
\mu &= x^{\mathrm{H}}Ex + x^{\mathrm{H}}AUp + x^{\mathrm{H}}EUp \\
&= x^{\mathrm{H}}Ex + h^{\mathrm{H}}(\lambda I - C)^{-1}U^{\mathrm{H}}Ex + f^{\mathrm{H}}(\lambda I - C)^{-1}g \\
&= (1, h^{\mathrm{H}}(\lambda I - C)^{-1})\begin{pmatrix} x^{\mathrm{H}} \\ U^{\mathrm{H}} \end{pmatrix} Ex + f^{\mathrm{H}}(\lambda I - C)^{-1}g \\
&= y^{\mathrm{H}}Ex + f^{\mathrm{H}}(\lambda I - C)^{-1}g,
\end{aligned} \tag{4.13}$$

where

$$y^{\mathrm{H}} = (1, h^{\mathrm{H}}(\lambda I - C)^{-1})\begin{pmatrix} x^{\mathrm{H}} \\ U^{\mathrm{H}} \end{pmatrix}, \tag{4.14}$$

and

$$f^{\mathrm{H}} = x^{\mathrm{H}}EU.$$

On comparing (4.14) with (3.11) of the last section, we see that y^{H} is simply the left eigenvector of A corresponding to λ, normalized so that $y^{\mathrm{H}}x = 1$.

This completes the perturbation analysis. The results give us reason to expect that $A + E$ will have an eigenvalue $\lambda + \mu$ with eigenvector $x + Up$, where μ is given by (4.13) and p is given by (4.12). However, it is important to remember that we have proved nothing. To turn these estimates into rigorous bounds is beyond the scope of this book. Instead we merely state the following theorem.

THEOREM 4.2. Let λ be a simple eigenvalue of $A \in \mathbb{C}^{n \times n}$ with eigenvector x and left eigenvector y. Suppose x has been scaled so that $\| x \|_2 = 1$ and y has been scaled so that $y^{\mathrm{H}}x = 1$ (by Theorem 3.6 this is always possible). Let $U \in \mathbb{C}^{n \times (n-1)}$ be chosen so that (x, U) is unitary and set

$$(x, U)^{\mathrm{H}}A(x, U) = \begin{pmatrix} \lambda & h^{\mathrm{H}} \\ 0 & C \end{pmatrix}.$$

Let $E \in \mathbb{C}^{n \times n}$ be given and let

$$g = U^{\mathrm{H}} E x \tag{4.15}$$

and

$$f = U^{\mathrm{H}} E^{\mathrm{H}} x.$$

Let

$$\varepsilon = \| E \|_2, \qquad \eta = \| h \|_2, \qquad \gamma = \| g \|_2,$$

and

$$\delta = \| (\lambda I - C)^{-1} \|_2^{-1}.$$

Then if

$$\frac{\gamma(\eta + \varepsilon)}{(\delta - \varepsilon)^2} < \frac{1}{4},$$

there is an eigenvalue λ' of $A + E$ with eigenvector x' satisfying

$$\lambda' = \lambda + y^{\mathrm{H}} E x + f^{\mathrm{H}} (\lambda I - C)^{-1} g + \eta O(\varepsilon^2) + O(\varepsilon^3) \tag{4.16}$$

and

$$x' = x + U(\lambda I - C)^{-1} g + \eta O(\varepsilon^2). \tag{4.17}$$

It is worthwhile to study the results of Theorem 4.2 in detail. We shall first derive a condition number for the eigenvalue λ. If we ignore the $O(\varepsilon^3)$ term in (4.16), we have

$$\begin{aligned} | \lambda - \lambda' | &\leq | y^{\mathrm{H}} E x | + | f^{\mathrm{H}} (\lambda I - C)^{-1} g | + \eta O(\varepsilon^2) \\ &\leq \varepsilon \| y \|_2 \| x \|_2 + \frac{\| f \|_2 \| g \|_2}{\delta} + \eta O(\varepsilon^2). \end{aligned}$$

However, since $\| x \|_2 = 1$ and $\| U \|_2 = 1$,

$$\| f \|_2 \| g \|_2 \leq \varepsilon^2.$$

Hence

$$| \lambda - \lambda' | \leq \varepsilon \| y \|_2 + \frac{\varepsilon^2}{\delta} + \eta O(\varepsilon^2). \tag{4.18}$$

For small ε the last terms in (4.18) are insignificant. The first term says that a perturbation of size ε in A induces a perturbation of size $\| y \|_2 \varepsilon$ in the eigenvalue λ. Thus if $\| y \|_2$ is large, λ will be ill conditioned. To summarize, *if x and y are right and left eigenvectors of A with simple eigenvalue λ and $x^{\mathrm{H}} x = y^{\mathrm{H}} x = 1$, then $\| y \|_2$ is a condition number for λ.*

The following simple example illustrates these considerations.

EXAMPLE 4.3. Let

$$A = \begin{pmatrix} 1 & 100 \\ 0 & 2 \end{pmatrix}.$$

Then A has the simple eigenvalue $\lambda = 1$ with eigenvectors $x = (1, 0)^T$ and $y = (1, -100)^T$. Since $\| y \|_2 \cong 100$, we should expect λ to be moderately ill conditioned. And in fact the matrix

$$A + E = \begin{pmatrix} 1 & 100 \\ -10^{-5} & 2 \end{pmatrix}$$

has an eigenvalue $\lambda' \cong 1.001$. Thus a perturbation of order 10^{-5} in A generates a perturbation of order 10^{-3} in λ.

It should be noted that the approximate inequality (4.18) bounds only the absolute error in λ. If λ is small compared with ε, then the relative error in λ' will be large, even though $\| y \|_2$ may be near unity.

We can similarly derive a condition number for the eigenvector x. If we ignore the term $\eta O(\varepsilon^2) + O(\varepsilon^3)$ in (4.17), we get

$$\| x - x' \|_2 \leq \frac{\gamma}{\delta} \leq \frac{\varepsilon}{\delta}.$$

Thus the number δ^{-1} is a condition number for the eigenvector x.

To gain some insight into the nature of the number δ, let $\lambda_2, \lambda_3, \ldots, \lambda_n$ be the eigenvalues, other than λ, of A. As we have observed above, $\lambda_2, \lambda_3, \ldots, \lambda_n$ are also the eigenvalues of C. Thus the eigenvalues of $(\lambda I - C)^{-1}$ are $(\lambda - \lambda_i)^{-1}$, $i = 2, 3, \ldots, n$. From Theorem 2.15,

$$\max\{| \lambda - \lambda_i |^{-1}\} = \varrho[(\lambda I - C)^{-1}] \leq \| (\lambda I - C)^{-1} \|_2 = \delta^{-1}.$$

Thus

$$\delta \leq \min\{| \lambda - \lambda_i |\}. \tag{4.19}$$

The inequality (4.19) shows that δ is always less than the separation of λ from the other eigenvalues of A. In other words *if λ is near another eigenvalue of A, then its eigenvector will be ill conditioned.*

EXAMPLE 4.4. Let

$$A = \begin{pmatrix} 1 & 0 \\ 0 & \gamma \end{pmatrix}.$$

Then A has the eigenvalue $\lambda = 1$ with eigenvector $x = (1, 0)^{\mathrm{T}}$. In the notation above C is the 1×1 matrix (γ). Hence $\delta = |1 - \gamma|$, and if γ is near unity, we should expect x to be ill conditioned. In fact if

$$A + E = \begin{pmatrix} 1 & 0 \\ 10^{-5} & \gamma \end{pmatrix},$$

then $A + E$ has the eigenvector $x' = (1, 10^{-5}/(1 - \gamma))$ corresponding to the eigenvalue $\lambda' = 1$. If γ is near unity, then x' differs significantly from x.

It should not be thought that the only ill-conditioned eigenvectors are those corresponding to poorly separated eigenvalues. Rather innocuous appearing matrices with well separated eigenvalues can have very ill-conditioned eigenvectors. A horrible example is given in the exercises.

Theorem 4.2 not only gives us estimates for the accuracy of an eigenvalue or an eigenvector, but it gives us approximations to the eigenvalue and eigenvector of the perturbed matrix. This fact can be used to improve the accuracy of an approximate set of eigenvectors. For if the columns of $X \in \mathbb{C}^{n \times n}$ are approximate eigenvectors of A, then $X^{-1}AX$ has the form

$$X^{-1}AX = D + E,$$

where the entries of $D = \operatorname{diag}(\delta_1, \delta_2, \ldots, \delta_n)$ are approximate eigenvalues of A, and E is a small matrix with zero diagonal elements. If the diagonal elements of D are sufficiently well separated, we may apply Theorem 4.2 to compute approximate eigenvectors $y_1, y_2, \ldots, y_n$ of $D + E$. If we set $Y = (y_1, y_2, \ldots, y_n)$, then $Y^{-1}(D + E)Y = (XY)^{-1}A(XY)$ will be more nearly diagonal than $X^{-1}AX$. Hence XY will be an improved set of eigenvectors for A. Moreover, since the eigenvalues of $D + E$ are the eigenvalues of A we may also apply the theorem to estimate the accuracy of δ_i as an approximate eigenvalue of A.

EXAMPLE 4.5. Let

$$A = \begin{pmatrix} 2.5 & 1.0 & 0.5 \\ 0.5 & 2.0 & -0.5 \\ -0.5 & -1.0 & 1.5 \end{pmatrix},$$

and

$$X = \begin{pmatrix} -1 & 1 & 1 \\ 1 & -1 & 1 \\ 1 & 1 & -1 \end{pmatrix}.$$

Then

$$X^{-1}AX = \operatorname{diag}(1, 2, 3)$$

so that the columns of X are eigenvectors of A. Suppose we are given the matrix of approximate eigenvectors

$$\tilde{X} = \begin{pmatrix} -1.001 & 0.999 & 0.999 \\ 0.999 & -1.001 & 0.999 \\ 0.999 & 0.999 & -1.001 \end{pmatrix}.$$

Then, except for errors of order 10^{-6},

$$\tilde{X}^{-1}A\tilde{X} = \begin{pmatrix} 1.00000 & 1.000\times 10^{-3} & 2.003\times 10^{-3} \\ -1.003\times 10^{-3} & 2.00000 & 1.003\times 10^{-3} \\ -2.003\times 10^{-3} & -1.000\times 10^{-3} & 3.00000 \end{pmatrix}$$

Let $\tilde{X}^{-1}A\tilde{X} = D + E$, where $D = \operatorname{diag}(1, 2, 3)$. Then, as suggested above, we wish to find approximate eigenvectors of $D + E$. Starting with the eigenvector e_1 of D, the quantities in Theorem 4.2 are

$$U = (e_2, e_3),$$

$$g = \begin{pmatrix} -1.003\times 10^{-3} \\ -2.003\times 10^{-3} \end{pmatrix}, \tag{4.20}$$

and

$$C = \operatorname{diag}(2, 3). \tag{4.21}$$

Hence

$$p = (I - C)^{-1}g = \begin{pmatrix} 1.0030\times 10^{-3} \\ 1.0015\times 10^{-3} \end{pmatrix}$$

and the vector

$$y = e_1 + Up = \begin{pmatrix} 1.0000 \\ 1.0030\times 10^{-3} \\ 1.0015\times 10^{-3} \end{pmatrix}$$

is an approximate eigenvector of $\tilde{X}^{-1}A\tilde{X}$. The other approximate eigenvectors y_2 and y_3 of $\tilde{X}^{-1}A\tilde{X}$ may be determined in the same way. The result is the matrix

$$Y = \begin{pmatrix} 1.0000 & 1.0000\times 10^{-3} & 1.0015\times 10^{-3} \\ 1.0030\times 10^{-3} & 1.0000 & 1.0030\times 10^{-3} \\ 1.0015\times 10^{-3} & 1.0000\times 10^{-3} & 1.0000 \end{pmatrix}.$$

If we form $\tilde{X}Y$ and scale its columns so that its diagonal elements are unity, there results the new matrix

$$\tilde{\tilde{X}} = \begin{pmatrix} -1.000000 & 0.999996 & 1.000002 \\ 0.999999 & -1.000000 & 0.999997 \\ 1.000001 & 1.000000 & -1.000000 \end{pmatrix}$$

of approximate eigenvectors of A. The matrix $\tilde{\tilde{X}}$ is obviously a better approximation to X than $\tilde{X}$.

Except for rounding errors made in the computation of $D + E$, the eigenvalues of A are the same as the eigenvalues of $D + E$. Now e_1 is both a left and right eigenvector of D corresponding to the eigenvalue $\delta_1 = 1$. Hence by Theorem 4.2 there is an eigenvalue μ_1 of $D + E$ given by

$$\mu_1 \cong \delta_1 + e_1^{\mathrm{H}} E e_1 + f^{\mathrm{H}}(I - C)^{-1} g,$$

where g is given by (4.20), $f = (1.000 \times 10^{-3}, 2.003 \times 10^{-3})^{\mathrm{T}}$, and C is given by (4.21). Since the (1, 1)-element of E is zero, $e_1^{\mathrm{H}} E e_1 = 0$ and

$$\mu_1 = \delta_1 - 3 \times 10^{-6}.$$

Thus to about six figures δ_1 is an eigenvalue of $D + E$. Since errors of order 10^{-6} were made in computing $D + E$, δ_1 is also a five- or six-digit approximation to an eigenvalue of A (actually it is equal to the eigenvalue 1 of A). A similar argument shows that $\delta_2 = 2$ and $\delta_3 = 3$ may be expected to err in their sixth figures as approximations to eigenvalues of A.

Although the above example illustrates the power of Theorem 4.2 as a technique for refining a system of approximate eigenvectors, it is important to realize that the method does not lend itself well to automatic computation. In the first place, to obtain an accurate correction matrix Y, one must compute the elements of $D + E = \tilde{X}^{-1} A \tilde{X}$ with reasonable accuracy. Since the off-diagonal elements of $D + E$ are small, one must use higher-precision arithmetic to attain this accuracy (compare this situation with Algorithm 5.5.1 for refining an approximate solution of a linear system in which the residual must be computed in double precision). This difficulty is aggravated when X is ill conditioned with respect to inversion. Of course, if one desires only to estimate the accuracy of a system of approximate eigenvectors, one needs only the magnitudes of the elements of E, which may then be computed to lower precision.

A more substantial difficulty is that Theorem 4.2 does not apply to multiple eigenvalues or to clusters of poorly separated eigenvalues. Thus if in the above example D has turned out to be diag(1, 1, 3), we would have been unable to refine the approximate eigenvectors $\tilde{x}_1$ and $\tilde{x}_2$, or even estimate the accuracy of the approximate eigenvalues 1 and 1. We shall return to the question of multiple eigenvalues in a moment.

One final comment on Example 4.5 is in order. Note that although the elements of E are of order 10^{-3}, the eigenvalues of D and $D + E$ differ by quantities of order 10^{-6}. This is because the left and right eigenvectors y_i and x_i satisfy $y_i^{\mathrm{H}} E x_i = 0$, so that the error in the eigenvalues is given by $f^{\mathrm{H}}(\lambda I - C)^{-1}g$, which is proportional to $\| E \|^2$. Such *off-diagonal* perturbations occur frequently in practice, which is the reason we retained the term $f^{\mathrm{H}}(\lambda I - C)^{-1}g$ in (4.16).

We have already pointed out in connection with Example 4.5 that Theorem 4.2 does not apply to multiple eigenvalues. A satisfactory perturbation theory for eigenvectors corresponding to multiple eigenvalues requires the introduction of the notion of invariant subspace and is beyond the scope of this book. The situation regarding the condition of multiple eigenvalues is mixed; a multiple eigenvalue may or may not be ill conditioned, depending on a number of factors. We begin with an example of an ill-conditioned multiple eigenvalue.

EXAMPLE 4.6. Consider the matrix

$$J_1^{(n)} + E = \begin{pmatrix} 1 & 1 & 0 & \cdots & 0 \\ 0 & 1 & 1 & \cdots & 0 \\ \vdots & \vdots & \vdots & & \vdots \\ 0 & 0 & 0 & & 1 \\ \varepsilon & 0 & 0 & \cdots & 1 \end{pmatrix},$$

whose characteristic polynomial is

$$(\lambda - 1)^n + \varepsilon = 0.$$

Then all the eigenvalues of $J_1^{(n)} + E$ satisfy

$$| \lambda - 1 | = \sqrt[n]{| \varepsilon |}.$$

In other words the eigenvalues of $J_1^{(n)} + E$ differ from the single eigenvalue

1 of $J_1^{(n)}$ by $\sqrt[n]{|\varepsilon|}$. This difference can be significant, even when ε is not. For example, if $n = 10$, and $\varepsilon = 10^{-10}$, then $\sqrt[n]{|\varepsilon|} = 0.1$. In this case a perturbation of order 10^{-10} introduces an error of 0.1 in the eigenvalue.

In spite of Example 4.6 it should not be thought that all multiple eigenvalues are ill conditioned. We shall see in Section 5 that the eigenvalues of a Hermitian matrix are perfectly conditioned, whatever their multiplicity. In order to investigate the sensitivity of multiple eigenvalues, we prove the widely used Gerschgorin theorem.

THEOREM 4.7. Let $A \in \mathbb{C}^{n\times n}$. Then each eigenvalue of A lies in one of the disks in the complex plane

$$\mathfrak{D}_i = \left\{\lambda : |\lambda - \alpha_{ii}| \le \sum_{\substack{j=1\\ j\ne i}}^{n} |\alpha_{ij}|\right\} \qquad (i = 1, 2, \ldots, n).$$

PROOF. Let $\lambda \in \lambda(A)$ and let x be an eigenvector corresponding to λ. Suppose that $|\xi_i| \ge |\xi_j|$ $(j \ne i)$ is a largest component of x. Since $x \ne 0$, $\xi_i \ne 0$. Now from the equation $(\lambda I - A)x = 0$, we have

$$(\lambda - \alpha_{ii})\xi_i = \sum_{j\ne i} \alpha_{ij}\xi_j.$$

Hence, since $|\xi_j|/|\xi_i| \le 1$ for $j \ne i$,

$$|\lambda - \alpha_{ii}| \le \sum_{j\ne i} |\alpha_{ij}| \frac{|\xi_j|}{|\xi_i|} \le \sum_{j\ne i} |\alpha_{ij}|$$

and $\lambda \in \mathfrak{D}_i$. ■

The sets $\mathfrak{D}_i$ are disks in the complex plane centered about α_{ii} and of radius $\sum_{j\ne i} |\alpha_{ij}|$. They are called the Gerschgorin disks of the matrix A. The proof of the theorem shows not only that each eigenvalue of A must lie in a Gerschgorin disk, but also that if the ith component of an eigenvector is maximal, then the corresponding eigenvalue must lie in the ith disk.

EXAMPLE 4.8. Let

$$A = \begin{pmatrix} 1 & 10^{-4} & 10^{-4} \\ 10^{-4} & 1 & 10^{-4} \\ 10^{-4} & 10^{-4} & 2 \end{pmatrix}.$$

Then the Gerschgorin disks of A are

$$\mathfrak{D}_1 = \mathfrak{D}_2 = \{\lambda : |\lambda - 1| \leq 2\times 10^{-4}\}$$

and

$$\mathfrak{D}_3 = \{\lambda : |\lambda - 2| \leq 2\times 10^{-4}\}.$$

Thus all the eigenvalues of A lie in the union of two disks of radius 2×10^{-4} centered at 1 and 2.

Example 4.8 illustrates two inadequacies in the Gerschgorin theorem as given above. The first difficulty is that the theorem does not state in which disks the eigenvalues lie. In the above example, it is conceivable that all the eigenvalues of A lie in $\mathfrak{D}_3$. However, this is contrary to one's natural expectations, which demand that two eigenvalues be near 1 and the other near 2. The following theorem takes care of this problem. Its proof depends on the continuity of the eigenvalues of a matrix as functions of the elements of the matrix and is omitted.

THEOREM 4.9. If k Gerschgorin disks of the matrix A are disjoint from the other disks, then exactly k eigenvalues of A lie in the union of the k disks.

Thus in Example 4.8, the disks $\mathfrak{D}_1$ and $\mathfrak{D}_2$ are disjoint from $\mathfrak{D}_3$. Hence there are two eigenvalues of A in $\mathfrak{D}_1 \cup \mathfrak{D}_2$; that is two eigenvalues of A lie near 1. Likewise $\mathfrak{D}_3$ contains exactly one eigenvalue.

The second inadequacy in the Gerschgorin theorem is that it does not give a very good bound for the eigenvalue near 2. To see this, write A in the form $A = D + E$, where $D = \operatorname{diag}(1, 1, 2)$. Then from Theorem 4.2 it follows that there is an eigenvalue λ of A satisfying

$$\lambda = 2 + 2\times 10^{-8} + O(10^{-12}). \tag{4.22}$$

Thus we should expect to find an eigenvalue of A in, say, the disk $\{|\lambda - 2| \leq 3\times 10^{-8}\}$, a disk which is far smaller than $\mathfrak{D}_3$.

This difficulty can be circumvented as follows. Let $D = \operatorname{diag}(1, 1, 2\times 10^{-4})$. Then $A' = DAD^{-1}$ has the form

$$A' = \begin{pmatrix} 1 & 10^{-4} & .5 \\ 10^{-4} & 1 & .5 \\ 2\times 10^{-8} & 2\times 10^{-8} & 2 \end{pmatrix}.$$

The Gerschgorin disks for A' are

$$\mathfrak{D}_1' = \mathfrak{D}_2' = \{\lambda : |\lambda - 1| \le .5001\}$$

and

$$\mathfrak{D}_3' = \{\lambda : |\lambda - 2| \le 4 \times 10^{-8}\}.$$

The disk $\mathfrak{D}_3'$ is disjoint from $\mathfrak{D}_1'$ and $\mathfrak{D}_2'$ and therefore contains exactly one eigenvalue of A', and hence of A. The radius of the disk $\mathfrak{D}_3'$ compares favorably with the estimate (4.22).

The technique just illustrated above is not just a trick. It can be applied to reduce the size of any isolated Gerschgorin disk, although it is most effective when the original matrix is a small perturbation of a diagonal matrix. A more systematic treatment is given in the exercises.

We conclude this section with a general theorem that provides an overall condition number for the eigenvalues of a nondefective matrix.

THEOREM 4.10. Let A be nondefective and $X^{-1}AX = \Lambda \equiv \operatorname{diag}(\lambda_1, \lambda_2, \ldots, \lambda_n)$. Let $\|\cdot\| : \mathbb{C}^{n\times n} \to \mathbb{R}$ denote a consistent matrix norm such that $\|\operatorname{diag}(\gamma_1, \gamma_2, \ldots, \gamma_n)\| = \max\{|\gamma_i| : i = 1, 2, \ldots, n\}$. Then for any $E \in \mathbb{C}^{n\times n}$, the eigenvalues of $A + E$ lie in the union of the disks

$$\mathfrak{B}_i = \{\lambda : |\lambda - \lambda_i| \le \varkappa(X) \|E\|\},$$

where $\varkappa(X) = \|X\| \|X^{-1}\|$.

PROOF. Let $F = X^{-1}EX$. Then the eigenvalues of $A + E$ are the same as the eigenvalues of $X^{-1}(A + E)X = \Lambda + F$. Now let λ be an eigenvalue of $\Lambda + F$. Then $\lambda I - \Lambda - F$ is singular. If $\lambda = \lambda_i$ for any i, then $\lambda \in \mathfrak{B}_i$. Otherwise $\lambda I - \Lambda$ is nonsingular and

$$(\lambda I - \Lambda)^{-1}(\lambda I - \Lambda - F) = I - (\lambda I - \Lambda)^{-1}F$$

is singular. Since $\|I\| = 1$, it follows from Theorem 4.3.4 that

$$\begin{aligned} 1 \le \|(\lambda I - \Lambda)^{-1}F\| &\le \|(\lambda I - \Lambda)^{-1}\| \|F\| \\ &\le \max\{|\lambda - \lambda_i|^{-1} : i = 1, 2, \ldots, n\} \|X^{-1}EX\| \\ &\le \varkappa(X) \|E\| \max\{|\lambda - \lambda_i|^{-1} : i = 1, 2, \ldots, n\}. \end{aligned}$$

Hence

$$\min\{|\lambda - \lambda_i| : i = 1, 2, \ldots, n\} \le \varkappa(X) \|E\|,$$

which implies that λ lies in one of the disks $\mathfrak{B}_i$. ■

It follows from Theorem 4.10 that $\varkappa(X)$ is an upper bound on the condition of the individual eigenvalues of A.

EXERCISES

1. Let λ be a simple eigenvalue of A with right eigenvector x and left eigenvector y. Show that

$$\frac{\partial \lambda}{\partial \alpha_{ij}} = \frac{\bar{\eta}_i \xi_j}{y^{\mathrm{H}} x}.$$

   [*Hint*: Differentiate the relation $Ax = \lambda x$ implicitly with respect to α_{ij} and premultiply by y^{H}.]

2. Use the results of the last exercise to show that if E is a small perturbation matrix, then

$$\lambda + \frac{y^{\mathrm{H}} E x}{y^{\mathrm{H}} x}$$

   is, up to terms of $O(\| E \|^2)$, an eigenvalue of $A + E$.

3. Apply Theorem 4.2 to estimate the eigenvalues and eigenvectors of the matrix

$$\begin{pmatrix} 1.00 & 1\times 10^{-3} & -2\times 10^{-3} & -1\times 10^{-3} \\ -1\times 10^{-3} & 2.00 & -3\times 10^{-3} & 1\times 10^{-3} \\ 1\times 10^{-3} & -4\times 10^{-3} & 3.00 & 2\times 10^{-3} \\ 1\times 10^{-3} & -1\times 10^{-3} & 1\times 10^{-3} & 4.00 \end{pmatrix}.$$

   Approximately how accurate are the estimates?

4. Show that if U is unitary, the eigenvalues of A and $U^{\mathrm{H}} A U$ have the same condition numbers.

5. Let $A \in \mathbb{R}^{n\times n}$ be the matrix illustrated below for $n = 6$.

$$\begin{pmatrix} 0 & -1 & -1 & -1 & -1 & -1 \\ 0 & 1 & -1 & -1 & -1 & -1 \\ 0 & 0 & 1 & -1 & -1 & -1 \\ 0 & 0 & 0 & 1 & -1 & -1 \\ 0 & 0 & 0 & 0 & 1 & -1 \\ 0 & 0 & 0 & 0 & 0 & -1 \end{pmatrix}.$$

   Compute the left eigenvector corresponding to the eigenvalue 0. From Theorem 4.2 deduce the sensitivity of the eigenvalue 0 to a perturbation ε in the (i, j)-element.

6. For the matrix A_n of the last exercise what is the value of δ (in Theorem 4.2) corresponding to the eigenvalue 0.

7. Discuss the condition of the eigenvalues of the following matrix of order 20:

$$\begin{bmatrix} 20 & 20 & & & & \bigcirc \\ & 19 & 20 & & & \\ & & 18 & 20 & & \\ & & & \ddots & \ddots & \\ & \bigcirc & & & 2 & 20 \\ & & & & & 1 \end{bmatrix}.$$

8. Let $A \in \mathbb{C}^{n \times n}$ and let $\varepsilon = \max_{i \neq j} | \alpha_{ij} |$ be presumed small. It is desired to choose a diagonal matrix $R = (\varrho, 1, 1, \ldots, 1)$ so that the first Gerschgorin disk of RAR^{-1} is as small as possible, without overlapping the other disks. Show that as $\varepsilon \to 0$,

$$\varrho = \frac{\varepsilon}{\min_{i>1} | \alpha_{11} - \alpha_{ii} |} + O(\varepsilon^2) \equiv \frac{\varepsilon}{\delta} + O(\varepsilon^3)$$

and hence the radius of the first Gerschgorin disk is given approximately by $(n-1)\varepsilon^2/\delta$. Compare this results with the estimates obtained from Theorem 4.2.

9. Compute minimal isolated Gerschgorin disks for the matrix of Exercise 3.

10. Use the Gerschgorin theorem to show that a symmetric, diagonally dominant matrix with positive diagonal elements is positive definite.

11. Let λ be a simple eigenvalue of A with right and left eigenvectors x and y. Show that if $\| x \|_2 = \| y \|_2 = 1$, then $1/| y^{\mathrm{H}} x |$ is a condition number of λ. Show that this number is the secant of the angle between x and y.

NOTES AND REFERENCES

The study of the behavior of the eigenvalues and eigenvectors of a matrix under perturbations has been and continues to be the subject of many investigations. For a fuller treatment see Householder (1964) or Wilkinson (AEP, Chapter II).

First- (or higher-) order perturbation estimates have long been used by scientists and engineers. They have the advantage that they provide a quick, nonrigorous look at what might otherwise be a mathematically difficult problem. Moreover, they often suggest how to improve an approximate solution, in the spirit of Example 4.5.

Theorem 4.2 is an embellishment of a result of Stewart (1972). The fact that $\| y \|_2$ is a condition number for a simple eigenvalue was first observed by Wilkinson (AEP, Chapter II) (he takes $\| y \|_2 = 1$, and uses $| y^{\mathrm{H}}x |^{-1}$ as a condition number). It has long been known that if an eigenvalue is poorly separated from its neighbors the corresponding eigenvector will be ill conditioned. However the use of $\delta = \| (\lambda I - C)^{-1} \|_2^{-1}$ to measure the separation of λ from its neighbors appears to be new and is justified by the simplicity of the resulting perturbation bound.

Taussky (1949) gives a historical discussion of Gerschgorin's theorem. A proof of Theorem 4.9 may be found in Wilkinson (AEP, Chapter II). The technique for reducing the size of Gerschgorin disks by diagonal similarity transformations appeared in Gerschgorin's original paper and has been extensively investigated by Varga (1965).

Wilkinson (AEP, Chapter V) has developed a comprehensive perturbation theory based on the Jordan canonical form and Gerschgorin's theorem supplemented by diagonal similarity transformations. Where his theory and Theorem 4.2 overlap they give essentially the same results. Otherwise the two approaches complement one another. Theorem 4.2 fails for a multiple eigenvalue, which Wilkinson's theory handles readily. On the other hand Wilkinson's theory requires considerable work to keep nontrivial Jordan blocks from interfering with the perturbation bounds for a simple eigenvalue or eigenvector.

Theorem 4.10 is due to Bauer and Fike (1960). The technique used to prove it is due to Householder (1958a).

Although an eigenvector is ill conditioned when its eigenvalue is near other eigenvalues, the subspace spanned by the eigenvectors corresponding to a cluster of eigenvalues may be insensitive to perturbations in the elements of the matrix. For more on this subject see papers by Varah (1967, 1970), Davis and Kahan (1970), Ruhe (1970a,b), and Stewart (1972).

5. HERMITIAN MATRICES

In this section we shall be concerned with the properties of the eigenvalues of Hermitian matrices. Such matrices are frequently encountered in practice, usually in the form of real symmetric matrices. Of course, all the theory developed in the last four sections applies directly to Hermitian matrices; however, the special structure of Hermitian matrices is reflected in the particularly nice behavior of its eigenvalues and eigenvectors.

The algebraic properties of the eigenvalues and eigenvectors of a Hermitian matrix are summarized in the following theorem.

THEOREM 5.1. Let $A \in \mathbb{C}^{n \times n}$ be Hermitian. Then A has n real eigenvalues $\lambda_1, \lambda_2, \ldots, \lambda_n$ corresponding to a complete system of eigenvectors $x_1, x_2, \ldots, x_n$. The eigenvectors may be chosen so that $X = (x_1, x_2, \ldots, x_n)$ is unitary; that is the eigenvectors may be chosen to form an orthonormal basis for $\mathbb{C}^n$.

PROOF. By Theorem 3.7 there is a unitary matrix X such that $\Lambda = X^{\mathrm{H}}AX$ is upper triangular. But $X^{\mathrm{H}}AX$ is Hermitian. Hence Λ is diagonal with real diagonal elements $\lambda_1, \lambda_2, \ldots, \lambda_n$. By Theorem 3.2, λ_i is an eigenvalue of A with the ith column of X as an eigenvector. ■

As we pointed out in connection with Theorem 3.2, the decomposition $X^{\mathrm{H}}AX = \Lambda$ is essentially unique. The columns of X corresponding to an eigenvalue λ may be chosen from any orthonormal basis for $\mathfrak{N}(\lambda I - A)$, but this is the only freedom we have in forming X. However, this freedom is sufficient to allow us to include one prespecified eigenvector from $\mathfrak{N}(\lambda I - A)$ for each distinct eigenvalue λ of A in the columns of X.

Theorem 5.1 has two immediate important corollaries. The first expresses the 2-norm and the Frobenius norm of a Hermitian matrix in terms of its eigenvalues.

COROLLARY 5.2. Let A be Hermitian with eigenvalues $\lambda_1, \lambda_2, \ldots, \lambda_n$. Then

$$\| A \|_2 = \max\{| \lambda_i | : i = 1, 2, \ldots, n\},$$

and

$$\| A \|_{\mathrm{F}} = \sqrt{\sum_{i=1}^{n} \lambda_i^2}.$$

PROOF. As in Theorem 5.1, let $X^{\mathrm{H}}AX = \Lambda = \mathrm{diag}(\lambda_1, \ldots, \lambda_n)$, where X is unitary. By Theorem 5.1.7, $\| A \|_2 = \| \Lambda \|_2$ and $\| A \|_{\mathrm{F}} = \| \Lambda \|_{\mathrm{F}}$. The results then follow from the easy computation of $\| \Lambda \|_2$ and $\| \Lambda \|_{\mathrm{F}}$. ■

Thus the spectral norm of a Hermitian matrix is the absolute value of its largest eigenvalue. The square of its Frobenius norm is the sum of squares of its eigenvalues.

The next consequence of Theorem 5.1 is a useful characterization of positive definite matrices, which will be used in the next section.

COROLLARY 5.3. The Hermitian matrix A is positive definite (semidefinite) if and only if the eigenvalues of A are positive (nonnegative).

PROOF. Let X and Λ be as in the proof of Theorem 5.1. Then, since X is nonsingular, A is positive definite (semidefinite) if and only if $\Lambda = X^{\mathrm{H}}AX$ is. Since Λ is diagonal, Λ is positive definite (semidefinite) if and only if its eigenvalues are positive (nonnegative). ■

We turn now to perturbation theory for the eigenvalues and eigenvectors of a Hermitian matrix. We first examine the simplifications that obtain in Theorem 4.2 when the matrix A is Hermitian.

We have seen in the discussion following Theorem 4.2 that if y and x are left and right eigenvectors corresponding to the simple eigenvalue λ of A and $y^{\mathrm{H}}x = 1$, then $\| y \|_2$ is a condition number for λ. For Hermitian A, if $Ax = \lambda x$, then λ is real and

$$x^{\mathrm{H}}A = x^{\mathrm{H}}A^{\mathrm{H}} = \bar{\lambda}x^{\mathrm{H}} = \lambda x^{\mathrm{H}},$$

so that x is also a left eigenvector corresponding to λ. Thus if $\| x \|_2 = 1$, then $x^{\mathrm{H}}x = 1$, and the condition number for λ is unity. In other words, the simple eigenvalues of a Hermitian matrix are perfectly conditioned; a sufficiently small perturbation of size ε will induce an error essentially no greater than ε in λ.

EXAMPLE 5.4. The matrix

$$A = \begin{pmatrix} 1.005 & -.9950 \\ -.9950 & 1.005 \end{pmatrix}$$

has eigenvalues $\lambda_1 = 2.000$ and $\lambda_2 = .01$ while the matrix

$$A' = \begin{pmatrix} 1.007 & -1.013 \\ -1.013 & 1.007 \end{pmatrix}$$

has eigenvalues $\lambda_1' = 2.020$ and $\lambda_2' = -.006$. Notice that $\| A - A' \|_2$

$\cong 10^{-2}$ and, as predicted by our theory, $|\lambda_1 - \lambda_1'| \cong 10^{-2}$ and $|\lambda_2 - \lambda_2'| \cong 10^{-2}$.

It is important to realize that when we say that the eigenvectors of a Hermitian matrix are perfectly conditioned, we are referring to absolute errors. If an eigenvalue is small, a small change in the elements of the matrix may induce a large relative error in the eigenvalue. This point is illustrated by Example 5.4, in which the eigenvalue λ_2 actually changes sign under the relatively small perturbation in A.

Regarding the eigenvectors of a Hermitian matrix, we have seen that if x is an eigenvector of A corresponding to λ with $\| x \|_2 = 1$, then there is an eigenvector x' of $A + E$ satisfying

$$\| x - x' \|_2 \leq \frac{\| E \|_2}{\delta} + O(\| E \|_2^2), \tag{5.1}$$

where

$$\delta^{-1} = \| (\lambda I - C)^{-1} \|_2 .$$

Now since C has the form $U^{\mathrm{H}}AU$, C is also Hermitian, and by Corollary 5.2

$$\delta^{-1} = \max\{| \lambda - \lambda_i |^{-1}\},$$

where the λ_i are the eigenvalues of C. But the eigenvalues of C are just the other eigenvalues of A. Hence

$$\delta = \min\{| \lambda - \mu | : \mu \in \lambda(A), \mu \neq \lambda\}.$$

This proves that *if an eigenvector of a Hermitian matrix corresponds to an eigenvalue that is well separated from the other eigenvalues, then it is well conditioned.*

The proof of Theorem 4.2 does not exploit the special properties of Hermitian matrices. Hence it might be expected that the results discussed above concerning the condition of eigenvalues and eigenvectors of a Hermitian matrix are not as general as they might be. So far as eigenvectors are concerned, this is not so; a bound of the form (5.1) is about as much as can be expected. The eigenvalues of a Hermitian matrix, however, have properties that enable us to obtain much more informative perturbation bounds. Accordingly, the rest of this section is devoted to a discussion of these special properties.

We have seen that if y is an approximate eigenvector of A, then the Rayleigh quotient $y^{\mathrm{H}}Ay/y^{\mathrm{H}}y$ is an approximate eigenvalue of A. When A is Hermitian, however, $y^{\mathrm{H}}Ay/y^{\mathrm{H}}y$ is far more accurate as an approximate eigenvalue than is y as an approximate eigenvector.

THEOREM 5.5. Let x be an eigenvector of the Hermitian matrix A corresponding to the eigenvalue λ. If

$$y = x + O(\varepsilon), \tag{5.2}$$

then

$$\frac{y^{\mathrm{H}}Ay}{y^{\mathrm{H}}y} = \lambda + O(\varepsilon^2).$$

PROOF. We merely sketch the proof, leaving the details as an exercise. We may assume without loss of generality that $\| x \| = 1$. Let $X = (x, x_2, \ldots, x_n)$ be an orthonormal system of eigenvectors corresponding to the eigenvalues $\lambda, \lambda_2, \ldots, \lambda_n$ of A. Let $y = Xr$. Then

$$\frac{y^{\mathrm{H}}Ay}{y^{\mathrm{H}}y} = \frac{r^{\mathrm{H}}X^{\mathrm{H}}AXr}{r^{\mathrm{H}}X^{\mathrm{H}}Xr} = \frac{|\varrho_1|^2 \lambda + \sum_{i=2}^{n} |\varrho_i|^2 \lambda_i}{|\varrho_1|^2 + \sum_{i=2}^{n} |\varrho_i|^2}.$$

Now from (5.2) it follows that

$$\varrho_1 \cong \| x \|_2 = 1$$

and

$$\varrho_i = O(\varepsilon) \qquad (i = 2, 3, \ldots, n).$$

Hence

$$\frac{y^{\mathrm{H}}Ay}{y^{\mathrm{H}}y} = \frac{|\varrho_1|^2 \lambda}{|\varrho_1|^2} + O(\varepsilon^2) = \lambda + O(\varepsilon^2). \quad \blacksquare$$

EXAMPLE 5.6. The matrix A of Example 5.4 has an eigenvector $(1, -1)^{\mathrm{T}}$ corresponding to the eigenvalue 2. If we set

$$y = (1.01, -.99),$$

then $y^{\mathrm{H}}Ay/y^{\mathrm{H}}y = 1.9998$, which is accurate to four places.

Theorem 5.5 shows the importance of the Rayleigh quotient as a device for estimating eigenvalues. However, the properties of the Rayleigh quotient go far deeper than this; for the Rayleigh quotient plays a basic role in the well-known minimax theorems, which are of fundamental importance in the investigation of the eigenvalues of Hermitian matrices. In what follows we shall assume that A is a Hermitian matrix with eigenvalues $\lambda_1, \lambda_2, \ldots, \lambda_n$, ordered so that

$$\lambda_1 \leq \lambda_2 \leq \cdots \leq \lambda_{n-1} \leq \lambda_n, \tag{5.3}$$

and corresponding eigenvectors $x_1, x_2, \ldots, x_n$ which are taken to be orthonormal. We begin our development with a theorem which describes the range of values that can be assumed by a Rayleigh quotient.

THEOREM 5.7. If $x \neq 0$, then

$$\lambda_1 \leq \frac{x^{\mathrm{H}}Ax}{x^{\mathrm{H}}x} \leq \lambda_n. \tag{5.4}$$

PROOF. Let $X = (x_1, x_2, \ldots, x_n)$ and $x = Xr$. Then, as in the proof of Theorem 5.5,

$$\frac{x^{\mathrm{H}}Ax}{x^{\mathrm{H}}x} = \frac{\sum_{i=1}^{n} |\varrho_i|^2 \lambda_i}{\sum_{i=1}^{n} |\varrho_i|^2}. \tag{5.5}$$

Since $\sum |\varrho_i|^2 = \|x\|_2^2 \neq 0$, Equation (5.5) expresses the Rayleigh quotient as a weighted average of the numbers $\lambda_1, \lambda_2, \ldots, \lambda_n$. In view of (5.3), the Rayleigh quotient must satisfy (5.4). ■

Theorem 5.7 says that the Rayleigh quotient must lie between λ_1 and λ_n. Moreover, for $x = x_1$ the Rayleigh quotient is λ_1 and for $x = x_n$ it is λ_n. In other words

$$\lambda_1 = \min_{\substack{x \in \mathbb{C}^n \\ x \neq 0}} \frac{x^{\mathrm{H}}Ax}{x^{\mathrm{H}}x} \leq \max_{\substack{x \in \mathbb{C}^n \\ x \neq 0}} \frac{x^{\mathrm{H}}Ax}{x^{\mathrm{H}}x} = \lambda_n,$$

which expresses λ_1 and λ_n in terms of extrema of Rayleigh quotients.

Expressions for the other eigenvalues may be obtained by allowing x to vary over subspaces of $\mathbb{C}^n$. To see how this is done, let $\mathcal{S}$ be a two-dimensional subspace of $\mathbb{C}^n$. Then from elementary geometric considerations it follows that there is a vector x in $\mathcal{S}$ that is orthogonal to x_1. If we write x in the form $x = Xr$, where $X = (x_1, \ldots, x_n)$, then $r = X^{\mathrm{H}}x$, and since $x_1{}^{\mathrm{H}}x = 0$, $\varrho_1 = 0$. Thus

$$\frac{x^{\mathrm{H}}Ax}{x^{\mathrm{H}}x} = \frac{\sum_{i=2}^{n} |\varrho_i|^2 \lambda_i}{\sum_{i=2}^{n} |\varrho_i|^2}.$$

Reasoning as in the proof of Theorem 5.7, we find that

$$\lambda_2 \leq \frac{x^{\mathrm{H}}Ax}{x^{\mathrm{H}}x}.$$

Another way of putting this is to say that

$$\max_{\substack{x \in \mathcal{S} \\ x \neq 0}} \frac{x^{\mathrm{H}}Ax}{x^{\mathrm{H}}x} \geq \lambda_2. \tag{5.6}$$

Now if we let $\mathcal{S}'$ be the space spanned by x_1 and x_2, then any $x \in \mathcal{S}'$ can be written in the form $x = \varrho_1 x_1 + \varrho_2 x_2$. Then

$$\frac{x^{\mathrm{H}}Ax}{x^{\mathrm{H}}x} = \frac{|\varrho_1|^2 \lambda_1 + |\varrho_2|^2 \lambda_2}{|\varrho_1|^2 + |\varrho_2|^2} \leq \lambda_2.$$

In other words for this particular two-dimensional subspace $\mathcal{S}'$

$$\max_{\substack{x \in \mathcal{S}' \\ x \neq 0}} \frac{x^{\mathrm{H}}Ax}{x^{\mathrm{H}}x} \leq \lambda_2. \tag{5.7}$$

If we combine (5.6) and (5.7), we get

$$\min_{\dim(\mathcal{S})=2} \max_{\substack{x \in \mathcal{S} \\ x \neq 0}} \frac{x^{\mathrm{H}}Ax}{x^{\mathrm{H}}x} = \lambda_2. \tag{5.8}$$

Equation (5.8) is the minimax characterization of λ_2. It is quite general. To obtain an expression for λ_i one need only minimize over all subspaces of dimension i. However, before we can prove this result in general, we must first establish the generalization of the geometric result that we used to obtain a vector $x \in \mathcal{S}$ orthogonal to x_1.

LEMMA 5.8. Let $\mathcal{S}$ and $\mathcal{T}$ be subspaces of C^n. If $\dim(\mathcal{S}) > \dim(\mathcal{T})$, then there is a nonzero vector in $\mathcal{S}$ that is orthogonal to $\mathcal{T}$.

PROOF. Let the columns of $S = (s_1, s_2, \ldots, s_l)$ form a basis for $\mathcal{S}$ and the columns of $T = (t_1, t_2, \ldots, t_m)$ form a basis for $\mathcal{T}$. Then $l = \dim(\mathcal{S}) > \dim(\mathcal{T}) = m$. Now x is a nonzero member of $\mathcal{S}$ if and only if $x = Sz$, where $z \neq 0$. Such a vector Sz will be orthogonal to $\mathcal{T}$ if and only if

$$T^{\mathrm{H}}Sz = 0. \tag{5.9}$$

But, $T^{\mathrm{H}}S$ is an $m \times l$ matrix with $m < l$. Hence (5.9) is an underdetermined system of homogeneous equations and always has a nonzero solution z. The vector Sz is then the vector orthogonal to $\mathscr{T}$ which was to be constructed. ■

We are now in a position to prove the minimax theorem.

THEOREM 5.9. Let the Hermitian matrix A have eigenvalues $\lambda_1, \lambda_2, \ldots, \lambda_n$ ordered so that $\lambda_1 \leq \lambda_2 \leq \cdots \leq \lambda_n$. Then

$$\lambda_i = \min_{\dim(\mathbb{S})=i} \max_{\substack{x \in \mathbb{S} \\ x \neq 0}} \frac{x^{\mathrm{H}}Ax}{x^{\mathrm{H}}x} = \max_{\dim(\mathbb{S})=n-i+1} \min_{\substack{x \in \mathbb{S} \\ x \neq 0}} \frac{x^{\mathrm{H}}Ax}{x^{\mathrm{H}}x}.$$

PROOF. We shall establish the first equality, the second being proved similarly. Let $x_1, x_2, \ldots, x_n$ be a set of orthonormal eigenvectors corresponding to $\lambda_1, \lambda_2, \ldots, \lambda_n$. Let $\mathbb{S}$ be a subspace of $\mathbb{C}^n$ with $\dim(\mathbb{S}) = i$. Then by Lemma 5.8, there is a vector $x \neq 0$ in $\mathbb{S}$ that is orthogonal to $x_1, x_2, \ldots, x_{i-1}$. If we set $X = (x_1, \ldots, x_n)$ and express x in the form $x = Xr$, then $\varrho_1 = \varrho_2 = \cdots = \varrho_{i-1} = 0$. Hence

$$\frac{x^{\mathrm{H}}Ax}{x^{\mathrm{H}}x} = \frac{|\varrho_i|^2 \lambda_i + \cdots + |\varrho_n|^2 \lambda_n}{|\varrho_i|^2 + \cdots + |\varrho_n|^2} \geq \lambda_i.$$

This shows that

$$\max_{\substack{x \in \mathbb{S} \\ x \neq 0}} \frac{x^{\mathrm{H}}Ax}{x^{\mathrm{H}}x} \geq \lambda_i. \tag{5.10}$$

On the other hand if $\mathbb{S}'$ is the i-dimensional subspace spanned by $x_1, x_2, \ldots, x_i$, then any nonzero $x \in \mathbb{S}'$ may be written in the form $\varrho_1 x_1 + \cdots + \varrho_i x_i$ and

$$\frac{x^{\mathrm{H}}Ax}{x^{\mathrm{H}}x} = \frac{|\varrho_1|^2 \lambda_1 + \cdots + |\varrho_i|^2 \lambda_i}{|\varrho_1|^2 + \cdots + |\varrho_i|^2} \leq \lambda_i.$$

This shows that

$$\max_{\substack{x \in \mathbb{S}' \\ x \neq 0}} \frac{x^{\mathrm{H}}Ax}{x^{\mathrm{H}}x} \leq \lambda_i. \tag{5.11}$$

The result follows on combining (5.10) and (5.11). ■

The power of the minimax theorem is that it provides explicit inequalities involving the eigenvalues of a matrix that do not require any special knowledge of the eigenvalues and eigenvectors of the matrix. We illustrate this observation by proving the following perturbation theorem.

THEOREM 5.10. Let $\alpha_1 \leq \alpha_2 \leq \cdots \leq \alpha_n$, $\beta_1 \leq \beta_2 \leq \cdots \leq \beta_n$, and $\gamma_1 \leq \gamma_2 \leq \cdots \leq \gamma_n$ be the eigenvalues of the Hermitian matrices A, B, and

$$C = A + B.$$

Then

$$\alpha_i + \beta_1 \leq \gamma_i \leq \alpha_i + \beta_n \qquad (i = 1, 2, \ldots, n). \tag{5.12}$$

PROOF. Let $x_1, x_2, \ldots, x_n$ be an orthonormal system of eigenvectors of A corresponding to $\alpha_1, \alpha_2, \ldots, \alpha_n$, and let $\mathcal{S}$ be the space spanned by $x_1, x_2, \ldots, x_i$. Then from Theorem 5.9

$$\gamma_i \leq \max_{\substack{x \in \mathcal{S} \\ x \neq 0}} \frac{x^{\mathrm{H}}Cx}{x^{\mathrm{H}}x} \leq \max_{\substack{x \in \mathcal{S} \\ x \neq 0}} \frac{x^{\mathrm{H}}Ax}{x^{\mathrm{H}}x} + \max_{\substack{x \in \mathcal{S} \\ x \neq 0}} \frac{x^{\mathrm{H}}Bx}{x^{\mathrm{H}}x}$$

$$= \alpha_i + \max_{\substack{x \in \mathcal{S} \\ x \neq 0}} \frac{x^{\mathrm{H}}Bx}{x^{\mathrm{H}}x} \leq \alpha_i + \beta_n,$$

where the last inequality follows from Theorem 5.7. This establishes the right-hand inequality in (5.12). On the other hand, if $D = -B$, then

$$A = C + D,$$

and by the result just proved

$$\alpha_i \leq \gamma_i + \delta_n, \tag{5.13}$$

where δ_n is the largest eigenvalue of D. But $\delta_n = -\beta_1$; hence (5.13) is equivalent to the left-hand inequality in (5.12). ∎

In applications of this theorem one usually finds that B is small, or equivalently that β_1 and β_n are small, in which case the theorem states that the eigenvalues of C are near those of A. However, Theorem 5.10 is much more powerful than our previous perturbation theorems, for it associates with each eigenvalue of A a unique eigenvalue of C that lies in a well-defined interval near the eigenvalue of A. Even if these intervals overlap, the eigenvalue of C cannot migrate out of its own interval, in contrast to the case of overlapping Gerschgorin disks where one of the disks may fail to contain an eigenvalue. Moreover, the specific form of the intervals in terms of the greatest and least eigenvalues of B may be informative. For example, if B is positive definite, then $\beta_1 > 0$ and the eigenvalues of C satisfy $\gamma_i > \alpha_i$.

In practice we will not usually know the eigenvalues of B. However, from Theorem 2.15 we have the bounds

$$|\beta_1|, |\beta_n| \leq \|B\|_2.$$

Hence

$$\alpha_i - \|B\|_2 \leq \gamma_i \leq \alpha_i + \|B\|_2.$$

This proves the following corollary.

COROLLARY 5.11. Let the eigenvalues $\lambda_1, \lambda_2, \ldots, \lambda_n$ and $\lambda_1', \lambda_2', \ldots, \lambda_n'$ of the Hermitian matrices A and $A' = A + E$ be arranged in ascending order. Then $\lambda_i' \in [\lambda_i - \|E\|_2, \lambda_i + \|E\|_2]$.

Corollary 5.11 is another, more complete, statement of the perfect conditioning of the eigenvalues of a Hermitian matrix. It says that a Hermitian perturbation E can change no eigenvalue of A by more than $\|E\|_2$, even if it is a multiple eigenvalue.

EXERCISES

1. What are the eigenvalues and eigenvectors of an elementary reflector?

2. Let

$$B = \begin{pmatrix} \alpha & a^{\mathrm{H}} \\ a & A \end{pmatrix}$$

be Hermitian. Show that there is an eigenvalue of B in the interval $\{\lambda : |\alpha - \lambda| \leq \|a\|_2\}$.

3. Give a formal proof of Theorem 5.5.

4. Let $A \in \mathbb{C}^{n \times n}$ be Hermitian and let $X \in \mathbb{C}^{n \times p}$ have orthonormal columns. The matrix $B = X^{\mathrm{H}}AX$ is called a *p-section* of A. Show that if $\lambda_1 \leq \lambda_2 \leq \cdots \leq \lambda_n$ are the eigenvalues of A and $\mu_1 \leq \mu_2 \leq \cdots$

$\leq \mu_p$ are the eigenvalues of B, then

$$\lambda_i \leq \mu_i \qquad (i = 1, 2, \ldots, p)$$

and

$$\mu_{p-i+1} \leq \lambda_{n-i+1} \qquad (i = 1, 2, \ldots, p).$$

[*Hint*: Use the minimax theorem on B to obtain an expression for μ_i and then interpret the expression in terms of A.]

5. Let $A \in \mathbb{C}^{n\times n}$ be Hermitian and let $B \in \mathbb{C}^{n-1\times n-1}$ be a principal submatrix of A. Show that the eigenvalues $\lambda_1, \lambda_2, \ldots, \lambda_n$ of A and $\mu_1, \mu_2, \ldots, \mu_{n-1}$ of B can be ordered so that

$$\lambda_1 \leq \mu_1 \leq \lambda_2 \leq \mu_2 \leq \cdots \leq \lambda_{n-1} \leq \mu_{n-1} \leq \lambda_n.$$

6. Let $A, B \in \mathbb{C}^{n\times n}$ be Hermitian. Show that if A is positive definite and $\| A^{-1} \|_2 \| B \|_2 < 1$, then $A + B$ is positive definite.
7. Let $A = \mathrm{diag}(1, 1, 0)$ and $E \in \mathbb{R}^{3\times 3}$ be symmetric and presumed small. Show that the space spanned by the eigenvectors corresponding to eigenvalues near unity of $A + E$ is in some sense near the space spanned by e_1 and e_2. [*Hint*: This space is orthogonal to the space spanned by the eigenvector corresponding to the eigenvalue near zero.]

NOTES AND REFERENCES

The material in this section is classical. For extensions of the minimax theorem see Householder (1964).

Since $\| E \|_F \geq \| E \|_2$ the spectral norm may be replaced by the Frobenius norm in Corollary 5.11. However, this is not the strongest possible result. The following theorem, due to Hoffman and Wielandt (1953), is useful in practical applications [an elementary, but lengthy, proof is given by Wilkinson (AEP, Chapter 2)]. *Under the hypotheses of Corollary 5.11,*

$$\sum_{i=1}^{n} (\lambda_i - \lambda_i')^2 \leq \| E \|_F^2.$$

6. THE SINGULAR VALUE DECOMPOSITION

In some applications it is necessary to determine the rank of a matrix, particularly to determine if the matrix is of less than full rank. In theory this presents no problems. One applies an algorithm such as Gaussian elimination to the matrix and reads off the rank from the final reduced form.

In practice the situation is more complicated. In the first place, the elements of a matrix are seldom given exactly, and, even if the original matrix is defective in rank, it is unlikely that its approximation will also be. Thus instead of asking if the given matrix is defective in rank, we must ask if it is near a matrix of defective rank. In the second place, the transformations of, say, Gaussian elimination may take a matrix that is very nearly defective in rank and turn it into one that is clearly of full rank. Finally, it is not always easy to recognize when even a triangular matrix, which is the end product of Gaussian elimination, is nearly defective in rank.

In this section we shall investigate a decomposition in which a general matrix is reduced to a diagonal form by premultiplying and postmultiplying it by unitary matrices. Observe that such a decomposition answers the last two difficulties raised in the last paragraph. Because only unitary matrices are used, the final result is near a matrix of defective rank (in the sense of $\|\cdot\|_2$ or $\|\cdot\|_F$) if and only if the original matrix is. Moreover, it is easy to recognize when a diagonal matrix is nearly defective in rank.

The diagonal elements of the decomposition are called the singular values of the matrix, and hence the decomposition itself is called the singular value decomposition. The singular values of a general matrix have many analogies with the eigenvalues of a Hermitian matrix; indeed for a positive definite matrix, they are the same. The decomposition has a number of practical applications in statistics. In this section we shall establish the existence and fundamental properties of the decomposition and show how it may be applied to solve degenerate least squares problems. In the next chapter we shall describe an algorithm for computing the singular value decomposition.

The existence of the singular value decomposition is established in the following theorem.

THEOREM 6.1. Let $A \in \mathbb{C}^{m \times n}$. Then there are unitary matrices U and V such that

$$V^{\mathrm{H}}AU = \begin{pmatrix} \Sigma & 0 \\ 0 & 0 \end{pmatrix}, \tag{6.1}$$

where $\Sigma = \mathrm{diag}(\sigma_1, \sigma_2, \ldots, \sigma_r)$ and $\sigma_1 \geq \sigma_2 \geq \cdots \geq \sigma_r > 0$.

PROOF. Since $A^{\mathrm{H}}A$ is positive semidefinite, its eigenvalues are nonnegative. Let them be $\sigma_1^2, \sigma_2^2, \ldots, \sigma_n^2$, where $\sigma_1 \geq \sigma_2 \geq \cdots \geq \sigma_r > 0 = \sigma_{r+1} = \sigma_{r+2} = \cdots = \sigma_n$. Let $u_1, u_2, \ldots, u_n$ be a set of orthonormal eigenvectors for $\sigma_1^2, \sigma_2^2, \ldots, \sigma_n^2$, and let $U_1 = (u_1, u_2, \ldots, u_r)$ and $U_2 = (u_{r+1}, u_{r+2}, \ldots, u_n)$. Then if $\Sigma = \mathrm{diag}(\sigma_1, \sigma_2, \ldots, \sigma_r)$, we have $U_1^{\mathrm{H}}A^{\mathrm{H}}AU_1 = \Sigma^2$ and consequently

$$\Sigma^{-1}U_1^{\mathrm{H}}A^{\mathrm{H}}AU_1\Sigma^{-1} = I. \tag{6.2}$$

Also $U_2^{\mathrm{H}}A^{\mathrm{H}}AU_2 = 0$, whence

$$AU_2 = 0.$$

Now let

$$V_1 = AU_1\Sigma^{-1}. \tag{6.3}$$

Then from (6.2), $V_1^{\mathrm{H}}V_1 = I$; that is the columns of V_1 are orthonormal. Let V_2 be chosen so that $V = (V_1, V_2)$ is unitary. Then

$$\begin{aligned} V^{\mathrm{H}}AU &= \begin{pmatrix} (V_1^{\mathrm{H}})AU_1 & V_1^{\mathrm{H}}(AU_2) \\ V_2^{\mathrm{H}}(AU_1) & V_2^{\mathrm{H}}(AU_2) \end{pmatrix} \\ &= \begin{pmatrix} (\Sigma^{-1}U_1^{\mathrm{H}}A^{\mathrm{H}})AU_1 & V_1^{\mathrm{H}}(0) \\ V_2^{\mathrm{H}}(V_1\Sigma) & V_2^{\mathrm{H}}(0) \end{pmatrix} \\ &= \begin{pmatrix} \Sigma & 0 \\ 0 & 0 \end{pmatrix}. \quad \blacksquare \end{aligned}$$

The decomposition of Theorem 6.1 is essentially unique. In the first place, from (6.1) we have

$$U^{\mathrm{H}}A^{\mathrm{H}}AU = \mathrm{diag}(\Sigma^2, 0). \tag{6.4}$$

Thus the numbers $\sigma_1^2, \sigma_2^2, \ldots, \sigma_r^2$ must be the nonzero eigenvalues of $A^{\mathrm{H}}A$ arranged in descending order. This, along with the requirement that the σ_i be nonnegative, completely determines the σ_i.

The matrices U and V are less well determined. Since the columns of U are eigenvectors of $A^{\mathrm{H}}A$, if $A^{\mathrm{H}}A$ has a multiple eigenvalue $\sigma^2 > 0$, the corresponding columns of U may be chosen as any orthonormal basis for the space spanned by the eigenvectors corresponding to σ^2. Once U_1 is chosen, however, V_1 is determined by (6.3). The matrices U_2 and V_2 may be any matrices with orthonormal columns spanning $\mathfrak{N}(A)$ and $\mathfrak{N}(A^{\mathrm{H}})$ respectively.

The numbers $\sigma_1, \sigma_2, \ldots, \sigma_r$ are called singular values of the matrix A. Conventions differ on what to do with the remaining zero eigenvalues of $A^{\mathrm{H}}A$. We shall make the following definition.

DEFINITION 6.2. Let $A \in \mathbb{C}^{m\times n}$ have the singular value decomposition (6.1). Then the numbers $\sigma_1, \sigma_2, \ldots, \sigma_n$, where $\sigma_{r+1} = \sigma_{r+2} = \cdots \sigma_n = 0$ are called the *singular values* of A. The vectors columns of U are *right singular vectors* of A and the columns of V are *left singular vectors* of A.

Thus if $A \in \mathbb{C}^{m\times n}$, A has n singular values, which are the square roots of the eigenvalues of $A^{\mathrm{H}}A$. The matrix A^{H} has m singular values, which are the square roots of the eigenvalues of AA^{H}. From (6.1) it is evident that the nonzero singular values of A are the same as those of A^{H}.

The right singular vectors $u_1, u_2, \ldots, u_n$ of A are eigenvectors of $A^{\mathrm{H}}A$. The left singular vectors $v_1, v_2, \ldots, v_m$ are eigenvectors of AA^{H}. From the equation

$$AU = V\begin{pmatrix} \Sigma & 0 \\ 0 & 0 \end{pmatrix},$$

it follows that

$$Au_i = \sigma_i v_i \qquad (i = 1, 2, \ldots, \min\{m, n\}), \tag{6.5}$$

and

$$Au_i = 0 \qquad (i = \min\{m, n\} + 1, \ldots, n). \tag{6.6}$$

We have mentioned that the singular values of a matrix have many appealing analogies with the eigenvalues of Hermitian matrices. In fact if A is Hermitian, then the singular values of A are just the absolute values of the eigenvalues of A.

THEOREM 6.3. Let A be Hermitian with eigenvalues $\lambda_1, \lambda_2, \ldots, \lambda_n$. Then the singular values of A are $|\lambda_1|, |\lambda_2|, \ldots, |\lambda_n|$.

PROOF. The singular values of A are the square roots of the eigenvalues of $A^{\mathrm{H}}A = A^2$. Thus the singular values of A are given by $\sqrt{\lambda_i^2} = |\lambda_i|$ $(i = 1, 2, \ldots, n)$. ■

The following analog of Corollary 5.2, is established by reducing A to diagonal form as in (6.1).

THEOREM 6.4. Let $A \in \mathbb{C}^{m\times n}$ have singular values $\sigma_1 \geq \sigma_2 \geq \cdots \geq \sigma_n$. Then

$$\| A \|_2 = \sigma_1$$

and

$$\| A \|_{\mathrm{F}}^2 = \sigma_1^2 + \sigma_2^2 + \cdots + \sigma_n^2.$$

Thus the 2-norm of a matrix A is simply its largest singular value, that is the square root of the largest eigenvalue of $A^{\mathrm{H}}A$ or AA^{H}. The square of the Frobenius norm of A is the sum of squares of the singular values of A.

Like the eigenvalues of a Hermitian matrix, the singular values of a general matrix have a minimax characterization. The characterization is established in almost the same way as the results of Theorem 5.9, except that the relations (6.5) and (6.6) take the place of the eigenvalue–eigenvector relation $X^{\mathrm{H}}AX = \Lambda$. Consequently, we leave the proof of the following theorem as an exercise.

THEOREM 6.5. Let $A \in \mathbb{C}^{m\times n}$ have singular values $\sigma_1 \geq \sigma_2 \geq \cdots \geq \sigma_n$. Then

$$\sigma_i = \min_{\dim(\mathbb{S})=n-i+1} \max_{\substack{x \in \mathbb{S} \\ x\neq 0}} \frac{\| Ax \|_2}{\| x \|_2}. \tag{6.7}$$

It was a consequence of Theorem 5.10 that the eigenvalues of a Hermitian matrix are in some sense perfectly conditioned. The same thing is true of the singular values of a matrix, as the following theorem shows.

THEOREM 6.6. Let $A, B \in \mathbb{C}^{m\times n}$ have singular values $\sigma_1 \geq \sigma_2 \geq \cdots \geq \sigma_n$ and $\tau_1 \geq \tau_2 \geq \cdots \geq \tau_n$ respectively. Then

$$| \sigma_i - \tau_i | \leq \| A - B \|_2 \qquad (i = 1, 2, \ldots, n). \tag{6.8}$$

PROOF. Let $A = B + E$, and let $\mathbb{S}$ be an $(n - i + 1)$-dimensional subspace for which the minimum in (6.7) is obtained. Then

$$\begin{aligned} \sigma_i &= \max_{\substack{x \in \mathbb{S} \\ x\neq 0}} \frac{\| Bx + Ex \|_2}{\| x \|_2} \\ &\leq \max_{\substack{x \in \mathbb{S} \\ x\neq 0}} \frac{\| Bx \|_2}{\| x \|_2} + \max_{\substack{x \in \mathbb{S} \\ x\neq 0}} \frac{\| Ex \|_2}{\| x \|_2} \\ &\leq \tau_i + \| E \|_2 = \tau_i + \| A - B \|_2. \end{aligned} \tag{6.9}$$

Similarly,

$$\tau_i \leq \sigma_i + \| A - B \|_2. \tag{6.10}$$

Inequalities (6.9) and (6.10) together imply (6.8). ■

In the introduction to this section we discussed the problem of determining the rank of a matrix. As we pointed out there, about the best we can expect to accomplish with a matrix whose elements are not exactly specified is to determine if it is near a matrix of defective rank. This can be accomplished as follows.

Suppose for definiteness that $A \in \mathbb{C}^{m\times n}$, with $m \geq n$ and that the elements of A are accurate to quantities of order ε. Let the singular values of A be $\sigma_1 \geq \sigma_2 \geq \cdots \geq \sigma_n$ and let

$$A = V\begin{pmatrix} \Sigma \\ 0 \end{pmatrix}U^{\mathrm{H}},$$

where $\Sigma = \operatorname{diag}(\sigma_1, \sigma_2, \ldots, \sigma_n)$ be the singular value decomposition of A. Suppose that, having calculated the singular value decomposition of A, we find that

$$\sigma_{r+1}^2 + \sigma_{r+2}^2 + \cdots + \sigma_n{}^2 < \varepsilon^2.$$

Then if we set $\Sigma' = \operatorname{diag}(\sigma_1, \ldots, \sigma_r, 0, \ldots, 0)$, and

$$A' = V\begin{pmatrix} \Sigma' \\ 0 \end{pmatrix}U^{\mathrm{H}},$$

we have

$$\| A - A' \|_{\mathrm{F}} = \sqrt{\sigma_{r+1}^2 + \cdots + \sigma_n{}^2} < \varepsilon.$$

Thus A' is a matrix of rank r lying within ε of A in the Frobenius norm, and we are justified in saying that A is near a matrix of rank r.

Actually the above construction has given a matrix A' of rank r that is nearest A in the Frobenius norm. Specifically, we have the following theorem.

THEOREM 6.7. Let A and A' be as above. Then

$$\| A - A' \|_{\mathrm{F}} = \min_{\operatorname{rank}(B)=r} \| A - B \|_{\mathrm{F}}. \tag{6.11}$$

PROOF. We prove the theorem under the assumption that the minimum in (6.11) actually exists (this assumption can easily be established by analytic considerations). Let B be a matrix of rank r such that $\| A - B \|_F$ is minimal. Let the singular value decomposition of B be

$$C = Q^H BP = \begin{pmatrix} C_{11} & 0 \\ 0 & 0 \end{pmatrix},$$

where $C_{11} = \text{diag}(\gamma_1, \gamma_2, \ldots, \gamma_r)$. Let

$$D = Q^H AP = \begin{pmatrix} D_{11} & D_{12} \\ D_{21} & D_{22} \end{pmatrix}$$

be partitioned conformally with C.

We claim that $D_{11} = C_{11}$, $D_{12} = 0$, and $D_{21} = 0$. Suppose, for example, that $D_{12} \neq 0$. Then the matrix

$$C' = \begin{pmatrix} C_{11} & D_{12} \\ 0 & 0 \end{pmatrix}$$

has rank r and $\| D - C' \|_F < \| D - C \|_F$. However, if we set $B' = QC'P^H$, then B' is also of rank r and

$$\| A - B' \|_F = \| D - C' \|_F < \| D - C \|_F = \| A - B \|_F,$$

contradicting the minimality of B. Similar arguments show that $D_{21} = 0$ and $D_{11} = C_{11}$.

It follows that D has the form

$$D = \begin{pmatrix} C_{11} & 0 \\ 0 & D_{22} \end{pmatrix},$$

and

$$\| A - B \|_F = \| D - C \|_F = \| D_{22} \|_F.$$

Since C_{11} is diagonal, it consists of singular values of A, and $\| D_{22} \|_F^2$ will be the sum of squares of those left over. Obviously, this is a minimum when

$$\| D_{22} \|_F^2 = \sigma_{r+1}^2 + \sigma_{r+2}^2 + \cdots + \sigma_n{}^2 = \| A - A' \|_F^2. \quad ■$$

The singular value decomposition also has important applications to the linear least squares problem. The reason for this is that the unitary matrices that transform A to diagonal form in (6.1) do not change the 2-norms of vectors. In fact this observation is the basis for the proof of the following theorem.

THEOREM 6.8. Let $A \in \mathbb{C}^{m\times n}$ have the singular value decomposition (6.1). If $b \in \mathbb{C}^n$, then the vector x given by

$$x = U\begin{pmatrix} \Sigma^{-1} & 0 \\ 0 & 0 \end{pmatrix} V^{\mathrm{H}} b \tag{6.12}$$

is a vector that minimizes $\| b - Ax \|_2^2$. Moreover, if $\| b - Ax' \|_2^2$ is minimal and $x' \neq x$, then $\| x \|_2 < \| x' \|_2$.

PROOF. Let $z = U^{\mathrm{H}}x$, and $c = V^{\mathrm{H}}b$. Partition z and c in the form

$$z = \begin{pmatrix} z_1 \\ z_2 \end{pmatrix}, \qquad c = \begin{pmatrix} c_1 \\ c_2 \end{pmatrix},$$

where $z_1, c_1 \in C^r$. Then

$$\begin{aligned} \| b - Ax \|_2^2 &= \| V^{\mathrm{H}}(b - AUU^{\mathrm{H}}x) \|_2^2 \\ &= \left\| \begin{pmatrix} c_1 \\ c_2 \end{pmatrix} - \begin{pmatrix} \Sigma & 0 \\ 0 & 0 \end{pmatrix} \begin{pmatrix} z_1 \\ z_2 \end{pmatrix} \right\|_2^2 \\ &= \left\| \begin{pmatrix} c_1 - \Sigma z_1 \\ c_2 \end{pmatrix} \right\|_2^2 . \end{aligned} \tag{6.13}$$

Obviously (6.13) will be minimal when $z_1 = \Sigma^{-1}c_1$. The value z_2 is arbitrary; however if we let

$$z = \begin{pmatrix} \Sigma^{-1}c_1 \\ 0 \end{pmatrix}$$

and let

$$z' = \begin{pmatrix} \Sigma^{-1}c_1 \\ z_2 \end{pmatrix} \qquad (z_2 \neq 0), \tag{6.14}$$

be a different minimizer of (6.13), then

$$\| z \|_2^2 = \| \Sigma^{-1}c_1 \|_2^2 < \| \Sigma^{-1}c_1 \|_2^2 + \| z_2 \|_2^2 = \| z' \|_2^2 .$$

Then $x = Uz$, which is given by (6.12), minimizes (6.13), and if $x' = Uz'$ is any different minimizer, $\| x \|_2 < \| x' \|_2$. ∎

Theorem 6.8 shows how the singular value decomposition of a matrix A may be used to solve the linear least squares problem, even when null(A) $\neq 0$. In fact the proof of the theorem shows us how to find the set of all solutions. They are given by Uz', where z' is any vector of the form (6.14). However, if we take $z_2 = 0$ we obtain the unique solution of smallest norm.

In the next chapter we shall give an algorithm for computing the singular value decomposition of a matrix. Thus the formula (6.12) represents a practical way of solving least squares problems when the least squares matrix is not of full rank. In practice, the computed singular value decomposition will never have singular values that are exactly zero. However, as we pointed out above, it is reasonable to set to zero any singular values that are smaller than the uncertainty in the elements of the matrix. In the following algorithm we assume that this has already been done.

ALGORITHM 6.9. Let $A \in \mathbb{C}^{m\times n}$ have the singular value decomposition (6.1) and let $b \in \mathbb{C}^m$. Let $U = (u_1, u_2, \ldots, u_n)$ and $V = (v_1, v_2, \ldots, v_m)$. This algorithm returns the unique solution x of smallest 2-norm that minimizes $\| b - Ax \|_2^2$.

$$1)\quad \delta_i = \sigma_i^{-1} v_i^{\mathrm{H}} b_i \qquad (i = 1, 2, \ldots, r)$$

$$2)\quad x = \delta_1 u_1 + \delta_2 u_2 + \cdots + \delta_r u_r$$

We conclude with a final observation. If we define the *pseudo-inverse* of A by

$$A^\dagger = U\begin{pmatrix} \Sigma^{-1} & 0 \\ 0 & 0 \end{pmatrix} V^{\mathrm{H}}, \tag{6.15}$$

then $x = A^\dagger b$ solves the linear least squares problem of minimizing $\| b - Ax \|_2^2$. It should be observed that this definition of $A^\dagger$ does not depend on the particular choice of U and V in the singular value decomposition of A; for, whatever U and V are chosen, $x = A^\dagger b$ is uniquely determined as the least squares solution of smallest 2-norm. It is easily verified that if null$(A) \neq 0$, then $A^\dagger$ as defined by (6.15) is equal to $(A^{\mathrm{H}}A)^{-1}A^{\mathrm{H}}$, so that the definition of $A^\dagger$ given in Section 5.2 coincides with (6.15).

EXERCISES

1. Let A be normal with eigenvalues $\lambda_1, \lambda_2, \ldots, \lambda_n$. Show that the singular values of A are $|\lambda_1|, |\lambda_2|, \ldots, |\lambda_n|$.

2. What are the singular values of a unitary matrix?

3. Prove Theorem 6.5.

4. Let $X \in \mathbb{C}^{m\times n}$ have orthonormal columns and be partitioned in the form

$$X = \begin{pmatrix} X_1 \\ X_2 \end{pmatrix}.$$

Show that if u is a singular vector of X_1 with singular value γ, then u is a singular vector of X_2 with singular value σ, where $\gamma^2 + \sigma^2 = 1$.

5. Show that if $X \in \mathbb{C}^{m\times n}$ has orthonormal columns and Y is obtained by deleting $n - 1$, or fewer, rows of X, then $\| Y \|_2 = 1$.

6. Let $A \in \mathbb{C}^{m\times n}$. Prove that $A^\dagger$ is the unique matrix satisfying Penrose's conditions:

   (a) $A^\dagger A A^\dagger = A^\dagger$,
   (b) $A A^\dagger A = A$,
   (c) $(AA^\dagger)^{\mathrm{H}} = AA^\dagger$,
   (d) $(A^\dagger A)^{\mathrm{H}} = A^\dagger A$.

7. Let $A \in \mathbb{C}^{m\times n}$. Show that for $x \in \mathbb{C}^m$, $AA^\dagger x$ is the projection of x onto $\mathbb{R}(A)$.

NOTES AND REFERENCES

For real square matrices the singular value decomposition was established by Sylvester (1889), for general matrices by Eckart and Young (1939). A thorough survey of the properties of singular values of completely continuous operators is given by Gohberg and Kreĭn (1969).

Theorem 6.7 is due to Eckart and Young (1936) [see also Householder and Young (1938)] and has important statistical applications. For other applications of the singular value decomposition see Golub (1969).

THE *QR* ALGORITHM

CHAPTER 7

In this chapter we shall consider methods for computing eigenvalues and eigenvectors. It is far beyond the scope of this book to treat all the important numerical methods for the algebraic eigenvalue problem in detail; hence we shall confine ourselves to describing a class of methods for reducing a real matrix to diagonal or quasi-triangular form by orthogonal similarity transformations. The heart of these methods is an iterative technique called the QR algorithm. Because of the remarkable numerical properties of this iteration, which are described in Section 3, QR methods have come to be the preferred way to compute all the eigenvalues and eigenvectors of a matrix whose elements can be contained in the high speed memory of a computer.

The QR algorithm is prohibitively expensive when applied to a full matrix, requiring a multiple of n^3 operations for each iteration. Consequently, the matrix must first be reduced to a simpler form, such as upper Hessenberg form. In Section 1 we shall describe one method for reducing a matrix to upper Hessenberg form by orthogonal similarity transformations. When the method is applied to a real symmetric matrix, the result is a tridiagonal matrix.

In Section 2 we shall consider three closely related techniques for finding a single eigenvector of a matrix: the power method, the inverse power method, and the Rayleigh quotient iteration. Each of these methods is of independent interest; however, their properties also serve to explain the success of the QR algorithm.

In Section 3 we shall describe the QR algorithm, sketch its properties, and consider the practical details of its implementation. One of the drawbacks of the algorithm as given in Section 3 is that it may require complex "origin shifts" and hence complex arithmetic. In Section 4 an implicit shift technique is described which circumvents the problem of complex shifts by effecting two conjugate shifts simultaneously in real arithmetic.

One justification for confining ourselves to the QR algorithm is that it can be extended to solve other problems. In Section 5 we shall show how the QR algorithm may be adapted to find the singular value decomposition of a matrix. In Section 6 we shall give an algorithm for the solution of the generalized eigenvalue problem $Ax = \lambda Bx$.

1. REDUCTION TO HESSENBERG AND TRIDIAGONAL FORMS

In this chapter we shall be concerned with algorithms for reducing a matrix to triangular or quasi-triangular form by unitary similarity transformations. The possibility of accomplishing such reductions was established in Theorems 6.3.7 and 6.3.9. In fact the proofs of these theorems were essentially algorithms for effecting these reductions and required only that one be able to find a single eigenvalue and eigenvector of any given matrix.

Since the proofs of Theorems 6.3.7 and 6.3.9 are so nearly constructive algorithms, it is not unreasonable to expect that they could be made completely so. The result would be a method which would reduce a matrix to triangular form by a finite number of similarity transformations. Unfortunately these expectations cannot be realized for the following reason. It is a famous classical result that there is no algorithm, involving only a finite number of additions, subtractions, multiplications, divisions, and root extractions, that will solve a general polynomial equation of degree five (or greater). We have seen (Exercise 6.1.18) that every polynomial has an easily constructed *companion matrix* whose eigenvalues are the zeros of the polynomial. It follows that any finite procedure for triangularizing a matrix by similarity transformations can be turned into a finite procedure for solving polynomial equations.

Reductions, such as Gaussian elimination, that terminate after a finite number of steps are called direct methods. For the reasons just cited we cannot achieve a direct reduction of a matrix to triangular form by similarity transformations; rather we shall construct iterative methods in which the subdiagonal elements of the matrix are made smaller and smaller by a sequence of similarity transformations. After a number of such transformations, the subdiagonal elements will be so small as to be negligible and the matrix will be effectively triangular. However, such iterative methods will not yield a matrix that is exactly triangular.

The process of computing a similarity transformation is usually expensive. For example, if A and U are real $n \times n$ matrices with U orthogonal, it requires about $2n^3$ multiplications to compute the matrix $A' = U^{\mathrm{H}}AU$. If an iterative process requires even as little as one similarity transformation per eigenvalue, the complete reduction will require about $2n^4$ multiplications (as opposed to, say, $\frac{2}{3}n^3$ multiplications for the Gaussian reduction). This operation count is prohibitively large, even for matrices of modest size.

One cure for this problem is to transform the given matrix to a simple form which has a large number of zero elements. Since such a preprocessing algorithm should be direct, the simple form will have to be more complex than a triangular matrix. In this section we shall describe an algorithm for reducing a matrix to upper Hessenberg form by unitary similarity transformations. The algorithm may be carried out using elementary reflectors and requires about $\frac{5}{3}n^3$ multiplications. Since a unitary similarity transformation leaves Hermitian matrices Hermitian, the algorithm when applied to a Hermitian matrix yields a tridiagonal matrix.

Since the matrices with complex elements occur rather infrequently, we shall limit our descriptions to real matrices, both in this section and in subsequent sections. Although the modifications required to adapt the resulting algorithms to complex matrices are sometimes tedious, they are usually straightforward.

A typical step of the reduction to Hessenberg form goes as follows. At the second step a matrix of order six will have the form

$$A_2 = \left(\begin{array}{c:c:cccc} x & x & x & x & x & x \\ x & x & x & x & x & x \\ \hdashline 0 & x & x & x & x & x \\ 0 & \textcircled{x} & x & x & x & x \\ 0 & \textcircled{x} & x & x & x & x \\ 0 & \textcircled{x} & x & x & x & x \end{array}\right). \tag{1.1}$$

An elementary reflector U_2 of the form

$$U_2 = \begin{pmatrix} I_2 & 0 \\ 0 & R_2 \end{pmatrix}$$

is chosen so that U_2A_2 has zeros in the position distinguished in (1.1). Moreover, the zero elements in the first column of A_2 are obviously undisturbed by the premultiplication by U_2. Also, postmultiplication of a matrix by U_2^{H} leaves its first two columns undisturbed. Hence $A_3 = U_2A_2U_2^{\mathrm{H}}$ has the form

$$A_3 = \begin{pmatrix} x & x & x & x & x & x \\ x & x & x & x & x & x \\ 0 & x & x & x & x & x \\ 0 & 0 & x & x & x & x \\ 0 & 0 & x & x & x & x \\ 0 & 0 & x & x & x & x \end{pmatrix},$$

which carries the reduction one step further.

More generally, let $A_1 = A \in \mathbb{R}^{n\times n}$. At the kth step, the reduced matrix A_k will have the form

$$A_k = \begin{pmatrix} A_{11}^{(k)} & a_{12}^{(k)} & A_{13}^{(k)} \\ 0 & a_{22}^{(k)} & A_{23}^{(k)} \end{pmatrix},$$

where $A_{11}^{(k)} \in \mathbb{R}^{k\times(k-1)}$. Let $R_k \in \mathbb{R}^{(n-k)\times(n-k)}$ be an elementary reflector such that

$$R_k a_{22}^{(k)} = \pm\| a_{22}^{(k)} \|_2 e_1,$$

and let

$$U_k = \begin{pmatrix} I_k & 0 \\ 0 & R_k \end{pmatrix}. \tag{1.2}$$

Then

$$A_{k+1} = U_kA_kU_k^{\mathrm{T}} = \begin{pmatrix} A_{11}^{(k)} & a_{12}^{(k)} & A_{13}^{(k)}R_k \\ 0 & \pm\| a_{22}^{(k)} \|_2 e_1 & R_kA_{23}^{(k)}R_k \end{pmatrix}, \tag{1.3}$$

which carries the reduction one step further.

As far as the practical details of the algorithm are concerned, the elementary reflector R_k may be determined in the form

$$R_k = I - \pi_k^{-1}u_ku_k^{\mathrm{T}}, \tag{1.4}$$

where

$$u_k = (v_{k+1,k}, v_{k+2,k}, \ldots, v_{nk})^{\mathrm{T}}. \tag{1.5}$$

We shall use the rather conservative Algorithm 5.3.6 to determine u_k. The elements $v_{k+2,k}, \ldots, v_{n,k}$ may be stored in positions $(k+2, k), \ldots, (n, k)$ of the array A, since it is precisely these elements that are annihilated by the transformation. Provisions must be made to store the number π_k and $v_{k+1,k}$. We shall store these in the $(n+1)$th and $(n+2)$th rows of the array A. Thus for $n = 6$ and $k = 3$, the array A will have the form

$$\begin{matrix} \alpha_{11} & \alpha_{12} & \alpha_{13} & \alpha_{14} & \alpha_{15} & \alpha_{16} \\ \alpha_{21} & \alpha_{22} & \alpha_{23} & \alpha_{24} & \alpha_{25} & \alpha_{26} \\ v_{31} & \alpha_{32} & \alpha_{33} & \alpha_{34} & \alpha_{35} & \alpha_{36} \\ v_{41} & v_{42} & \alpha_{43} & \alpha_{44} & \alpha_{45} & \alpha_{46} \\ v_{51} & v_{52} & \alpha_{53} & \alpha_{54} & \alpha_{55} & \alpha_{56} \\ v_{61} & v_{62} & \alpha_{63} & \alpha_{64} & \alpha_{65} & \alpha_{66} \\ \pi_1 & \pi_2 & & & & \\ v_{21} & v_{32} & & & & \end{matrix} .$$

The premultiplication by U_k amounts to overwriting $A_{23}^{(k)}$ by $R_k A_{23}^{(k)}$. This can be done columnwise by means of Algorithm 5.3.7. The post-multiplication by U_k^{H}, which amounts to overwriting

$$\begin{pmatrix} A_{13}^{(k)} \\ R_k A_{23}^{(k)} \end{pmatrix}$$

by

$$\begin{pmatrix} A_{13}^{(k)} \\ R_k A_{23}^{(k)} \end{pmatrix} R_k,$$

can be accomplished similarly.

ALGORITHM 1.1. Let $A \in \mathbb{R}^{n\times n}$. This algorithm determines elementary reflectors $U_1, U_2, \ldots, U_{n-2}$ such that $A_{n-1} = U_{n-2}U_{n-3} \cdots U_1 A U_1 \cdots U_{n-3}U_{n-2}$ is upper Hessenberg. The nonzero elements of A_{n-1} overwrite those of A. The numbers $v_{k+2,k}, v_{k+3,k}, \ldots, v_{nk}$ defining U_k [cf. (1.2), (1.4), and (1.5)] overwrite the corresponding elements of A, and the numbers π_k and $v_{k+1,k}$ are stored in positions $(n+1, k)$ and $(n+2, k)$ of the array A. If no transformation is required at the kth step, π_k is set to zero.

1) For $k = 1, 2, \ldots, n-2$
   1) Determine the transformation
      1) $\eta = \max\{|\alpha_{ik}| : i = k+1, k+2, \ldots, n\}$
      2) If $\eta = 0$
         1) $\alpha_{n+1,k} = 0$
         2) Step k
      3) $\alpha_{ik} \leftarrow v_{ik} = \alpha_{ik}/\eta \qquad (i = k+1, k+2, \ldots, n)$
      4) $\sigma = \operatorname{sign}(v_{k+1,k})\sqrt{v_{k+1,k}^2 + \cdots + v_{nk}^2}$
      5) $v_{k+1,k} \leftarrow v_{k+1,k} + \sigma$
      6) $\alpha_{k+1,k} \leftarrow \pi_k = \sigma v_{k+1,k}$
   2) Premultiply
      1) For $j = k+2, k+3, \ldots, n$
         1) $\varrho = \pi_k^{-1} \sum_{i=k+1}^{n} v_{ik}\alpha_{ij}$
         2) $\alpha_{ij} \leftarrow \alpha_{ij} - \varrho v_{ik} \qquad (i = k+1, k+2, \ldots, n)$
   3) Postmultiply
      1) For $i = 1, 2, \ldots, n$
         1) $\varrho = \pi_k^{-1} \sum_{j=k+1}^{n} \alpha_{ij}v_{jk}$
         2) $\alpha_{ij} \leftarrow \alpha_{ij} - \varrho v_{jk} \qquad (j = k+1, k+2, \ldots, n)$
   4) Compute $\alpha_{k+1,k}$ and save $v_{k+1,k}$
      1) $\alpha_{n+2,k} \leftarrow v_{k+1,k}$
      2) $\alpha_{k+1,k} \leftarrow -\eta\sigma$

Note that in Algorithm 1.1, we have let $v_{k+1,k}$ initially overwrite $\alpha_{k+1,k}$, and only after the premultiplication and postmultiplication have we placed it in its final position.

The bulk of the computational work is in statement 1.2, which requires about $2(n-k)^2$ multiplications, and in statement 1.3, which requires about $2n(n-k)$ multiplications. Thus the algorithm as a whole requires approximately $\frac{5}{3}n^3$ multiplications.

The algorithm is quite stable. It can be shown that if Algorithm 1.1 is executed in t-digit arithmetic, then the computed A_{n-1} is unitarily similar to $A + E$, where

$$\| E \|_F \leq \gamma n^2 10^{-t} \| A \|_F , \tag{1.6}$$

and γ is a constant of order unity. Moreover, the orthogonal matrix U that transforms $A + E$ into the computed A_{n-1} is just the product of the orthogonal matrices $U_{n-2} U_{n-3} \cdots U_1$, where U_i corresponds to the exact transformation for the computed A_i. It follows that if we use the information stored in the lower part of the array A to generate the product $U_{n-2} U_{n-3} \cdots U_1$, then the computed result is very near the matrix U that reduces $A + E$.

If inner products are accumulated in double precision at the appropriate places in Algorithm 1.1, the bound (1.6) becomes

$$\| E \|_F \leq \gamma n 10^{-t} \| A \|_F ,$$

a very satisfactory result.

The reader may reasonably ask why we have confined ourselves to reduction by unitary matrices. The critical step in the derivation of Algorithm 1.1 was the construction of R_k. However, all that is really required of R_k is that it be easily invertible and that it satisfy the equation $R_k a_{22}^{(k)} = \sigma e_1$ for some constant σ. Obviously we can choose R_k in the form

$$R_k = T_k P_k ,$$

where P_k is an elementary permutation and T_k is an elementary lower triangular matrix. Such a choice leads to an algorithm that is related to Gaussian elimination as Algorithm 1.1 is related to the algorithm of orthogonal triangularization.

Actually there is a strong case to be made for preferring such an algorithm. In the first place, it requires about half the number of operations. In the second place, it is usually just as stable. The qualification "usually" is necessary here; for the rounding-error bounds for the algorithm contain growth factors analogous to those found in the bounds for Gaussian elimination. However, an untoward growth seldom occurs in practice. In the third place, the algorithm admits of a variant, analogous to the Crout algorithm, in which inner products may be accumulated in double precision. If this is done, the rounding-error bounds become about as small as possible (when growth factors are ignored).

We have chosen to describe the orthogonal reduction because of its simplicity. Moreover, it is never a bad algorithm; it is always stable in the sense of the error bound (1.6), and, when inner products cannot be accumulated, it is almost as good as the nonorthogonal algorithm. However, the reader should bear in mind that it is not necessarily the best algorithm for general use.

In the case where A is symmetric, the orthogonal algorithm enjoys a clear superiority over the nonorthogonal algorithm. This is because the upper Hessenberg matrix A_{n-1} is also symmetric and hence tridiagonal. Thus for symmetric A, Algorithm 1.1 is a method for reducing A to tridiagonal form. However, Algorithm 1.1 can be modified to further take advantage of the symmetry of A.

In the first place A_{n-1} has the form

$$A_{n-1} = \begin{pmatrix} \delta_1 & \gamma_1 & & & & \bigcirc \\ \gamma_1 & \delta_2 & \gamma_2 & & & \\ & \gamma_2 & \delta_2 & \ddots & & \\ & & \ddots & \ddots & \ddots & \\ & & & \ddots & \delta_{n-1} & \gamma_{n-1} \\ \bigcirc & & & & \gamma_{n-1} & \delta_n \end{pmatrix},$$

and for later applications it is convenient to store the δ_i and γ_i separately as they are generated. When this is done, we need not save the elements α_{kk} and $\alpha_{k+1,k}$, which may then be overwritten by π_k and $v_{k+1,k}$. Finally because all the matrices A_k are symmetric, we need only store, say, the lower halves of the matrices. These storage considerations are illustrated in the following array, in which $n = 6$ and $k = 3$:

$$\begin{array}{llllll} \pi_1 & & & & & \\ v_{21} & \pi_2 & & & & \\ v_{31} & v_{32} & \alpha_{33} & & & \\ v_{41} & v_{42} & \alpha_{43} & \alpha_{44} & & \\ v_{51} & v_{52} & \alpha_{53} & \alpha_{54} & \alpha_{55} & \\ v_{61} & v_{62} & \alpha_{63} & \alpha_{64} & \alpha_{65} & \alpha_{66} \end{array}.$$

It is important to take advantage of the symmetry of A_k in computing $U_k A_k U_k^{\mathrm{H}}$. From (1.3) and the symmetry of A_k, it follows that all we must compute is

$$R_k A_{23}^{(k)} R_k = (I - \pi_k^{-1} u_k u_k^{\mathrm{T}}) A_{23}^{(k)} (I - \pi_k^{-1} u_k u_k^{\mathrm{T}}).$$

If we set

$$r_k = \pi_k^{-1} A_{23}^{(k)} u_k,$$

then from the symmetry of $A_{23}^{(k)}$

$$R_k A_{23}^{(k)} R_k = A_{23}^{(k)} - u_k r_k^{\mathrm{T}} - r_k u_k^{\mathrm{T}} + \pi^{-1} u_k u_k^{\mathrm{T}} r_k u_k^{\mathrm{T}}.$$

Hence if we set

$$t_k = r_k - \frac{\pi^{-1}}{2} (u_k^{\mathrm{T}} r_k) u_k,$$

then

$$R_k A_{23}^{(k)} R_k = A_{23}^{(k)} - u_k t_k^{\mathrm{T}} - t_k u_k^{\mathrm{T}}.$$

Note that the unused locations $\gamma_{k+1}, \gamma_{k+2}, \ldots, \gamma_n$ are sufficient to store first the components of r_k and then the components of t_k.

The modified Algorithm 1.1 may be summarized as follows.

ALGORITHM 1.2. Let $A \in \mathbb{R}^{n\times n}$ be symmetric. This algorithm determined elementary reflectors $U_1, U_2, \ldots, U_{n-2}$ such that $A_{n-1} = U_{n-2} U_{n-3} \cdots U_1 A U_1 \cdots U_{n-3} U_{n-2}$ is tridiagonal. The diagonal elements of A_{n-1} are placed in $\delta_1, \delta_2, \ldots, \delta_n$, and the off-diagonal elements in $\gamma_1, \gamma_2, \cdots, \gamma_{n-1}$. The numbers $v_{k+1,k}, v_{k+2}, \ldots, v_{nk}$ defining U_k overwrite the corresponding elements of A, and the number π_k overwrites α_{kk}. The elements in the upper part of the array A are left undisturbed. If no transformation is required at the kth step, π_k is set to zero.

1) For $k = 1, 2, \ldots, n-2$
   1) $\delta_k = \alpha_{kk}$
   2) Determine the transformation
      1) $\eta = \max\{|\alpha_{ik}| : i = k+1, k+2, \ldots, n\}$
      2) If $\eta = 0$
         1) $\alpha_{kk} \leftarrow \pi_k = 0$
         2) $\gamma_k = 0$
         3) Step k
      3) $\alpha_{ik} \leftarrow v_{ik} = \alpha_{ik}/\eta \qquad (i = k+1, k+2, \ldots, n)$
      4) $\sigma = \operatorname{sign}(v_{k+1,k}) \sqrt{v_{k+1,k}^2 + \cdots + v_{nk}^2}$

5) $v_{k+1,k} \leftarrow v_{k+1,k} + \sigma$

6) $\alpha_{kk} \leftarrow \pi_k = \sigma v_{k+1,k}$

7) $\gamma_k = -\sigma\eta$

3) Apply the transformation

1) $\sigma = 0$

2) For $i = k+1, k+2, \ldots, n$

1) $\gamma_i \leftarrow \varrho_i = \sum_{j=k+1}^{i} \alpha_{ij}v_{jk} + \sum_{j=i+1}^{n} \alpha_{ji}v_{jk}$

2) $\sigma = \sigma + \varrho_i v_{ik}$

3) $\gamma_i \leftarrow \tau_i = \pi^{-1}[\varrho_i - (\sigma/2)v_{ik}] \quad (i = k+1, k+2, \ldots, n)$

4) For $i = k+1, k+2, \ldots, n$

1) For $j = k+1, k+2, \ldots, i$

1) $\alpha_{ij} \leftarrow \alpha_{ij} - v_{ik}\tau_j - v_{jk}\tau$

2) $\delta_{n-1} = \alpha_{n-1,n-1}$

3) $\delta_n = \alpha_{nn}$

4) $\gamma_{n-1} = \alpha_{n,n-1}$

Algorithm 1.2 has essentially the same numerical properties as Algorithm 1.1, except that it requires only $\frac{2}{3}n^3$ multiplications. If $U_1, U_2, \ldots, U_{n-2}$ are the exact transformations corresponding to the computed $A_1, A_2, \ldots, A_{n-2}$, then

$$A_{n+1} = U_{n-2}U_{n-3} \cdots U_1(A + E)U_1 \cdots U_{n-3}U_{n-2},$$

where $\| E \|_F$ satisfies a bound of the form (1.6). If inner products are accumulated in double precision, the factor n^2 in (1.6) may be reduced to n.

We have already seen in connection with Gaussian elimination that our rounding-error bounds are truly satisfactory only when the elements of the matrix are in some sense balanced. In the eigenvalue problem one unbalanced case occurs frequently enough to merit special attention. This is the case of "graded" matrices whose elements show a progressive decrease in size as one proceeds diagonally from left to right. For example, such a matrix might have elements with orders of magnitudes illustrated below:

$$A = \begin{pmatrix} 1 & 10^{-2} & 10^{-4} & 10^{-6} \\ 10^{-2} & 10^{-4} & 10^{-6} & 10^{-8} \\ 10^{-4} & 10^{-6} & 10^{-8} & 10^{-10} \\ 10^{-6} & 10^{-8} & 10^{-10} & 10^{-12} \end{pmatrix}.$$

Such a matrix will have very small eigenvalues, which, however, are quite insensitive to small relative perturbations in the elements of the matrix. If, as is often the case, the elements of the matrix are known to high relative accuracy, it makes sense to attempt to compute the small eigenvalues to high relative accuracy.

The error bounds cited above are not applicable to this case, for the bound (1.6) ensures only that the elements of E are small compared with the largest element of A, and such an error could overwhelm the smaller elements of A. Fortunately in practice the matrix E will usually have the same graded structure as A itself, and Algorithms 1.1 and 1.2 may be used with safety. It is important, however, that the elements of A be graded in descending order from left to right. If, for example, A is presented in the form

$$A = \begin{pmatrix} 10^{-12} & 10^{-10} & 10^{-8} & 10^{-6} \\ 10^{-10} & 10^{-8} & 10^{-6} & 10^{-4} \\ 10^{-8} & 10^{-6} & 10^{-4} & 10^{-2} \\ 10^{-6} & 10^{-4} & 10^{-2} & 1 \end{pmatrix},$$

the application of Algorithm 1.1 or Algorithm 1.2 will obliterate the smaller eigenvalues.

EXERCISES

1. Write an INFL program that takes the output of Algorithm 1.1 and generates in the array U the orthogonal matrix that transforms A into upper Hessenberg form.

2. Write an INFL program that takes the output of Algorithm 1.2 and generates in the array A the orthogonal matrix that transforms the original matrix to tridiagonal form.

3. Let $A \in \mathbb{R}^{n \times n}$. Show that there exists an elementary permutation P and an elementary lower triangular matrix M of index 2 such that

$(MP)A(MP)^{-1}$ has the form

$$A_2 \equiv (MP)A(MP)^{-1} = \begin{pmatrix} \alpha_{11}^{(2)} & \alpha_{12}^{(2)} & \cdots & \alpha_{1n}^{(2)} \\ \alpha_{21}^{(2)} & \alpha_{22}^{(2)} & \cdots & \alpha_{2n}^{(2)} \\ 0 & \alpha_{32}^{(2)} & \cdots & \alpha_{3n}^{(2)} \\ \vdots & \vdots & & \vdots \\ 0 & \alpha_{n2}^{(n)} & \cdots & \alpha_{nn}^{(2)} \end{pmatrix}.$$

What is a good choice of P?

4. Using the results of Exercise 3, devise an algorithm for reducing a square matrix A to upper Hessenberg form by nonorthogonal similarity transformations. Give an INFL code for this algorithm.

5. If no interchanges are made in the algorithm of Exercise 4, the final result is a lower triangular matrix M of the form

$$M = \begin{pmatrix} 1 & 0 \\ 0 & M' \end{pmatrix}$$

such that $B = M^{-1}AM$ is upper Hessenberg. By examining the equation $AM = MB$ deduce an algorithm, of the same character as the Crout algorithm, for determining M and B directly. Discuss the efficient use of storage. Give an INFL code.

6. Show how interchanges may be incorporated into the algorithm of Exercise 1.5.

7. Let $A \in \mathbb{C}^{n \times n}$, $x \in \mathbb{C}^n$, and $X = (x, Ax, \ldots, A^{n-1}x)$ (the columns of X are said to form a Krylov sequence). Show that if X is nonsingular, then $X^{-1}AX$ is upper Hessenberg and, in fact, is a companion matrix for the characteristic polynomial of A.

8. Let A and X be as in Exercise 1.7, and suppose R is nonsingular and upper triangular. Show that $(XR)^{-1}AXR$ is upper Hessenberg.

9. Let A and X be as in Exercise 1.8. Let $y \in \mathbb{C}^n$ and let $Y = (y, A^{\mathrm{H}}y, \ldots, (A^{\mathrm{H}})^{n-1}y)$ be nonsingular. Show that if $Y^{\mathrm{H}}X$ has an LDU decomposition then there is a unit upper triangular matrix R and an upper triangular matrix S such that $(XR)^{-1} = S^{\mathrm{H}}Y^{\mathrm{H}}$. Show further that $S^{\mathrm{H}}Y^{\mathrm{H}}AXR$ is a tridiagonal matrix that is similar to A.

10. In Exercise 1.9 let $U = XR$ and $V = YS$. Show that the columns of U

may be taken in the form

$$u = x, \qquad u_{i+1} = Au_i - \beta_{i,i+1}u_i - \beta_{i-1,i+1}u_{i-1} - \cdots - \beta_{1,i+1}u_i,$$

and the columns of V be taken in the form

$$v_1 = \gamma_{11}y, \quad v_{i+1} = \gamma_{i+1,i+1}A^{\mathrm{H}}v_i - \gamma_{i,i+1}v_i - \gamma_{i-1,i+1}v_{i-1} - \cdots - \gamma_{1,i+1}v_1.$$

11. In Exercise 10 show that

$$\beta_{i,i+1} = v_i^{\mathrm{H}}Au_i, \qquad \beta_{i-1,i} = v_{i-1}^{\mathrm{H}}Au_i,$$

and $\beta_{j,i+1} = 0$ $(j = 1, 2, \ldots, i-2)$. Develop similar expressions for the γ's. [*Hint*: Remember that $V^{\mathrm{H}}U = I$ and that $V^{\mathrm{H}}AU$ is tridiagonal.]

12. On the basis of the last five exercises, write an INFL program to reduce a matrix to tridiagonal form (this is the Lanczos tridiagonalization algorithm). Discuss conditions that can cause the algorithm to fail.

13. Discuss the simplifications that obtain in the Lanczos tridiagonalization algorithm when A is symmetric and $y = x$.

NOTES AND REFERENCES

The use of elementary reflectors to reduce a matrix to condensed form was first suggested by Householder and is mentioned in a paper by Householder and Bauer (1958). The reduction to tridiagonal form superseded an earlier reduction of Givens (1954) based on plane rotations (see Definition 3.3).

The reductions described in this section are two of many reductions by similarity transformations. Some of these are related to methods for computing the characteristic polynomial as a first step toward finding eigenvalues. For theoretical treatments see Householder (1964) and Bauer (1959). Wilkinson (AEP, Chapter VI) gives practical details and describes the numerical properties of the methods. Because of its historical importance and its connection with the method of conjugate gradients, special mention should be made of the method of Lanczos (1950) for reducing a general matrix to tridiagonal form.

Algorithms for reducing a matrix to Hessenberg form have been given by Martin and Wilkinson (1968a, HACLA/II/13), and for reducing a real symmetric matrix to tridiagonal form by Martin, Reinsch, and Wilkinson (1968, HACLA/II/2).

2. THE POWER AND INVERSE POWER METHODS

In this section we shall consider two methods for computing an eigenvector, and incidentally the associated eigenvalue, of a matrix. Both methods are in general use. The power method and its variants may be used to find a few eigenvectors corresponding to the largest eigenvalues of a matrix and is particularly applicable to large sparse matrices (i.e. matrices with a large number of zero elements). The inverse power method is one of the best methods for computing an eigenvector of a Hessenberg or tridiagonal matrix corresponding to a given approximate eigenvalue. We are interested in these methods primarily because they explain the remarkable numerical properties of the QR algorithm. Consequently, in this section we shall only sketch some of the properties of these methods and refer the reader to the literature for details.

In Example 6.2.2 we saw that if the matrix A has a largest eigenvalue λ_1, then for almost any vector q, the vectors $A^\nu q$ tend to lie along an eigenvector corresponding to λ_1. More specifically, suppose that A has a complete system of eigenvectors $x_1, x_2, \ldots, x_n$ corresponding to the eigenvalues $\lambda_1, \lambda_2, \ldots, \lambda_n$, which satisfy

$$|\lambda_1| > |\lambda_2| \geq |\lambda_3| \geq \cdots \geq |\lambda_n|. \tag{2.1}$$

Since the vectors $x_1, x_2, \ldots, x_n$ form a basis for C^n, any vector q_0 may be expressed in the form

$$q_0 = \gamma_1 x_1 + \gamma_2 x_2 + \cdots + \gamma_n x_n.$$

It follows that if we define

$$q_\nu = \frac{A^\nu q_0}{\lambda_1{}^\nu}, \tag{2.2}$$

then

$$q_\nu = \gamma_1 x_1 + \left(\frac{\lambda_2}{\lambda_1}\right)^\nu \gamma_2 x_2 + \cdots + \left(\frac{\lambda_n}{\lambda_1}\right)^\nu \gamma_n x_n. \tag{2.3}$$

Since $|\lambda_i/\lambda_1| < 1$ for $i = 2, 3, \ldots, n$,

$$\lim_{\nu\to\infty} \left(\frac{\lambda_i}{\lambda_1}\right)^\nu = 0 \qquad (i = 2, 3, \ldots, n).$$

Hence if $\gamma_1 \neq 0$,

$$\lim_{\nu\to\infty} q_\nu = \gamma_1 x_1.$$

Thus the sequence $q_1, q_2, q_3, \ldots$ approximates an eigenvector of A with increasing accuracy.

The above considerations suggest that we attempt to approximate the eigenvector x_1 by computing the vectors $q_1, q_2, q_3, \ldots$ and stopping when they appear to be sufficiently near a limit. However, formula (2.2) is not suitable for practical computations for two reasons. In the first place, we shall not usually know ahead of time the eigenvalue λ_1, so that we cannot compute the denominator $\lambda_1{}^\nu$ in (2.2). In the second place, the computation of A^ν is prohibitively expensive.

As far as the first problem is concerned, note that the factor $\lambda_1^{-\nu}$ in (2.2) serves only as a scaling factor. Since we are interested only in the directions of the q_ν, and not their sizes, we may use any other convenient scaling factor. For example, we may scale q_ν so that its largest component is unity, in which case the sequence $q_1, q_2, q_3, \ldots$ will converge to x_1, similarly scaled.

Concerning the second problem, the formation of A^ν may be avoided by observing that the vectors q_ν defined by (2.2) satisfy the recursion

$$q_{\nu+1} = \frac{Aq_\nu}{\lambda_1} \qquad (\nu = 0, 1, 2, \ldots). \tag{2.4}$$

If we replace λ_1 by a suitable scale factor σ_ν, the recursion (2.4) becomes

$$q_{\nu+1} = \frac{Aq_\nu}{\sigma_\nu} \qquad (\nu = 0, 1, 2, \ldots), \tag{2.5}$$

which only involves matrix–vector multiplications.

Formula (2.5) is the basis of the power method for finding an eigenvector of the matrix A. If the largest eigenvalue λ_1 of A satisfies (2.1) (we say that λ_1 is a *dominant eigenvalue*) and q_0 is chosen so that $\gamma_1 \neq 0$, then the sequence $q_0, q_1, q_2, \ldots,$ suitably scaled, will converge to an eigenvector corresponding to λ_1. Since the largest of the ratios $|\lambda_i/\lambda_1|$ is $|\lambda_2/\lambda_1|$, the second term in (2.3) will converge most slowly to zero. Thus the rate of convergence of the method is controlled by the degree of dominance of λ_1 as measured by the ratio $|\lambda_2/\lambda_1|$.

The condition $\gamma_1 \neq 0$ is not a severe restriction. Of course what is required in practice is not that γ_1 be merely nonzero, but that it be reasonably large. This condition will usually be satisfied by almost any randomly chosen vector q_0. However, if A has special structure, certain apparently reasonable choices of q_0 may go wrong. As a trivial example, if A is the triangular matrix

$$A = \begin{pmatrix} 3 & -1 \\ 0 & 2 \end{pmatrix},$$

then the vector $q_0 = (1, 1)^{\mathrm{T}}$ has no component along the dominant eigenvector $(1, 0)^{\mathrm{T}}$.

As a practical algorithm, the power method has some attractive features and some severe limitations. Its strongest point is the simplicity of the recursion (2.5). For example, only the nonzero elements of A are required to form the product Aq_ν; hence if A is sparse it is only necessary to store the nonzero elements of A. This makes it possible to calculate the dominant eigenvalue and eigenvector of a sparse matrix which is too large to store as a full square array. The method can also be used in connection with the deflation techniques of Section 6.3 to find some of the smaller eigenvalues.

A drawback to the method is that it will converge slowly when λ_1 is not strongly dominant. For example, if $|\lambda_1/\lambda_2| = .9$, then about 20 iterations will be required to reduce the second term in (2.3) by a factor of ten. One cure for this problem is to transform the matrix A so that the transformed matrix has a strongly dominant eigenvalue. Unfortunately, the only readily computable transformation that also preserves the sparseness of the matrix is an origin shift of the form $A - \pi I$. While it is possible to obtain improved convergence by such a shift (sometimes a striking improvement) the technique does not lend itself well to automatic computation.

The matrix A may even fail to have a dominant eigenvalue, in which case the power method will not converge. This occurs, for example, whenever a real matrix has a complex eigenvalue λ_1 that is equal to its spectral radius. For then $\bar{\lambda}_1 \neq \lambda_1$ is also an eigenvalue, and $|\lambda_1| = |\bar{\lambda}_1|$. Extensions of the power method which find several eigenvectors simultaneously have been developed, and these methods to some extent circumvent the convergence problems just mentioned. However, their description and analysis is beyond the scope of this book.

In discussing the power method we have analyzed only the special case where A is nondefective. The general case may be analyzed by expressing q_0 as a linear combination of the principal vectors associated with the Jordan canonical form of A (cf. Theorem 6.3.10 and the following discussion). The analysis shows that the method will converge even when λ_1 is a multiple eigenvalue, provided no other eigenvalue has as large an absolute value. When λ_1 has as many linearly independent eigenvectors as its multiplicity (i.e., when λ_1 appears only in 1×1 Jordan blocks), the rate of convergence depends on the ratio of λ_1 to the next largest eigenvalue. Otherwise the convergence is excessively slow. It should be appreciated that in practice the sharp distinction between defective and nondefective matrices is blurred owing to rounding errors. For example, if the eigenvectors corresponding to λ_1 and λ_2 are very nearly parallel, the power method may behave as if λ_1 were a defective multiple eigenvalue.

The fact that the rate of convergence of the power method depends on the size of the largest eigenvalue compared with the second largest suggests that one initially transform the matrix so that it has one very large eigenvalue. A very few iterations of the power method should then suffice to find an eigenvector of the transformed matrix. This idea is the basis of the inverse power method. To derive it, suppose again that A has eigenvalues $\lambda_1, \lambda_2, \ldots, \lambda_n$ (not necessarily in any special order) corresponding to the linearly independent eigenvectors $x_1, x_2, \ldots, x_n$, and suppose that λ is a close approximation to λ_1. Then by Theorem 6.2.5 the matrix $(\lambda I - A)^{-1}$ has eigenvalues $(\lambda - \lambda_1)^{-1}, (\lambda - \lambda_2)^{-1}, \ldots, (\lambda_1 - \lambda_n)^{-1}$ corresponding to eigenvectors $x_1, x_2, \ldots, x_n$. Consequently if we choose a starting vector x and express it in the form

$$x = \gamma_1 x_1 + \gamma_2 x_2 + \cdots + \gamma_n x_n, \tag{2.6}$$

we have

$$x' = (\lambda I - A)^{-1} x = \frac{\gamma_1 x_1}{\lambda - \lambda_1} + \frac{\gamma_2 x_2}{\lambda - \lambda_2} + \cdots + \frac{\gamma_n x_n}{\lambda - \lambda_n}. \tag{2.7}$$

Now if λ is much nearer λ_1 than to $\lambda_2, \lambda_3, \ldots, \lambda_n$, then $(\lambda - \lambda_1)^{-1}$ will be much larger than $(\lambda - \lambda_2)^{-1}, (\lambda - \lambda_3)^{-1}, \ldots, (\lambda - \lambda_n)^{-1}$. It follows that if γ_1 in (2.6) is not unreasonably small, the term $\gamma_1(\lambda - \lambda_1)^{-1} x_1$ in (2.7) will dominate all the others. In other words, if we start with almost any vector x and generate x' according to (2.7), then x' will be a good approximation to x_1. Of course the process may be repeated, starting with x', to improve the approximation further.

The rapidity with which this method converges can be quite striking. Suppose, for example,

$$\| x_i \|_2 = 1 \qquad (i = 1, 2, \ldots, n),$$

$$\gamma_1 = 10^{-2}, \qquad \gamma_i = 1 \qquad (i = 2, 3, \ldots, n),$$

and

$$\lambda - \lambda_1 = 10^{-4}, \qquad \lambda - \lambda_i = 1 \qquad (i = 2, 3, \ldots, n),$$

so that the starting vector x is deficient in its component along x_1. Then

$$x' = 10^2 x_1 + x_2 + \cdots + x_n,$$

and

$$x'' = (\lambda I - A)^{-1} x' = 10^6 x_1 + x_2 + \cdots + x_n.$$

Thus one application of the method gives a two-digit approximation to x_1, in spite of the unfortunate choice of x. A second application gives a six-digit approximation, and each subsequent application adds four more digits.

From the foregoing it is evident that the inverse power method is potentially a very effective method for computing eigenvectors. However, a number of practical points must be considered before the method can be accepted as a working technique. We shall only sketch these points, referring the interested reader to the literature for details.

In the first place, the method requires an accurate approximation λ to the eigenvalue λ_1. This is usually obtained by some other method. It should be noted that if λ_1 is well conditioned, λ will usually agree with λ_1 to almost the number of digits used in the calculations. In this case, the method may be expected to converge very rapidly indeed.

Of course one does not form $(\lambda I - A)^{-1}$ to compute x'; rather one solves the system

$$(\lambda I - A)x' = x \tag{2.8}$$

and then scales x' so that, say, its largest component is unity. It may be objected that the solution of (2.8) by the techniques of Chapter 3 requires approximately $\frac{2}{3}n^3$ multiplications, which is a high price to pay for a single eigenvector. Fortunately, the computation of the eigenvalues of A, which is a prerequisite for the application of the inverse power method, is usually accomplished via a preliminary reduction to Hessenberg form, say by Algorithm 1.1. Thus one may compute an eigenvector of the Hessenberg matrix and use the history of transformations of the reduction to transform the result back to an eigenvector of A. If we assume that A is Hessenberg in (2.8), then the computation of x' requires only a small multiple of n^2 operations, which is perfectly satisfactory. If A is tridiagonal, which will happen whenever the original matrix was symmetric, the solution of (2.8) requires only a small multiple of n operations.

The method will converge faster as λ approaches the eigenvalue λ_1. However, the nearer λ is to λ_1, the more nearly singular is $\lambda I - A$ and the more ill conditioned is the system (2.8). This raises serious questions concerning the stability of the method, since x' may be expected to be computed quite inaccurately. These questions may be answered as follows. The system (2.8) will be solved by first computing an LU decomposition of $\lambda I - A$. If, for simplicity, we ignore interchanges the computed matrices L and U will satisfy

$$LU = \lambda I - A - E,$$

where E is a small matrix. We have observed in Section 3.5 that the solution of the triangular systems $LUx' = x$ does not usually generate much error. Thus the computed x' will nearly satisfy the equation

$$[\lambda I - (A + E)]x' = x.$$

In other words, the inverse power method will converge to an eigenvector of the slightly perturbed matrix $A + E$, which is all that we require of a stable algorithm.

Another way of looking at this phenomenon is the following. The computed solution of (2.8) will indeed be inaccurate; however, if x_1 is well conditioned, most of the error will lie in the direction of x_1. Since it is only the direction of the eigenvector x_1 that we are trying to compute, this error has no effect on the final results.

The choice of the initial vector x and the treatment of multiple eigenvalues and ill-conditioned eigenvectors are problems that require close attention. Generally speaking, their solutions involve arbitrary decisions based on numerical experience. One problem of particular importance is to preserve the orthogonality of the eigenvectors of a symmetric matrix corresponding to a cluster of poorly separated eigenvalues.

An important variant of the inverse power method is the Rayleigh quotient iteration. Suppose that we have no good approximation to an eigenvalue of A, but that the starting vector x is known to be a fairly accurate approximate eigenvector. Then a reasonable choice for λ is the Rayleigh quotient (cf. Theorem 6.5.5)

$$\lambda = \frac{x^{\mathrm{H}}Ax}{x^{\mathrm{H}}x}. \tag{2.9}$$

If the inverse power method is applied with this value of λ, there results a new approximate eigenvector. The process may be iterated by computing a new Rayleigh quotient and applying the inverse power method again. More precisely, given the starting vector x, one computes λ from (2.9) and then solves the system

$$(\lambda I - A)x' = \sigma x, \tag{2.10}$$

where σ is any convenient scaling factor. The process may be repeated as many times as is desired.

In the next section we shall wish to know how fast this Rayleigh quotient iteration converges. To analyze the method, it is convenient to make a transformation of variables. If R is any unitary matrix and

$$y = R^{\mathrm{H}}x, \qquad y' = R^{\mathrm{H}}x',$$

then from (2.9)

$$\lambda = \frac{y^{\mathrm{H}}R^{\mathrm{H}}ARy}{y^{\mathrm{H}}y}$$

and from (2.10)

$$(\lambda I - R^{\mathrm{H}}AR)y' = \sigma y.$$

In other words y and y' are Rayleigh quotient iterates for the matrix $R^{\mathrm{H}}AR$.

In particular, if we assume that $\| x \|_2 = 1$, and $R = (x, U)$ is unitary, then $y = e_1$ and

$$R^{\mathrm{H}}AR = \begin{pmatrix} \lambda & h^{\mathrm{H}} \\ g & C \end{pmatrix},$$

where λ is just the Rayleigh quotient (2.9). It follows that y' is the solution of the system

$$\begin{pmatrix} 0 & -h^{\mathrm{H}} \\ -g & \lambda I - C \end{pmatrix} y' = \sigma e_1.$$

If we choose σ so that the first component of y' is unity and seek y' in the form $(1, p^{\mathrm{T}})^{\mathrm{T}}$, where $p = (\pi_2, \ldots, \pi_n)^{\mathrm{T}}$, then

$$\begin{pmatrix} 0 & -h^{\mathrm{H}} \\ -g & \lambda I - C \end{pmatrix} \begin{pmatrix} 1 \\ p \end{pmatrix} = \begin{pmatrix} \sigma \\ 0 \end{pmatrix},$$

or

$$p = (\lambda I - C)^{-1}g.$$

Now a measure of the accuracy of y as an eigenvector of $R^{\mathrm{H}}AR$ is the number

$$\varrho = \frac{\| \lambda y - R^{\mathrm{H}}ARy \|_2}{\| y \|_2},$$

which we have seen in Theorem 6.3.3 is the size of the smallest residual of the form $\mu y - R^{\mathrm{H}}ARy$. Likewise if λ' denotes the Rayleigh quotient for y', then

$$\varrho' = \frac{\| \lambda' y' - R^{\mathrm{H}}ARy' \|_2}{\| y' \|_2}$$

is a measure of the accuracy of y. Thus to determine how much of an improvement y' is over y, we must find expressions for the numbers ϱ and ϱ'.

Since $y = e_1$, it is easily verified that

$$\varrho = \| g \|_2 .$$

To find an expression for ϱ', suppose that Q is a unitary matrix such that $Q^{\mathrm{H}}y' = \| y' \|_2 e_1$. Then from Theorem 6.3.4 it follows that if

$$Q^{\mathrm{H}}R^{\mathrm{H}}ARQ = \begin{pmatrix} \lambda' & h'^{\mathrm{H}} \\ g' & C' \end{pmatrix},$$

then

$$\varrho' = \| g' \|_2 .$$

Now it is easily verified that Q may be taken in the form

$$Q = \begin{pmatrix} 1 & -p^{\mathrm{H}} \\ p & I \end{pmatrix} D,$$

where

$$D^2 = I + \mathrm{diag}(\| p \|_2^2, \pi_2^2, \pi_3^2, \ldots, \pi_n^2).$$

Obviously if g, and hence p, is small, D differs only negligibly from an identity matrix. Hence we may ignore it and compute g' approximately as

$$\begin{aligned} g' &\cong (-p, I) \begin{pmatrix} \lambda & h^{\mathrm{H}} \\ g & C \end{pmatrix} \begin{pmatrix} 1 \\ p \end{pmatrix} \\ &= -\lambda p - p h^{\mathrm{H}} p + g + Cp, \end{aligned}$$

or in view of the equality $(\lambda I - C)p = g$,

$$g' \cong -p h^{\mathrm{H}} p.$$

Thus we have approximately

$$\varrho' \leq \| p \|_2^2 \| h \|_2 \leq \| (\lambda I - C)^{-1} \|_2^2 \| h \|_2 \| g \|_2^2 = \varkappa_1 \varrho^2, \tag{2.11}$$

where

$$\varkappa_1 = \| (\lambda I - C)^{-1} \|_2^2 \| h \|_2 .$$

Inequality (2.11) represents a particularly nice state of affairs. Suppose that the Rayleigh quotient iteration is converging to an eigenvector corresponding to a simple eigenvalue. Then λ will be very near to that eigenvalue, and C will be very near a matrix containing the other eigenvalues. Hence $\lambda I - C$ will be nonsingular. Moreover $\| (\lambda I - C)^{-1} \|_2$ will not change much from iteration to iteration. Inequality (2.11) then says that the size ϱ' of the residual of one iterate is bounded by a constant times the square of the size of the residual of the preceding iterate. This behavior is called *quadratic convergence.* Once a quadratically convergent process starts converging, it approaches its limit very swiftly. For example, if $\varkappa_1 = 1$ and $\varrho = 10^{-2}$, then the subsequent residuals will have sizes 10^{-4}, 10^{-8}, $10^{-16}, \ldots$.

When A is Hermitian, we have $g = h$ in the above derivation. Hence, approximately,

$$\varrho' \leq \| (\lambda I - C)^{-1} \|_2^2 \, \| g \|_2^3 = \varkappa_2 \varrho^3,$$

where

$$\varkappa_2 = \| (\lambda I - C)^{-1} \|_2^2 .$$

This very swift *cubic convergence* of the Rayleigh quotient iteration for Hermitian matrices might have been anticipated on the grounds that the Rayleigh quotient for a symmetric matrix is more accurate as an approximate eigenvalue than the generating vector is as an approximate eigenvector (cf. Theorem 6.5.5).

Although we have had to assume that the iteration was converging to an eigenvector corresponding to a simple eigenvalue in order to justify the above informal analysis of the Rayleigh quotient iteration, all that is really necessary is that the eigenvalue have a full complement of eigenvectors. In particular, since a Hermitian matrix is never defective, the Rayleigh quotient iteration for a Hermitian matrix always converges cubically.

EXERCISES

1. Let $x \in \mathbb{C}^n$ be nonzero. Show that the vectors $x_k = (J_\lambda^{(n)})^k x$ tend to lie in direction of the eigenvector e_1 of $J_\lambda^{(n)}$. How fast is the limiting direction approached?

2. In Exercise 2.1 let $x = e_n$ and $\lambda = 1$. How many iterations will be required to produce an eigenvector that is accurate to 5 figures?

3. Describe the simplifications that obtain when the power method is applied to the companion matrix of Exercise 6.2.18 (when the eigenvalue is estimated by the ratios of the components of the approximate eigenvectors, this is simply Bernoulli's method for finding a zero of a polynomial).

4. Let $A \in \mathbb{C}^{n \times n}$ have real eigenvalues satisfying $\lambda_1 > \lambda_2 \geq \cdots \geq \lambda_{n-1} > \lambda_n$. Consider the power method for the matrix $A - \pi I$. Show that the choice of π that gives the fastest convergence to the eigenvector corresponding to λ_1 is $\pi = (\lambda_2 + \lambda_n)/2$. What is the rate of convergence?

5. Let $A \in \mathbb{R}^{n \times n}$ be symmetric with eigenvalues $\lambda_1, \lambda_2, \ldots, \lambda_n$ corresponding to the orthonormal system of eigenvector $x_1, x_2, \ldots, x_n$. Suppose that the eigenvalues of A can be ordered so that $|\lambda_1| > |\lambda_2| > |\lambda_3| \geq \cdots \geq |\lambda_n| > 0$, and let $Q_0 = (q_1^{(0)}, q_2^{(0)})$ have orthonormal columns. Consider the iteration

$$Q_{\nu+1} = AQ_\nu R_\nu,$$

where R_ν is an upper triangular matrix chosen so that $Q_{\nu+1}^{\mathrm{H}} Q_{\nu+1} = I$. Let Q_0 be represented in the form $Q_0 = XP_0$. Show that if $P_0^{[1]}$ and $P_0^{[2]}$ are nonsingular, then (except perhaps for the signs of the columns) $Q_\nu \to (x_1, x_2)$. [*Hint*: First show that $Q_\nu = A^\nu Q_0 \tilde{R}_\nu$, where $\tilde{R}_\nu = R_{\nu-1} R_{\nu-2} \cdots R_1$. Thus Q_ν is the matrix obtained by orthogonalizing the columns of $A^\nu Q_0$.]

6. Generalize the method described in Exercise 5 and give an INFL program implementing it. Use orthogonal triangularization to accomplish the orthogonalization.

7. In the inverse power method, let $A = J_\mu^{(n)}$ and $x = e_n$. Show that if λ is near μ, then x' will be an accurate eigenvector of $J_\mu^{(n)}$. Show further that the accuracy of the subsequent iterates $x'', x''', \ldots$ improves very slowly.

8. Let $A \in \mathbb{C}^{n \times n}$, and let λ and $u \in \mathbb{C}^n$ be given. Show that the vector $v = (\lambda I - A)^{-1} u$ is an eigenvector of $A + E$ where the matrix E may be chosen to satisfy

$$\| E \|_{\mathrm{F}} = \frac{\| u \|_2}{\| v \|_2}.$$

[*Hint*: Use Exercise 6.3.14.]

9. Let $A, E \in \mathbb{C}^{n\times n}$ and let λ be an eigenvalue of $A + E$ but not of A. Show there are vectors u and v such that $v = (\lambda I - A)^{-1}u$ and

$$\frac{\| u \|_2}{\| v \|_2} \leq \| E \|_2 .$$

10. The result of Exercise 8 indicates that if, in the inverse power method, $x' = (\lambda I - A)^{-1}x$ is sufficiently larger than x, then x' will be an acceptable approximate eigenvector. If x' is not sufficiently large, Exercise 9 suggests that it is reasonable to restart with a new vector, say one orthogonal to the old. Write an INFL program that starts with the vector $x_0 = (1, 1, \ldots, 1)^{\mathrm{T}}$ and produces a sequence x_0, $x_1, \ldots, x_{n-1}$ of orthonormal vectors for use in the inverse power method. [*Hint*: Take x_i as a multiple of the $(i - 1)$th column of the elementary reflector that annihilates the last $n - 1$ components of x_0.]

11. Let $A \in \mathbb{R}^{n\times n}$ be upper Hessenberg and suppose that λ is an eigenvalue of $A + E$ where E is a small matrix whose norm is bounded by the known number ε. Combine Exercises 3.4.1 and 2.10 to give an INFL algorithm for finding an approximate eigenvector of A corresponding to the approximate eigenvalue λ.

NOTES AND REFERENCES

Until recently the power method was the only alternative for finding the eigenvalues of a general matrix to computing the characteristic polynomial and then the eigenvalues. For many large sparse problems it remains the only practical method. Wilkinson (AEP, Chapter 9) has developed this technique, which does not automate easily, to a fine point.

Methods employing the power method to find several eigenvectors simultaneously were introduced by Bauer (1959) under the names "treppen (staircase) iteration" and "bi-iteration." Wilkinson (AEP, Chapter 9) describes an orthogonal variant. For symmetric matrices the orthogonal variant may be accelerated by a procedure discovered independently by Clint and Jennings (1970), Stewart (1969b), and Rutishauser (1969). A program has been published by Rutishauser (1970, HACLA/II/9).

The method of inverse iteration is due to Wielandt (1944). The practical method is treated by Wilkinson (AEP, Chapter 9) and by Peters and Wilkinson (HACLA/II/18), who also give program listings.

The Rayleigh quotient iteration and its variants have been exhaustively analyzed in a series of papers by Ostrowski (1958–1959).

3. THE EXPLICITLY SHIFTED QR ALGORITHM

In this and the following sections we shall be concerned with an iterative method, called the QR algorithm, for reducing a matrix to triangular or quasi-triangular form by orthogonal similarity transformations. When it is used in connection with a preliminary reduction to Hessenberg or tridiagonal form, the algorithm has proven to be one of the most effective methods for finding all the eigenvalues of a matrix whose elements can be fit into the high-speed storage of a computer. Eigenvectors may be found by calculating the corresponding eigenvectors of the final triangular matrix and transforming them back. In addition to its application to the eigenvalue problem, the QR algorithm may be adapted to find the singular values of a matrix and to solve the generalized eigenvalue problem $Ax = \lambda Bx$.

We shall introduce the QR algorithm without motivation, and then explain the properties, mathematical and numerical, that make it so good. Let $A = A_0 \in \mathbb{C}^{n \times n}$. The QR algorithm produces a sequence $A_0, A_1, A_2, \ldots$ of similar matrices as follows. Given A_ν, a scalar $\varkappa_\nu$, called an *origin shift*, is determined from the elements of A_ν (as the iteration converges, $\varkappa_\nu$ will approach an eigenvalue of A). The matrix $A_\nu - \varkappa_\nu I$ is then factored in the form

$$A_\nu - \varkappa_\nu I = Q_\nu R_\nu, \tag{3.1}$$

where Q_ν is unitary and R_ν is upper triangular. By Theorem 5.1.9 this factorization exists and is essentially unique, provided $A - \varkappa_\nu I$ is nonsingular. Finally $A_{\nu+1}$ is computed as

$$A_{\nu+1} = R_\nu Q_\nu + \varkappa_\nu I. \tag{3.2}$$

Note that from (3.1), we have $R_\nu = Q_\nu^{\mathrm{H}}(A_\nu - \varkappa_\nu I)$, and hence from (3.2)

$$A_{\nu+1} = Q_\nu^{\mathrm{H}}(A_\nu - \varkappa_\nu I)Q_\nu + \varkappa_\nu I = Q_\nu^{\mathrm{H}} A_\nu Q_\nu,$$

so that $A_{\nu+1}$ is indeed unitarily similar to A_ν.

On the face of it there is no reason for expecting this algorithm to do anything but produce a sequence of similar matrices at considerable computational expense. It turns out, however, that with the proper choice of the shifts $\varkappa_\nu$, the off-diagonal elements in the last row of the A_ν can be made to approach zero very swiftly. Moreover, all of the subdiagonal elements of the A_ν may approach zero somewhat more slowly. The reason is twofold.

1. The method is a variant of the inverse power method. In fact with proper choice of the origin shifts $\varkappa_\nu$, the method is a variant of the Rayleigh quotient iteration and converges quadratically.
2. The method is simultaneously an extension of the power method. Thus while the method is converging to one eigenvalue, it is also busy finding the others, albeit slowly.

The method is also computationally efficient. If A_0 is upper Hessenberg, then so are the subsequent A_ν, and the algorithm can be arranged so that it requires only a multiple of n^2 operations for each step. If A_0 is symmetric and tridiagonal, the number of operations is a multiple of n. After an eigenvalue has been found, the method automatically deflates the problem, so that the orders of the matrices involved decrease as more and more eigenvalues are found. Finally, the algorithm admits an implicitly shifted variant in which two complex conjugate shifts can be performed simultaneously in real arithmetic. This permits the efficient processing of real matrices with complex eigenvalues.

This section and the next are devoted to substantiating the above assertions. We begin with the relation of the method to the inverse power method. From (3.1),

$$Q_\nu = (A_\nu - \varkappa_\nu I)^{-\mathrm{H}} R_\nu^{\mathrm{H}}.$$

Since R_ν^{H} is lower triangular, $R_\nu^{\mathrm{H}} e_n = \bar{\varrho}_{nn} e_n$. Hence if Q_ν is partitioned in the form $Q_\nu = (q_1^{(\nu)}, q_2^{(\nu)}, \ldots, q_n^{(\nu)})$,

$$q_n^{(\nu)} = Q_\nu e_n = \bar{\varrho}_{nn} (A_\nu^{\mathrm{H}} - \bar{\varkappa}_\nu I)^{-1} e_n. \tag{3.3}$$

Thus $q_n^{(\nu)}$ is the approximate eigenvector of A_ν^{H} that results from applying one step of the inverse power method with shift $\bar{\varkappa}_\nu$ to the vector e_n.

To see what this means in terms of the elements of $A_{\nu+1}$, let A_ν be partitioned in the form

$$A_\nu = \begin{pmatrix} C_\nu & h_\nu \\ g_\nu^{\mathrm{H}} & \alpha_{nn}^{(\nu)} \end{pmatrix}, \tag{3.4}$$

with $A_{\nu+1}$ partitioned similarly.* Then $\| g_\nu \|_2$ is the size of the minimum residual of e_n regarded as an approximate eigenvector of A_ν^{H}. Since $A_{\nu+1}$

* The notation here is consistent with what has gone before. The submatrices C, g, and h occupy different positions because the last column of Q is an approximate eigenvector of A^{H} rather than the first column of Q being an approximate eigenvector of A.

$= Q^{\mathrm{H}} A_\nu Q$, $\| g_{\nu+1} \|_2$ is the size of the minimum residual of $q_n^{(\nu)}$ regarded as an approximate eigenvector of $A_\nu{}^{\mathrm{H}}$. If $\bar{\varkappa}_\nu$ is near an eigenvalue of A^{H}, then $q_n^{(\nu)}$ will be a more accurate approximate eigenvector than e_n, and $\| g_{\nu+1} \|_2$ will be smaller than $\| g_\nu \|_2$. Thus if the $\varkappa_\nu$ ever approximate an eigenvalue of A, the vectors g_ν may be expected to approach zero rapidly.

The natural choice for $\bar{\varkappa}_\nu$ in (3.3) is the Rayleigh quotient $e_n{}^{\mathrm{H}} A^{\mathrm{H}} e_n$; that is

$$\varkappa_\nu = \alpha_{nn}^{(\nu)}.$$

In this case the numbers $\| g_\nu \|_2$ are the sizes of the residuals corresponding to the Rayleigh quotient iteration. From the results of Section 2, we know that if the process converges, then the g_ν will generally approach zero quadratically. If A is Hermitian the convergence is cubic. Once g_ν is sufficiently small, we may neglect it and restart the process with the smaller matrix C_ν. We shall return to this automatic deflation later.

To elucidate the connection of the QR algorithm with the power method, we must introduce some additional notation. Let

$$\tilde{Q}_\nu = Q_0 Q_1 \cdots Q_\nu \tag{3.5}$$

and

$$\tilde{R}_\nu = R_\nu R_{\nu-1} \cdots R_0.$$

Then from the fact that $A_{\nu+1} = Q_\nu{}^{\mathrm{H}} A_\nu Q_\nu$, it follows that

$$A_{\nu+1} = \tilde{Q}_\nu{}^{\mathrm{H}} A \tilde{Q}_\nu, \tag{3.6}$$

or since $\tilde{Q}_\nu$ is unitary

$$A_{\nu+1} - \varkappa_\nu I = \tilde{Q}_\nu{}^{\mathrm{H}} (A - \varkappa_\nu I) \tilde{Q}_\nu. \tag{3.7}$$

The basic relation between the QR algorithm and the power method is contained in the following theorem.

THEOREM 3.1.

$$\tilde{Q}_\nu \tilde{R}_\nu = (A - \varkappa_0 I)(A - \varkappa_1 I) \cdots (A - \varkappa_\nu I). \tag{3.8}$$

PROOF. The proof is by induction. For $\nu = 0$, Equation (3.8) is just the defining relation for Q_0 and R_0. Assume the relation holds for $\tilde{Q}_{\nu-1} \tilde{R}_{\nu-1}$.

From (3.2) and (3.6) we have

$$R_\nu = (A_{\nu+1} - \varkappa_\nu I)Q_\nu{}^{\mathrm{H}} = \tilde{Q}_\nu{}^{\mathrm{H}}(A - \varkappa_\nu I)\tilde{Q}_\nu Q_\nu{}^{\mathrm{H}}$$
$$= \tilde{Q}_\nu{}^{\mathrm{H}}(A - \varkappa_\nu I)\tilde{Q}_{\nu-1}. \tag{3.9}$$

Postmultiplying (3.9) by $\tilde{R}_{\nu-1}$, we get

$$\tilde{R}_\nu = \tilde{Q}_\nu{}^{\mathrm{H}}(A - \varkappa_\nu I)\tilde{Q}_{\nu-1}\tilde{R}_{\nu-1}.$$

Hence

$$\tilde{Q}_\nu \tilde{R}_\nu = (A - \varkappa_\nu I)\tilde{Q}_{\nu-1}\tilde{R}_{\nu-1},$$

which in view of the inductive hypothesis is simply (3.8). ■

The significance of Theorem 3.1 becomes apparent when we consider the unshifted *QR* algorithm; that is, when we set $0 = \varkappa_0 = \varkappa_1 = \varkappa_2 = \cdots$. In this case, Equation (3.8) becomes

$$A^{\nu+1} = \tilde{Q}_\nu \tilde{R}_\nu.$$

In other words $\tilde{Q}_\nu \tilde{R}_\nu$ is the *QR* decomposition of $A^{\nu+1}$. In particular, since $\tilde{R}_\nu e_1 = \tilde{\varrho}_{11}^{(\nu)} e_1$, it follows that the first column of $\tilde{Q}_\nu$ is given by

$$\tilde{\varrho}_{11}^{(\nu)} \tilde{q}_1^{(\nu)} = \tilde{Q}_\nu \tilde{R}_\nu e_1 = A^{\nu+1} e_1.$$

Thus $q_1^{(\nu)}$ is the vector that would be obtained by applying ν steps of the power method to e_1. If A has a dominant eigenvalue λ_1 and e_1 is not deficient in the corresponding eigenvector, then $\tilde{q}_1^{(\nu)}$ approaches that eigenvector.

To see what this means in terms of the elements of A_ν, let A_ν be partitioned in the form

$$A_\nu = \begin{pmatrix} \alpha_{11}^{(\nu)} & h_\nu{}^{\mathrm{H}} \\ g_\nu & C_\nu \end{pmatrix}$$

[the matrices g_ν, h_ν, and C_ν are different from those in (3.4)]. Then from (3.6) and Theorem 6.3.4 follows that $\| g_{\nu+1} \|_2$ is the size of the minimum residual of $\tilde{q}_\nu$ regarded as an approximate eigenvector of A. Since $\tilde{q}_\nu$ is approaching an eigenvector of A, g_ν must converge to the zero vector. Thus under the conditions stated in the last paragraph, the subdiagonal elements of the first column of A_ν approach zero. The rate of convergence depends on the ratio of the second largest eigenvalue to the largest.

Actually, under suitable conditions, all or most of the subdiagonal elements of the A_ν tend to zero. In the following theorem we state only the simplest conditions under which this happens. The proof is not important to our development and is omitted.

THEOREM 3.2. Let A have eigenvalues $\lambda_1, \lambda_2, \ldots, \lambda_n$ satisfying $|\lambda_1| > |\lambda_2| > \cdots > |\lambda_n|$, and let the ith row of Y be a left eigenvector of A corresponding to λ_i. If Y has an LU decomposition, then the subdiagonal elements of the matrices A_ν of the unshifted QR algorithm tend to zero, and, for $i = 1, 2, \ldots, n$, $\alpha_{ii}^{(\nu)}$ tends to λ_i.

Theorem 3.2 and its variants account for some of the observed numerical properties of the QR algorithm. In the first place, it suggests that the QR algorithm will tend to find the eigenvalues of a matrix in descending order along the diagonal. Secondly, it accounts for the fact that the shifted form of the algorithm not only reduces the off-diagonal elements in the last row to zero but may simultaneously reduce the other subdiagonal elements, somewhat more slowly. To see how this can happen, suppose that A has real positive eigenvalues and $\alpha_{nn}^{(\nu)}$ is converging to λ_n, the smallest eigenvalue of A. Then, after a point, the shifts are very nearly equal to λ_n, and we are effectively performing the unshifted QR algorithm with matrix $A - \lambda_n I$. Hence by Theorem 3.2, the QR iteration will tend to reduce all the subdiagonal elements, while driving the ith diagonal element toward λ_i.

Although the above reasoning is quite informal, the phenomenon does occur in practice. While the off-diagonal elements in the last row of A_ν are converging to zero quadratically, the other subdiagonal elements are often observed to be decreasing. The decrease may not be much in any one iteration, but over many iterations it can be significant. In fact, some subdiagonal elements may become negligible. When A is Hessenberg, this means that the original problem may be decomposed into smaller problems. Even when this does not happen, the diagonal element $\alpha_{ii}^{(\nu)}$ may become an increasingly good approximate eigenvalue, so that when it comes time to use it as a shift, the convergence will be very fast indeed.

When applied to a full matrix, the QR algorithm has the defect that it is computationally expensive. If the QR factorization of $A_\nu - \varkappa_\nu I$ is obtained by means of Algorithm 5.3.8 and the product $R_\nu Q_\nu$ formed from the factored form of Q_ν, a single iteration will require a multiple of n^3 multiplications. However, if A_0 is upper Hessenberg, then all subsequent iterates A_ν are also upper Hessenberg. Moreover, the calculations can be arranged so that each iteration requires only a multiple of n^2 multiplications. The proof of these facts is best accomplished by describing the actual algorithm for upper Hessenberg matrices, which we shall now do.

We could, of course, take Algorithm 5.3.8 as a starting point and observe the simplifications that obtain in the reduction when the matrix is upper Hessenberg. Instead we shall describe a different reduction based on a class of orthogonal matrices called plane rotations. These easily constructed matrices can be used to introduce a single zero element into a vector or a matrix.

DEFINITION 3.3. A plane rotation in the (i, j)-plane is a matrix P_{ij} of the form

$$P_{ij} = \begin{matrix} \\ \\ \\ \\ i \\ \\ j \\ \\ \end{matrix} \begin{pmatrix} 1 & \cdots & 0 & \cdots & 0 & \cdots & 0 & \cdots & 0 \\ \vdots & & \vdots & & \vdots & & \vdots & & \vdots \\ 0 & \cdots & 1 & \cdots & 0 & \cdots & 0 & \cdots & 0 \\ \vdots & & \vdots & & \vdots & & \vdots & & \vdots \\ 0 & \cdots & 0 & \cdots & \gamma & \cdots & \sigma & \cdots & 0 \\ \vdots & & \vdots & & \vdots & & \vdots & & \vdots \\ 0 & \cdots & 0 & \cdots & -\sigma & \cdots & \gamma & \cdots & 0 \\ \vdots & & \vdots & & \vdots & & \vdots & & \vdots \\ 0 & \cdots & 0 & \cdots & 0 & \cdots & 0 & \cdots & 1 \end{pmatrix},$$

(columns i and j are those containing γ, $-\sigma$ and σ, γ respectively)

where $\gamma^2 + \sigma^2 = 1$.

It is easily verified that P_{ij} in Definition 3.3 is orthogonal. The effect of multiplying a vector x by P_{ij} is easily determined. In fact if $y = P_{ij}x$, then

$$\begin{aligned} \eta_k &= \xi_k \qquad (k \neq i, j), \\ \eta_i &= \gamma\xi_i + \sigma\xi_j, \\ \eta_j &= -\sigma\xi_i + \gamma\xi_j. \end{aligned} \tag{3.10}$$

These formulas have two important consequences. First, if A is any matrix, $B = P_{ij}A$ differs from A only in its ith and jth rows, whose elements are given by

$$\beta_{ik} = \gamma\alpha_{ik} + \sigma\alpha_{jk} \qquad \text{and} \qquad \beta_{jk} = -\sigma\alpha_{ik} + \gamma\alpha_{jk}.$$

Similarly the matrix AP_{ij} differs from A only in its ith and jth columns.

The second consequence of (3.10) is that γ and σ may be chosen so that η_j is zero; that is given any vector x we may construct a plane rotation in the (i, j)-plane that annihilates the jth component of x. In fact, if we set

$$\gamma = \frac{\xi_i}{\sqrt{\xi_i^2 + \xi_j^2}} \qquad \text{and} \qquad \sigma = \frac{\xi_j}{\sqrt{\xi_i^2 + \xi_j^2}},$$

then $\gamma^2 + \sigma^2 = 1$ and $\eta_j = 0$ (if $\xi_i = \xi_j = 0$, set $\gamma = 1$ and $\sigma = 0$).

As in the determination of elementary reflectors, underflows and overflows can cause difficulties in the computation of γ and σ. The following algorithm circumvents this difficulty. Note that the algorithm provides a means for other programs to call it.

ALGORITHM 3.4. Given the scalars α and β, this algorithm produces scalars γ, σ, and ν such that $\gamma^2 + \sigma^2 = 1$, and

$$\begin{pmatrix} \gamma & \sigma \\ -\sigma & \gamma \end{pmatrix} \begin{pmatrix} \alpha \\ \beta \end{pmatrix} = \begin{pmatrix} \nu \\ 0 \end{pmatrix}.$$

The algorithm may be referenced by the INFL statement

$$(\gamma, \sigma, \nu) = \text{rot}(\alpha, \beta)$$

1) $\eta = \max\{|\alpha|, |\beta|\}$

2) $\alpha' = \alpha/\eta$

3) $\beta' = \beta/\eta$

4) $\delta = \sqrt{\alpha'^2 + \beta'^2}$

5) $\gamma = \alpha'/\delta$

6) $\sigma = \beta'/\delta$

7) $\nu = \eta\delta$

Turning now to the QR algorithm, we consider the problem of effecting one unshifted QR step with the upper Hessenberg matrix $A \in \mathbb{R}^{n\times n}$. The first problem is to determine the QR factorization of A. This is done by reducing A to triangular form by means of premultiplications by plane rotations. The process is sufficiently well illustrated by the third step of the reduction for a matrix of order 6. At this stage, two plane rotations P_{12} and P_{23} have been determined so that $P_{23}P_{12}A$ has the form

$$\begin{matrix} \nu_1 & x & x & x & x & x \\ 0 & \nu_2 & x & x & x & x \\ 0 & 0 & \alpha_{33} & x & x & x \\ 0 & 0 & \textcircled{\alpha_{43}} & x & x & x \\ 0 & 0 & 0 & x & x & x \\ 0 & 0 & 0 & 0 & x & x \end{matrix}. \tag{3.11}$$

A plane rotation P_{34} in the (3, 4)-plane is then chosen so that the (4, 3)-element of $P_{34}P_{23}P_{12}A$ is zero. Specifically, the constants γ_3 and σ_3 determining P_{34} are calculated using Algorithm 3.4 via the INFL statement

$$(\gamma_3, \sigma_3, \nu_3) = \mathrm{rot}(\alpha_{33}, \alpha_{43}). \tag{3.12}$$

The matrix $P_{34}P_{23}P_{12}A$ then has the form

$$\begin{matrix} \nu_1 & x & x & x & x & x \\ 0 & \nu_2 & x & x & x & x \\ 0 & 0 & \nu_3 & x & x & x \\ 0 & 0 & 0 & x & x & x \\ 0 & 0 & 0 & x & x & x \\ 0 & 0 & 0 & 0 & x & x \end{matrix}, \tag{3.13}$$

which carries the reduction one step further.

At the end of the reduction we have determined plane rotations P_{12}, $P_{23}, \ldots, P_{n-1,n}$ such that

$$P_{n-1,n} \cdots P_{23}P_{12}A = R$$

is upper triangular. If we set $Q = P_{12}^{\mathrm{T}}P_{23}^{\mathrm{T}} \cdots P_{n-1,n}^{\mathrm{T}}$, then

$$A = QR,$$

and Q is the orthogonal part of the QR factorization of A. Thus to form the next QR iterate, we must compute

$$RQ = RP_{12}^{\mathrm{T}}P_{23}^{\mathrm{T}} \cdots P_{n-1,n}^{\mathrm{T}}.$$

Since P_{12} is a rotation in the (1, 2)-plane, the product RP_{12}^{T} differs from R only in its first and second columns, which are linear combinations of one another. Since R is upper triangular, RP_{12}^{T} must have the form

$$\begin{matrix} x & x & x & x & x & x \\ x & x & x & x & x & x \\ 0 & 0 & x & x & x & x \\ 0 & 0 & 0 & x & x & x \\ 0 & 0 & 0 & 0 & x & x \\ 0 & 0 & 0 & 0 & 0 & x \end{matrix}. \tag{3.14}$$

Likewise P_{23} is a rotation in the (2, 3)-plane. Hence $RP_{12}^{\mathrm{T}}P_{23}^{\mathrm{T}}$ differs from RP_{12}^{T} only in its second and third columns, which are linear combinations of one another. Since RP_{12}^{T} has the form (3.14), $RP_{12}^{\mathrm{T}}P_{23}^{\mathrm{T}}$ must have the form

$$\begin{matrix} x & x & x & x & x & x \\ x & x & x & x & x & x \\ 0 & x & x & x & x & x \\ 0 & 0 & 0 & x & x & x \\ 0 & 0 & 0 & 0 & x & x \\ 0 & 0 & 0 & 0 & 0 & x \end{matrix}.$$

In general, postmultiplication of $RP_{12}^{\mathrm{T}} \cdots P_{k-1,k}^{\mathrm{T}}$ by $P_{k,k+1}^{\mathrm{T}}$ affects only columns k and $k+1$ and introduces a single new nonzero element in the $(k+1, k)$-position. The final result RQ is upper Hessenberg.

The above construction, which shows that the upper Hessenberg form is invariant under the QR transformations, is effectively an algorithm for accomplishing one QR step. The INFL statement (3.12) gives an explicit means for obtaining the plane rotations. The premultiplications, each of which alters only two rows of the matrix, can obviously be accomplished in the array that originally contained the matrix, at the cost of destroying the original matrix. Likewise the postmultiplication can be accomplished with no more storage.

In fact, we can avoid the necessity of storing the quantities γ_i and σ_i that determine the rotations by performing the postmultiplications simultaneously with the premultiplications. An examination of (3.11) and (3.12) shows that once $P_{23}P_{12}A$ has been formed the first two columns of the array never enter into the remaining steps of the reduction to triangular form.

Since postmultiplication by P_{12}^{T} affects only the first two columns of the array, we may postmultiply before proceeding with the reduction. The process may be continued by alternating pre- and postmultiplications, thus:

$$P_{23}P_{12}A \to P_{23}P_{12}AP_{12}^{\mathrm{T}} \to P_{34}P_{23}P_{12}AP_{12}^{\mathrm{T}} \to P_{34}P_{23}P_{12}AP_{12}^{\mathrm{T}}P_{23}^{\mathrm{T}} \to \cdots.$$

The incorporation of shifts into the algorithm presents no problems. The shift may be subtracted from the diagonal elements of A before the QR step and added back afterward. Or, as is done in the following algorithm, the shift may be subtracted from the $(k+1)$th diagonal element just before $P_{k,k+1}$ is premultiplied and added back to the kth diagonal element after the postmultiplication by $P_{k,k+1}^{\mathrm{T}}$.

The above considerations are summed up in the following algorithm.

ALGORITHM 3.5. Given the upper Hessenberg matrix A of order n and the scalar $\varkappa$, this algorithm overwrites A with $Q^{\mathrm{H}}AQ$, where Q is the orthogonal matrix of the QR algorithm with shift $\varkappa$. The matrix Q is expressed as the product $P_{12}^{\mathrm{T}}P_{23}^{\mathrm{T}} \cdots P_{n-1,n}^{\mathrm{T}}$, where $P_{k,k+1}$ is a plane rotation in the $(k, k+1)$-plane. The quantities γ_k and σ_k that determine $P_{k,k+1}$ are computed by Algorithm 3.4.

1) $\alpha_{11} \leftarrow \alpha_{11} - \varkappa$

2) For $k = 1, 2, \ldots, n$

   1) If $k = n$, go to 2.4

   2) Determine $P_{k,k+1}$ and premultiply

      1) $(\gamma_k, \sigma_k, \nu_k) = \mathrm{rot}(\alpha_{kk}, \alpha_{k+1,k})$

      2) $\alpha_{kk} \leftarrow \nu_k$

      3) $\alpha_{k+1,k} \leftarrow 0$

      4) $\alpha_{k+1,k+1} \leftarrow \alpha_{k+1,k+1} - \varkappa$

      5) $\begin{pmatrix} \alpha_{kj} \\ \alpha_{k+1,j} \end{pmatrix} \leftarrow \begin{pmatrix} \gamma_k & \sigma_k \\ -\sigma_k & \gamma_k \end{pmatrix} \begin{pmatrix} \alpha_{kj} \\ \alpha_{k+1,j} \end{pmatrix} \quad (j = k+1, k+2, \ldots, n)$

   3) If $k = 1$, step k

   4) Postmultiply by $P_{k-1,k}^{\mathrm{T}}$ and restore the shift

      1) $(\alpha_{i,k-1}, \alpha_{ik}) \leftarrow (\alpha_{i,k-1}, \alpha_{ik}) \begin{pmatrix} \gamma_{k-1} & -\sigma_{k-1} \\ \sigma_{k-1} & \gamma_{k-1} \end{pmatrix} \quad (i = 1, 2, \ldots, k)$

      2) $\alpha_{k-1,k-1} \leftarrow \alpha_{k-1,k-1} + \varkappa$

3) $\alpha_{nn} \leftarrow \alpha_{nn} + \varkappa$

Algorithm 3.5 is quite efficient. For large n, the bulk of the work is concentrated in statements 2.2.5 and 2.4.1. An elementary operations count on these statements shows that they require about $4n^2$ multiplications. Thus if approximately k QR steps are required to find an eigenvalue, then about $4kn^3$ multiplications will be required to find all the eigenvalues of A. If k is not too large, this is well within the practical limitations of a modern computer. However, the general algorithm can be made far more efficient than this.

In practice, Algorithm 3.5 will be applied repeatedly with different shifts $\varkappa_0, \varkappa_1, \varkappa_2, \ldots$ to generate a sequence of unitarily similar upper Hessenberg matrices $A_0, A_1, A_2, \ldots$. We have seen above that if we take $\varkappa_\nu = \alpha_{nn}^{(\nu)}$, we may expect the off-diagonal elements in the last row of A_ν, in this case $\alpha_{n,n-1}^{(\nu)}$ since A_ν is upper Hessenberg, to converge at least quadratically to zero. Once $\alpha_{n,n-1}^{(\nu)}$ is small enough, it may be set to zero to yield a matrix of the form illustrated below:

$$\left(\begin{array}{ccccc|c} X & X & X & X & X & x \\ X & X & X & X & X & x \\ 0 & X & X & X & X & x \\ 0 & 0 & X & X & X & x \\ 0 & 0 & 0 & X & X & x \\ \hline 0 & 0 & 0 & 0 & 0 & \lambda_n \end{array}\right). \tag{3.15}$$

The number $\lambda_n = \alpha_{nn}^{(\nu)}$ is an approximate eigenvalue of A. Moreover the remaining eigenvalues of A are approximately the eigenvalues of the leading principal submatrix of A_ν of order $n-1$ [distinguished by the upper case X's in (3.15)]. Thus we may continue by applying the algorithm to this smaller matrix. In general, once m eigenvalues have been found, we need only work with the leading principal submatrix of order $n-m$. Obviously toward the end of the process we shall be working with very small matrices.

However, this is not the only savings that can be effected. Because of Theorem 3.2, we may expect that some of the subdiagonal elements of A_ν will approach zero also. While this convergence will be relatively slow, if n is sufficiently large, enough iterations may be performed to make one of these subdiagonal elements effectively zero. In this case A_ν will have the form illustrated for a typical case below [in which the (3, 2)-element has become zero]:

$$\begin{array}{cc:cccc:cc}
x & x & x & x & x & x & x & x \\
x & x & x & x & x & x & x & x \\
\hdashline
0 & 0 & X & X & X & X & x & x \\
0 & 0 & X & X & X & X & x & x \\
0 & 0 & 0 & X & X & X & x & x \\
0 & 0 & 0 & 0 & X & X & x & x \\
\hdashline
0 & 0 & 0 & 0 & 0 & 0 & \lambda_{n-1} & x \\
0 & 0 & 0 & 0 & 0 & 0 & 0 & \lambda_n
\end{array} \qquad (3.16)$$

In this case one need work with only the 4×4 submatrix distinguished by the upper case X's. After the eigenvalues of this submatrix have been found, the remaining eigenvalues of A are found in the leading 2×2 principal submatrix. This illustration is perfectly general. If m eigenvalues have been found and $\alpha_{l,l-1}$ is effectively zero, one iterates with the submatrix lying in the intersection of rows and columns l through $n - m$, and after that with the leading principal submatrix of order $l - 1$. One step of the restricted iteration may be accomplished by replacing 1 with l and n with $n - m$ in the appropriate statements in Algorithm 3.5.

To recapitulate, our analysis of the QR algorithm leads us to expect that the subdiagonal elements of the A_ν may approach zero, the last one quadratically, the others more slowly. Each time a subdiagonal element becomes effectively zero we may take advantage of the fact by working with a submatrix of A. These two observations account for the remarkable effectiveness of the QR algorithm.

We have used the term "sufficiently small" in connection with the subdiagonal elements of A without specifying precisely what we meant. It is difficult to give an entirely satisfactory convergence criterion for a method such as the QR algorithm, and none of the popular criteria is completely without objection. One test is easily motivated. Suppose that we are willing to tolerate an error of $\varepsilon \| A \|$ in the elements of our original matrix, and that at some stage

$$| \alpha_{k+1,k}^{(\nu)} | \leq \varepsilon \| A \|. \qquad (3.17)$$

Then because A is unitarily similar to A_ν, the result of setting $\alpha_{k+1,k}^{(\nu)}$ to zero corresponds to a perturbation in A that is on the order of $\varepsilon \| A \|$ or less. Of course one uses an easily computed norm in (3.17), say $\| \cdot \|_\infty$. Generally ε is taken to be on the order of 10^{-t}, where t is the number of significant digits carried in the calculations.

The convergence criterion (3.17) is unsatisfactory for graded matrices of the kind discussed in Section 1, for some of the eigenvalues of such a matrix may be smaller than $\varepsilon \| A \|$. If $\alpha_{nn}^{(\nu)}$, say, is converging to such an eigenvalue, the above convergence criterion may result in $\alpha_{n,n-1}^{(\nu)}$ being set to zero when it is larger than $\alpha_{nn}^{(\nu)}$; that is, before $\alpha_{nn}^{(\nu)}$ has actually converged. This suggests that the subdiagonal elements of A_ν be compared with their neighboring diagonal elements. One very cautious criterion is to regard $\alpha_{k+1,k}^{(\nu)}$ as effectively zero when

$$| \alpha_{k+1,k}^{(\nu)} | \leq \varepsilon \min\{| \alpha_{kk}^{(\nu)} |, | \alpha_{k+1,k+1}^{(\nu)} |\}. \tag{3.18}$$

Another less stringent criterion is to set $\alpha_{k+1,k}^{(\nu)}$ to zero when

$$| \alpha_{k+1,k}^{(\nu)} | \leq \varepsilon(| \alpha_{kk}^{(\nu)} | + | \alpha_{k+1,k+1}^{(\nu)} |). \tag{3.19}$$

Both criteria have the disadvantage that they may result in extra iterations. However, because of the swift convergence of the *QR* algorithm, the extra work is not usually substantial and is more than justified by the added safety of (3.18) or (3.19).

Algorithm 1.1, Algorithm 3.5, the deflation techniques, and the convergence criteria together constitute the basis of a general algorithm for finding all the eigenvalues of a matrix, with one important exception. Because Algorithm 3.5 is performed in real arithmetic, the shift $\varkappa = \alpha_{nn}$ cannot approximate a complex eigenvalue. Since the swift convergence of the shifted *QR* algorithm requires that the shift be near an eigenvalue, it follows that Algorithm 3.5 cannot be used efficiently to find a complex eigenvalue. One cure for this difficulty is to convert Algorithm 3.5 to complex arithmetic and use a different shifting strategy. When the original matrix A is complex, this is perfectly satisfactory. However, when A is real, we know from Theorem 6.3.9 that A can be reduced to quasi-triangular form by orthogonal similarity transformations, and it is natural to attempt to adapt the *QR* algorithm to compute this form without using complex arithmetic. In the next section we shall show how this may be done.

When A is symmetric, the problem of complex shifts does not arise, for all the eigenvalues of A are real. In this case, moreover, the matrix A is tridiagonal, and so are all subsequent A_ν. Algorithm 3.5 can be adapted to take advantage of this fact, and the resulting algorithm requires only a small multiple of n operations. If this adapted algorithm is combined with initial reduction to tridiagonal form, the result is an efficient algorithm for finding all the eigenvalues of a symmetric matrix.

In the symmetric case, if we choose the shift $\varkappa$ to be equal to α_{nn}, then we know from the theory of the Rayleigh quotient iteration that once the process starts converging it must converge cubically. In practice a somewhat different shift strategy is used. The shift $\varkappa$ is taken to be the eigenvalue of

$$\begin{pmatrix} \alpha_{n-1,n-1} & \alpha_{n-1,n} \\ \alpha_{n,n-1} & \alpha_{nn} \end{pmatrix}$$

that is nearest α_{nn}. It can be shown that for this choice of shifts the iteration always converges and that the convergence is usually cubic. The details of adapting Algorithm 3.5 to real symmetric matrices are left as an exercise.

In general, Algorithm 3.5 is quite stable. It can be shown that if it is applied with shift $\varkappa$ to the matrix A, the resulting matrix is orthogonally similar to $A + E$, where

$$\| E \|_{\mathrm{F}} \leq \phi(n) 10^{-t} \max\{\| A \|_{\mathrm{F}}, | \varkappa |\}. \tag{3.20}$$

Here $\phi(n)$ is a slowly growing function of n. If the elements of A are reasonably balanced, the well-conditioned eigenvalues of A will not be much affected. However, this result is unsatisfactory for graded matrices, since it is possible that the perturbation E will overwhelm the smaller elements of A. Fortunately, the algorithm is usually stable, even for graded matrices, provided the elements decrease as one proceeds along the diagonal. In this case the small eigenvalues will be found first, so that $| \varkappa |$ in (3.20) can become large only after the small eigenvalues have been deflated from the problem.

The *QR* algorithm can also be adapted to compute the eigenvectors of a general matrix. Such a calculation might go as follows. First Algorithm 1.1 is applied to A to yield an orthogonal matrix U such that

$$A_0 = U^{\mathrm{T}} A U$$

is upper Hessenberg. Algorithm 3.5 is then applied iteratively to yield a sequence of similar matrices $A_1, A_2, \ldots$. The sequence will terminate with a matrix A_ν that is effectively upper triangular (or, if the algorithm of the next section is used, quasi-triangular). If $\tilde{Q}_\nu$ is defined by (3.5), then

$$A_\nu = \tilde{Q}_\nu^{\mathrm{T}} A_0 \tilde{Q}_\nu = \tilde{Q}_\nu^{\mathrm{T}} U^{\mathrm{T}} A U \tilde{Q}_\nu. \tag{3.21}$$

Now the kth eigenvector y_k of the upper triangular matrix A_ν may be found by setting the kth component of y_k to unity and solving the upper triangular system

$$(A_\nu - \alpha_{kk}^{(\nu)} I)y_k = 0.$$

The last $n - k$ components of y_k will be zero; hence the matrix $Y = (y_1, y_2, \ldots, y_k)$ of eigenvectors will be upper triangular. In view of (3.21), the eigenvectors of A will be given by

$$x_i = U\tilde{Q}_\nu y_i.$$

There is one problem. If the deflation strategy suggested by (3.16) is used, the similarity transformation is performed only on the current submatrix. This means that from one step to the next the array A does not contain similar matrices. To ensure that successive iterates are similar, we must apply the transformations to the entire array. In the illustration (3.16) this is equivalent to altering the elements denoted by upper case X's below.

$$\begin{array}{cc|cccc|cc}
x & x & X & X & X & X & x & x \\
x & x & X & X & X & X & x & x \\
\hline
0 & 0 & X & X & X & X & X & X \\
0 & 0 & X & X & X & X & X & X \\
0 & 0 & 0 & X & X & X & X & X \\
0 & 0 & 0 & 0 & X & X & X & X \\
\hline
0 & 0 & 0 & 0 & 0 & 0 & \lambda_{n-1} & x \\
0 & 0 & 0 & 0 & 0 & 0 & 0 & \lambda_n
\end{array}.$$

Algorithm 3.5 can be modified in an obvious way to accomplish this.

Practically, the matrix U may be formed from the information provided by Algorithm 1.1 and stored in an array X. As each plane rotation P is computed, X is postmultiplied by P^{T}, an operation which changes only two columns of X. When the final matrix A_ν is reached X will contain the product $U\tilde{Q}_0$. If the eigenvectors of A_ν are computed in the order $y_n, y_{n-1}, \ldots, y_1$ they may overwrite the corresponding columns of the array A (the diagonal elements of A_ν, which are the approximate eigenvalues, are stored elsewhere). Finally, the product XY, whose columns are eigenvectors of A, can overwrite X.

When A is real and symmetric, the final matrix A_ν is diagonal. Hence $Y = I$, and the eigenvectors of A are just the columns of X.

EXERCISES

1. Give INFL code describing one step of the QR algorithm for a symmetric tridiagonal matrix with shift $\varkappa$.

2. Let $A \in \mathbb{R}^{n \times n}$ be symmetric and tridiagonal. Write an INFL program to find all the eigenvalues and eigenvectors of A by the QR algorithm. Take full advantage of negligable elements. Use the shift strategy described on page 353.

3. Every matrix $A \in \mathbb{R}^{n \times n}$ can be written in the form $A = QL$, where Q is unitary and L is lower triangular (see Exercise 5.3.5). Describe a "QL algorithm" for finding the eigenvalues of a matrix. Where do the eigenvalues appear and in what order? What should be used as a shift? Give INFL code for the shifted algorithm for symmetric tridiagonal matrices.

4. Let $A \in \mathbb{R}^{n \times n}$ be upper Hessenberg and have real eigenvalues. Write an INFL code to reduce A to triangular form by the QR algorithm. Take full advantage of negligable elements.

5. Let $A \in \mathbb{R}^{n \times n}$ be upper triangular with distinct diagonal elements. Write an INFL program to find all the eigenvectors of A (cf. Exercise 6.2.29).

6. Combine Exercises 3.4 and 3.5 to give an INFL algorithm for finding all the eigenvalues and eigenvectors of a real upper Hessenberg matrix with real distinct eigenvalues.

7. Consider one step of the explicitly shifted QR algorithm applied to the matrix A with shift α_{nn}. After premultiplication by the first $n-2$ rotations (cf. page 358) the matrix $A - \alpha_{nn}I$ will have the form

$$\begin{matrix} x & x & x & x & x & x \\ 0 & x & x & x & x & x \\ 0 & 0 & x & x & x & x \\ 0 & 0 & 0 & x & x & x \\ 0 & 0 & 0 & 0 & a & b \\ 0 & 0 & 0 & 0 & \varepsilon & 0 \end{matrix},$$

where ε will be small if the algorithm is converging. Show that after the complete QR step, the $(n, n-1)$-element of A will be given by

$$\frac{-\varepsilon^2 b}{a^2 + \varepsilon^2}.$$

Conclude that the shifted QR algorithm will tend to converge quadratically.

8. In the spirit of Exercise 3.7, show that the QR algorithm applied to a symmetric tridiagonal matrix will tend to converge cubically.

9. Fill in the details of the following proof of Theorem 3.2. From Theorem 3.1 it is known that $A^\nu = \tilde{Q}_\nu \tilde{R}_\nu$. It will be shown that $\tilde{Q}_\nu \to \tilde{Q}$, where $\tilde{Q}$ is the unitary part of the QR factorization of the matrix $X = Y^{-1}$ of right eigenvectors of A. (How will this prove the theorem?) Let $\Lambda = \operatorname{diag}(\lambda_1, \lambda_2, \ldots, \lambda_n)$ and $Y = LU$ where L is unit lower triangular and U is upper triangular. Since $A = X\Lambda Y$,

$$A^\nu = X\Lambda^\nu Y = (X\Lambda^\nu L\Lambda^{-\nu})\Lambda^\nu U.$$

Let $P_\nu S_\nu$ be the QR factorization for $X\Lambda^\nu L\Lambda^\nu$. Then since $A^\nu = P_\nu(S_\nu \Lambda^\nu U)$ and $S_\nu \Lambda^\nu U$ is upper triangular, $P_\nu = \tilde{Q}_\nu$. Show that $\Lambda^\nu L\Lambda^{-\nu} \to I$, and conclude that $\tilde{Q}_\nu \to \tilde{Q}$.

10. Let $A = A_0 \in \mathbb{C}^{n\times n}$. Consider the sequence of matrices A generated as follows. Given A_ν, factor it in the form $A_\nu = L_\nu R_\nu$ where L_ν is unit lower triangular and R_ν is upper triangular (note that this factorization may not exist, in which case the sequence cannot be formed). Set $A_{\nu+1} = R_\nu L_\nu$. Show that $A_{\nu+1}$ is similar to A_ν. Prove an analog of Theorem 3.1. Discuss the relation of this "LR algorithm" to the power method.

11. In Exercise 3.10 suppose that A_0 is upper Hessenberg. Show that all the matrices A_ν are upper Hessenberg.

12. Shifts may be incorporated in the LR algorithm just as in the QR algorithm. Give INFL code to perform one step of the shifted LR algorithm with shift R on the upper Hessenberg matrix A.

13. Show how interchanges for numerical stability may be incorporated in the algorithm of Exercise 3.12. Note that the final result is upper Hessenberg. (The reader is warned that there is no satisfactory theory for this LR algorithm with interchanges, although it works well enough in practice.)

14. Show, in the spirit of Exercise 3.7, that if the last diagonal element is used as a shift in the LR algorithm, the convergence will tend to be quadratic.

4. THE IMPLICITLY SHIFTED QR ALGORITHM

In this section we shall consider a variant of the QR algorithm in which the accelerated convergence associated with an origin shift is achieved without having to subtract the shift from the diagonal and later restore it. The main application of this *implicit shift technique* is to the computation of the complex eigenvalues of a real matrix; however, it will also be used as a basis for the algorithms of the next two sections.

The technique rests on an elementary property of *unreduced* Hessenberg matrices.

DEFINITION 4.1. Let B be an upper Hessenberg matrix of order n. Then B is *unreduced* if $\beta_{i+1,i} \neq 0$ $(i = 1, 2, \ldots, n-1)$.

Thus an unreduced upper Hessenberg matrix is an upper Hessenberg matrix whose subdiagonal elements are all nonzero. The significance of unreduced Hessenberg matrices is that the QR algorithm works exclusively with them; for whenever a subdiagonal element is zero, the QR algorithm may be applied to an unreduced submatrix, as was described in the last section.

The property of unreduced Hessenberg matrices that we require is stated in the following theorem.

THEOREM 4.2. Let $A, B, Q \in \mathbb{C}^{n\times n}$ with Q unitary and B an unreduced upper Hessenberg matrix with positive subdiagonal elements. If $B = Q^{\mathrm{H}}AQ$, then both B and Q are uniquely determined by the first column of Q (or equivalently by the first row of Q^{H}).

PROOF. The proof of the theorem amounts to an algorithm for computing B and Q. Let $Q = (q_1, q_2, \ldots, q_n)$ be partitioned by columns, and suppose that we have already computed $q_1, q_2, \ldots, q_k$ and the first $k-1$ columns of B (this process is well started, since we know q_1). We shall show how to compute the kth column of B and q_{k+1}. From the relation

$$QB = AQ$$

and the fact that B is upper Hessenberg, we obtain

$$\beta_{k+1,k}q_{k+1} + \beta_{k,k}q_k + \cdots + \beta_{1k}q_1 = Aq_k. \tag{4.1}$$

Since Q is unitary, $q_i^{\mathrm{H}}q_j = 0$ except when $i = j$, in which case $q_i^{\mathrm{H}}q_j = 1$.

Hence, upon multiplying (4.1) by q_i^{H}, we obtain

$$\beta_{ik} = q_i^{\mathrm{H}} A q_k \qquad (i = 1, 2, \ldots, k),$$

which determines the kth column of B, except for $\beta_{k+1,k}$. However, since B is unreduced, $\beta_{k+1,k} \neq 0$. Hence q_{k+1} is given by

$$q_{k+1} = \beta_{k+1,k}^{-1}\left(Aq_k - \sum_{i=1}^{k} \beta_{ik} q_i\right),$$

and the requirement that $q_{k+1}^{\mathrm{H}} q_{k+1} = 1$, uniquely determines the positive number $\beta_{k+1,k}$. ■

The requirement that $\beta_{k+1,k}$ be positive in the statement of the theorem was necessary only to tie down the uniqueness completely. Actually $\beta_{k+1,k}$ and q_{k+1} are determined up to a constant factor of absolute value unity simply by the requirement that $\beta_{k+1,k}$ be nonzero. It is this essential uniqueness of B and Q that we shall actually use.

To see how Theorem 4.2 may be applied to the QR algorithm, suppose that one step of the QR algorithm with shift $\varkappa$ has been applied to the unreduced upper Hessenberg matrix A to yield an unreduced upper Hessenberg matrix B. Then

$$B = Q^{\mathrm{H}} A Q$$

where Q is the unitary part of the QR factorization of A. We claim that the following algorithm is an alternative way of computing B.

1) Find a unitary matrix P^{H} whose first column is the same as Q
2) Apply Algorithm 1.1 to reduce PAP^{H} to the upper Hessenberg matrix B'

We must show that $B = B'$. To see this, note that if we set

$$Q'^{\mathrm{H}} = U_{n-2} U_{n-1} \cdots U_1 P,$$

where $U_1, U_2, \ldots, U_{n-2}$ are the transformations of Algorithm 1.1, then $B' = Q' A Q'^{\mathrm{H}}$. But because of their special form, premultiplying a matrix by one of the U_i does not change the first row of the matrix. Thus Q'^{H} has the same first row as P, which by construction has the same first row as Q^{H}. Hence by Theorem 4.2, $Q = Q'$ and $B = B'$.

The above algorithm concentrates the effect of the shift $\varkappa$ in the matrix P, which will change as $\varkappa$ changes. After PAP^{H} has been formed, the rest of the algorithm consists of a routine reduction to Hessenberg form. Before we can regard this proposed algorithm as a working technique, however, we must consider two problems. First, we must show how to determine P. Second, we must refine step 2 in the proposed algorithm to take advantage of the case where A is upper Hessenberg or tridiagonal.

To determine P, we must find the first column of Q. Now

$$A - \varkappa I = QR,$$

where R is upper triangular. Hence, if $Q = (q_1, q_2, \ldots, q_n)$,

$$\varrho_{11} q_1 = QRe_1 = (A - \varkappa I)e_1;$$

in other words, the first column of Q is a multiple of the first column of $A - \varkappa I$. In particular, if we denote this column by $a = (\alpha_{11} - \varkappa, \alpha_{21}, 0, \ldots, 0)^{\mathrm{T}}$ and choose P so that

$$Pa = \pm \| a \|_2 e_1,$$

then

$$P^{\mathrm{H}} e_1 = \pm \frac{a}{\| a \|_2},$$

so that the first column of P^{H} is the same as first column of Q, except for the sign. If A is real and upper Hessenberg, then only the first two elements of a are nonzero, and we may take P to be the plane rotation in the (1, 2)-plane that annihilates the second element in a. Specifically, the constants γ_1, σ_1, and ν_1 that determine P are given by the INFL statement

$$(\gamma_1, \sigma_1, \nu_1) = \mathrm{rot}(\alpha_{11} - \varkappa, \alpha_{21}). \tag{4.2}$$

We shall describe the simplification of step 2 of the proposed algorithm for the case where A is a real symmetric tridiagonal matrix. Suppose A has the form illustrated below for $n = 5$:

$$A = \begin{pmatrix} \alpha_1 & \beta_1 & & & \bigcirc \\ \beta_1 & \alpha_2 & \beta_2 & & \\ & \beta_2 & \alpha_3 & \beta_3 & \\ & & \beta_3 & \alpha_4 & \beta_4 \\ \bigcirc & & & \beta_4 & \alpha_5 \end{pmatrix}.$$

If $P_1 = P$ is determined by (4.2), then it is easily verified that $P_1AP_1^{\mathrm{T}}$ has the form

$$\begin{pmatrix} \alpha_1' & \pi_2 & \varrho_2 & & \bigcirc \\ \pi_2 & \tau_2 & \omega_2 & & \\ \varrho_2 & \omega_2 & \alpha_2 & \beta_3 & \\ & & \beta_3 & \alpha_4 & \beta_4 \\ & \bigcirc & & \beta_4 & \alpha_5 \end{pmatrix},$$

where

$$\begin{aligned} \alpha_1' &= \gamma_1^2\alpha_1 + 2\gamma_1\sigma_1\beta_1 + \sigma_1^2\alpha_2, \\ \pi_2 &= (\gamma_1^2 - \sigma_1^2)\beta_1 + \gamma_1\sigma_1(\alpha_2 - \alpha_1), \\ \tau_2 &= \sigma_1^2\alpha_1 - 2\gamma_1\sigma_1\beta_1 + \gamma^2\alpha_2, \\ \varrho_2 &= \sigma_1\beta_2, \\ \omega_2 &= \gamma_1\beta_2. \end{aligned}$$

To reduce $P_1AP_1^{\mathrm{T}}$ to tridiagonal form, we first choose a plane rotation in the (2, 3)-plane to introduce a zero into the position occupied by ϱ_2. The numbers γ_2 and σ_2 appearing in P_2 are given by the INFL statement

$$(\gamma_2, \sigma_2, \nu_2) = \mathrm{rot}(\pi_2, \varrho_2).$$

The matrix $P_2P_1AP_1^{\mathrm{T}}P_2^{\mathrm{T}}$ will then have the form

$$\begin{pmatrix} \alpha_1' & \beta_1' & & & \bigcirc \\ \beta_1' & \alpha_2' & \pi_3 & \varrho_3 & \\ & \pi_3 & \tau_3 & \omega_3 & \\ & \varrho_3 & \omega_3 & \alpha_4 & \beta_4 \\ \bigcirc & & & \beta_4 & \alpha_5 \end{pmatrix},$$

where

$$\begin{aligned} \beta_1' &= \nu_2, \\ \alpha_2' &= \gamma_2^2\tau_2 + 2\gamma_2\sigma_2\omega_2 + \sigma_2^2\alpha_3, \\ \pi_3 &= (\gamma_2^2 - \sigma_2^2)\omega_2 + \gamma_2\sigma_2(\alpha_3 - \tau_2), \\ \tau_3 &= \gamma_2^2\tau_2 - 2\gamma_2\sigma_2\omega_2 + \gamma_2^2\alpha_3, \\ \varrho_3 &= \sigma_2\beta_3, \\ \omega_3 &= \gamma_2\beta_3. \end{aligned} \tag{4.3}$$

The process may now be repeated. A rotation P_3 in the (3, 4)-plane is chosen to annihilate ϱ_3. Then $P_3P_2P_1AP_1^{\mathrm{T}}P_2^{\mathrm{T}}P_3^{\mathrm{T}}$ has the form

$$\begin{pmatrix} \alpha_1' & \beta_1' & & & \bigcirc \\ \beta_1' & \alpha_2' & \beta_2' & & \\ & \beta_2' & \alpha_3' & \pi_4 & \varrho_4 \\ & & \pi_4 & \tau_4 & \omega_4 \\ \bigcirc & & \varrho_4 & \omega_4 & \alpha_5 \end{pmatrix}. \tag{4.4}$$

The elements in (4.4) are given by (4.3) in which every subscript is increased by 1. An application of a rotation in the (4, 5)-plane to annihilate ϱ_4 completes the job.

Obviously the reduction described above is completely general. At the kth stage, the matrix $P_{k-1} \cdots P_1AP_1 \cdots P_{k-1}^{\mathrm{T}}$ is tridiagonal, except for the element ϱ_k in the $(k+1, k-1)$-position. This element is "chased" from the $(k+1, k-1)$-position to the $(k+2, k)$-position by a rotation in the $(k, k+1)$-plane. The elements of $P_k \cdots P_1AP_1^{\mathrm{T}} \cdots P_k$ are given by (4.3), in which every subscript is increased by $k-2$. In fact, if we set

$$\begin{aligned} \pi_1 &= \alpha_1 - \varkappa, \\ \varrho_1 &= \beta_1, \\ \tau_1 &= \alpha_1, \\ \omega_1 &= \beta_1, \end{aligned}$$

then the above formulas serve also for the first step.

These considerations are summarized in the following algorithm.

ALGORITHM 4.3. Let $A \in \mathbb{R}^{n \times n}$ be symmetric and tridiagonal with diagonal elements $\alpha_1, \alpha_2, \ldots, \alpha_n$ and subdiagonal elements $\beta_1, \beta_2, \ldots, \beta_{n-1}$. This algorithm overwrites α_i and β_i with the corresponding elements of $A' = Q^{\mathrm{T}}AQ$, where Q is the QR transformation associated with the shift $\varkappa$.

1) $\pi = \alpha_1 - \varkappa$

2) $\varrho = \beta_1$

3) $\tau = \alpha_1$

4) $\omega = \beta_1$

5) For $k = 1, 2, \ldots, n-1$

   1) $(\gamma, \sigma, \nu) = \text{rot}(\pi, \varrho)$
   2) If $k \neq 1$, $\beta_{k-1} \leftarrow \beta'_{k-1} = \nu$
   3) $\alpha_k \leftarrow \alpha_k{}' = \gamma^2\tau + 2\gamma\sigma\omega + \sigma^2\alpha_{k+1}$
   4) $\pi \leftarrow (\gamma^2 - \sigma^2)\omega + \gamma\sigma(\alpha_{k+1} - \tau)$
   5) $\tau \leftarrow \sigma^2\tau - 2\gamma\sigma\omega + \gamma^2\alpha_{k+1}$
   6) $\varrho \leftarrow \sigma\beta_{k+1}$
   7) $\omega \leftarrow \gamma\beta_{k+1}$

6) $\beta_{n-1} \leftarrow \beta'_{n-1} = \pi$

7) $\alpha_n \leftarrow \alpha_n{}' = \tau$

In the above algorithm we have dropped the subscripts on the intermediate quantities. The computations in statement 5 may be arranged more economically, at the cost of obscuring the relation of the algorithm to its derivation. However the computations are arranged, though, the algorithm will require only a multiple of n operations. The algorithm is stable; the computed matrix A' is similar to $A + E$, where

$$\| E \|_{\mathrm{F}} \leq \phi(n) \| A \|_{\mathrm{F}} 10^{-t}. \tag{4.5}$$

Here t is the number of digits carried in the computations, and ϕ is a slowly growing function of n. This bound differs from the bound (3.20) for the explicitly shifted algorithm by the absence of the term involving $|\varkappa|$. Otherwise put, the size of the shift has no influence on the stability of Algorithm 4.3.

Algorithm 4.3 will of course be applied iteratively to determine all the eigenvalues of A. As with the explicitly shifted algorithm, the shift $\varkappa$ is chosen as the eigenvalue of

$$\begin{pmatrix} \alpha_{n-1} & \beta_{n-1} \\ \beta_{n-1} & \alpha_n \end{pmatrix}$$

that lies nearest α_n. The element β_{n-1} will always converge to zero, usually cubically. When any β_i becomes effectively zero, the problem may be reduced in the usual way. Eigenvectors may be found by accumulating the products of the transformations. In general, algorithms based on Algorithm 1.2 and Algorithm 4.3 (or the explicitly shifted form) are the most efficient means of computing all the eigenvalues and eigenvectors of a real symmetric matrix.

We shall next describe how the implicit shift technique may be used to accomplish two shifted QR steps in one pass. Suppose that we have applied one QR step with shift $\varkappa_0$ to the matrix A to obtain the matrix

$$A_1 = Q_0^{\mathrm{H}} A Q_0,$$

and suppose a second QR step with shift $\varkappa_1$ is performed on the matrix A_1 to obtain the matrix

$$A_2 = Q_1^{\mathrm{H}} A_1 Q_1 = Q_1^{\mathrm{H}} Q_0^{\mathrm{H}} A Q_0 Q_1.$$

Consider the following algorithm:

1) Find a unitary matrix U with the same first row as $Q_1^{\mathrm{H}} Q_0^{\mathrm{H}}$
2) Apply Algorithm 1.1 to reduce UAU^{H} to an upper Hessenberg matrix A'

The same reasoning that we used above in connection with a single shift shows that the unitary matrix $U_{n-2}U_{n-3} \cdots U_1 U$, which transforms A into A', same first row as $Q_1^{\mathrm{H}} Q_0^{\mathrm{H}}$. Hence by Theorem 4.2, $U_{n-2} \cdots U_1 U = Q_1^{\mathrm{H}} Q_0^{\mathrm{H}}$ and $A' = A_2$. Thus the above algorithm accomplishes the transformation of A into A_2 without explicitly using the shifts $\varkappa_0$ and $\varkappa_1$.

As in the development of the single shift algorithm, we must first show how to determine U and then we must simplify the reduction to Hessenberg form. To obtain U, let R_0 and R_1 be the upper triangular parts of the QR factorization of A and A_1. Then by Theorem 3.1,

$$Q_1 Q_0 R_0 R_1 = (A - \varkappa_0 I)(A - \varkappa_1 I).$$

Since R_0 and R_1 are upper triangular the first column of $Q_1 Q_0$ is a multiple of the first column of $(A - \varkappa_0 I)(A - \varkappa_1 I)$. Hence we may take U to be the elementary reflector that reduces the first column of $(A - \varkappa_0 I)(A - \varkappa_1 I)$ to a multiple of e_1.

If A is upper Hessenberg, only the first three elements of the first column of $(A - \varkappa_0 I)(A - \varkappa_1 I)$ are nonzero. These elements are

$$\begin{aligned} \alpha_{10} &= \alpha_{11}^2 - (\varkappa_0 + \varkappa_1)\alpha_{11} + \varkappa_0\varkappa_1 + \alpha_{12}\alpha_{21}, \\ \alpha_{20} &= \alpha_{21}[\alpha_{11} + \alpha_{22} - (\varkappa_0 + \varkappa_1)] \\ \alpha_{30} &= \alpha_{21}\alpha_{32}. \end{aligned}$$

Now the conventional choice of shifts in the doubly shifted algorithm is to take $\varkappa_0$ and $\varkappa_1$ as the eigenvalues of the matrix

$$\begin{pmatrix} \alpha_{n-1,n-1} & \alpha_{n-1,n} \\ \alpha_{n,n-1} & \alpha_{nn} \end{pmatrix}. \tag{4.6}$$

By examining the characteristic polynomial of (4.6), it can be verified that

$$\varkappa_0 + \varkappa_1 = \alpha_{n-1,n-1} + \alpha_{nn},$$

and

$$\varkappa_0 \varkappa_1 = \alpha_{nn}\alpha_{n-1,n-1} - \alpha_{n-1,n}\alpha_{n,n-1}.$$

Using these values, we can verify that

$$\begin{aligned} \alpha_{10} &= \alpha_{21}\left(\frac{(\alpha_{nn} - \alpha_{11})(\alpha_{n-1,n-1} - \alpha_{11}) - \alpha_{n,n-1}\alpha_{n-1,n}}{\alpha_{21}} + \alpha_{12}\right), \\ \alpha_{20} &= \alpha_{21}[\alpha_{22} - \alpha_{11} - (\alpha_{nn} - \alpha_{11}) - (\alpha_{n-1,n-1} - \alpha_{11})], \\ \alpha_{30} &= \alpha_{21}\alpha_{32}. \end{aligned} \tag{4.7}$$

Since we are interested only in the direction of the first column of $(A - \varkappa_0 I)(A - \varkappa_1 I)$, the common factor α_{21} in (4.7) may be ignored in the determination of U.

When A is real, the numbers α_{10}, α_{20}, and α_{30} will also be real. It follows that U is an orthogonal matrix and that the subsequent reduction of UAU^{H} can be accomplished in real arithmetic. Of course nothing much is saved by this double shift technique if $\varkappa_0$ and $\varkappa_1$ are real. However, the eigenvalues of (4.6) may be complex conjugates; that is $\varkappa_0 = \bar{\varkappa}_1$. In this case the double shift technique accomplishes in real arithmetic the effect of performing two single shift steps, one with $\varkappa_0$ and the other with $\bar{\varkappa}_0$. Since performed separately each of these QR steps would have to be done in complex arithmetic, the double shift technique represents a substantial savings in computation.

The reduction of UAU^{T} to Hessenberg form is done in much the same way as it was done for tridiagonal matrices. Because only the first three elements of the first column of $(A - \varkappa_0 I)(A - \varkappa_1 I)$ are nonzero, U has the form

$$U = \begin{pmatrix} R & 0 \\ 0 & I_{n-3} \end{pmatrix},$$

where R is an elementary reflector of order 3. It follows that UAU has the form illustrated below for $n = 6$.

$$\begin{array}{cccccc} x & x & x & x & x & x \\ x & x & x & x & x & x \\ \boxed{x} & x & x & x & x & x \\ \boxed{x} & x & x & x & x & x \\ 0 & 0 & 0 & x & x & x \\ 0 & 0 & 0 & 0 & x & x \end{array}. \tag{4.8}$$

To reduce UAU to Hessenberg form, choose an elementary reflector U_2 of the form

$$U_2 = \begin{pmatrix} 1 & 0 & 0 \\ 0 & R_2 & 0 \\ 0 & 0 & I_{n-4} \end{pmatrix}$$

such that the elements of U_2UAU distinguished in (4.8) are zero. Then U_2UAUU_2 will have the form

$$\begin{array}{cccccc} x & x & x & x & x & x \\ x & x & x & x & x & x \\ 0 & x & x & x & x & x \\ 0 & \boxed{x} & x & x & x & x \\ 0 & \boxed{x} & x & x & x & x \\ 0 & 0 & 0 & 0 & x & x \end{array}. \tag{4.9}$$

The process is continued by choosing U_3 to annihilate the distinguished elements in (4.9), and so on. At each stage the pattern of three nonzero elements below the subdiagonal is chased one position further toward the lower right-hand corner of the array. After the last step, the matrix has the form

$$\begin{array}{cccccc} x & x & x & x & x & x \\ x & x & x & x & x & x \\ 0 & x & x & x & x & x \\ 0 & 0 & x & x & x & x \\ 0 & 0 & 0 & x & x & x \\ 0 & 0 & 0 & \boxed{x} & x & x \end{array}.$$

A rotation in the $(n-1, n)$-plane may then be used to eliminate the last nonzero element below the subdiagonal.

Some additional savings in the reduction may be achieved by taking advantage of the special form of the elementary reflectors U_k. At the kth step, U_k has the form

$$U_k = \begin{pmatrix} I_{k-1} & 0 & 0 \\ 0 & R_k & 0 \\ 0 & 0 & I_{n-k-2} \end{pmatrix},$$

where R_k is the elementary reflector that annihilates the last two elements of $(\alpha_{k,k-1}, \alpha_{k+1,k-1}, \alpha_{k+2,k-1})^{\mathrm{T}}$. Rather than representing R in the form $I - \pi^{-1}uu^{\mathrm{T}}$, it is more convenient to take R in the form $I - vw^{\mathrm{T}}$, where the first component of w is unity. This amounts to taking $w = u/v_1$ and $v = v_1 u/\pi = u/\sigma$, where u, π, and σ are the output of Algorithm 5.3.5. Hence we have the following algorithm for computing R, in which we scale to avoid underflows and overflows.

ALGORITHM 4.4. Given the vector $(\alpha_1, \alpha_2, \alpha_3)^{\mathrm{T}}$, this algorithm produces vectors v, w, $\in \mathbb{R}^3$ and a scalar σ such that $\omega_1 = 1$, $I - vw^{\mathrm{T}}$ is an elementary reflector, and

$$(I - vw^{\mathrm{T}})\begin{pmatrix} \alpha_1 \\ \alpha_2 \\ \alpha_3 \end{pmatrix} = \begin{pmatrix} -\sigma \\ 0 \\ 0 \end{pmatrix}.$$

The algorithm may be referenced by the INFL statement

$$(v, w, \sigma) = \mathrm{reflect}_3(\alpha_1, \alpha_2, \alpha_3)$$

1) $\eta = \max\{|\alpha_1|, |\alpha_2|, |\alpha_3|\}$
2) If $\eta = 0$
   1) $v = 0;\ w = 0;\ \sigma = 0;$ quit
3) $v = (\alpha_1, \alpha_2, \alpha_3)^{\mathrm{T}}/\eta$
4) $\sigma = \mathrm{sign}(v_1)\sqrt{v_1^2 + v_2^2 + v_3^2}$
5) $v_1 \leftarrow v_1 + \sigma$
6) $w = v/v_1$
7) $v \leftarrow v/\sigma$
8) $\sigma \leftarrow \eta\sigma$

We are now in a position to give an INFL description of the implicit double shift QR step.

ALGORITHM 4.5. Let $A \in \mathbb{R}^{n\times n}$ be upper Hessenberg. This algorithm overwrites A with $Q^{\mathrm{T}}AQ$, where Q is the transformation corresponding to two consecutive QR steps, the first with shift $\varkappa_0$ and the second with shift $\varkappa_1$. The shifts are the eigenvalues of the trailing 2×2 submatrix of A.

1) Compute α_{10}, α_{20}, and α_{30} from (4.7), ignoring the common factor α_{21}

2) For $k = 0, 1, 2, \ldots, n-3$
   1) $(v, w, \sigma) = \text{reflect}_3(\alpha_{k+1,k}, \alpha_{k+2,k}, \alpha_{k+3,k})$
   2) If $k \neq 0$
      1) $\alpha_{k+1,k} = -\sigma$
      2) $\alpha_{k+2,k} = 0$
      3) $\alpha_{k+3,k} = 0$
   3) Premultiply
      1) For $j = k+1, k+2, \ldots, n$
         1) $\tau = \alpha_{k+1,j} + \omega_2\alpha_{k+2,j} + \omega_3\alpha_{k+3,j}$
         2) $(\alpha_{k+1,j}, \alpha_{k+3,j})^{\mathrm{T}} \leftarrow (\alpha_{k+1,j}, \alpha_{k+2,j}, \alpha_{k+3,j})^{\mathrm{T}} - \tau(\nu_1, \nu_2, \nu_3)^{\mathrm{T}}$
   4) Postmultiply
      1) For $i = 1, 2, \ldots, \min\{k+4, n\}$
         1) $\tau = \alpha_{i,k+1} + \omega_2\alpha_{i,k+2} + \omega_3\alpha_{i,k+3}$
         2) $(\alpha_{i,k+1}, \alpha_{i,k+2}, \alpha_{i,k+3}) \leftarrow (\alpha_{i,k+1}, \alpha_{i,k+2}, \alpha_{i,k+3}) - \tau(\nu_1, \nu_2, \nu_3)$

3) $(\gamma, \sigma, \nu) = \text{rot}(\alpha_{n-1,n-2}, \alpha_{n,n-2})$

4) $\alpha_{n-1,n-2} = \nu$

5) $\alpha_{n,n-2} = 0$

6) $\begin{pmatrix}\alpha_{n-1,j}\\ \alpha_{nj}\end{pmatrix} \leftarrow \begin{pmatrix}\gamma & \sigma\\ -\sigma & \gamma\end{pmatrix}\begin{pmatrix}\alpha_{n-1,j}\\ \alpha_{nj}\end{pmatrix} \qquad (j = n-1, n)$

7) $(\alpha_{i,n-1}, \alpha_{in}) \leftarrow (\alpha_{i,n-1}, \alpha_{in})\begin{pmatrix}\gamma & -\sigma\\ \sigma & \gamma\end{pmatrix} \qquad (i = 1, 2, \ldots, n)$

Algorithm 4.5 requires about $5n^2$ multiplications. It is stable; the matrix produced by it is exactly similar to the original matrix plus an error that satisfies (4.5).

Algorithm 4.5 can be applied iteratively to find all the eigenvalues of a matrix. The comments made in connection with Algorithm 3.5 apply here. When a subdiagonal element of A becomes effectively zero, the problem is deflated in the usual way. Eigenvectors may be found by accumulating the product of the transformations and using this product to transform back the eigenvectors of the final matrix, which are easily computed.

We have noted that with the shift strategy $\varkappa_\nu = \alpha_{nn}^{(\nu)}$, a method based on Algorithm 3.5 cannot converge to a complex eigenvalue. No such problem exists for Algorithm 4.5, since the two shifts implicitly effected by the algorithm may be complex conjugates. Of course since the elements of the matrices A_ν are all real, the element $\alpha_{nn}^{(\nu)}$ cannot converge to a complex eigenvalue. Instead, what happens is that $\alpha_{n-1,n-2}^{(\nu)}$ converges quadratically to zero. After a time, the problem may be deflated, leaving a 2×2 trailing submatrix whose eigenvalues approximate a pair of complex conjugate eigenvalues of the original matrix. The final matrix obtained in this way will be in quasi-triangular form rather than triangular form.

We cannot guarantee that Algorithm 4.5 applied iteratively will always converge. In fact the matrix

$$\begin{pmatrix} 0 & 0 & 0 & 0 & 1 \\ 1 & 0 & 0 & 0 & 0 \\ 0 & 1 & 0 & 0 & 0 \\ 0 & 0 & 1 & 0 & 0 \\ 0 & 0 & 0 & 1 & 0 \end{pmatrix}$$

is left unchanged by Algorithm 4.5, and hence none of its subdiagonal elements can converge to zero. In practice, when the method fails to converge, one makes a single application of Algorithm 4.3 with a pair of randomly chosen shifts and thereafter proceeds as usual. This *ad hoc* shift is usually sufficient to destroy any special relation among the elements of A that may prevent convergence.

EXERCISES

1. The proof of Theorem 4.2 gives an algorithm for computing Q and B. Show that numerically it can yield a matrix Q whose columns are far from orthogonal. [*Hint*: Suppose some $\beta_{i+1,i}$ is very small.]

2. Let $A \in \mathbb{R}^{2\times 2}$ have real eigenvalues. Find a plane rotation R such that RAR^{T} is upper triangular, and give INFL code for accomplishing the reduction. [*Hint*: Find an eigenvalue λ of A (Exercise 6.2.17) and choose R so that the (2, 1)-element of $R(A - \lambda I)$ is zero. For numerical stability λ should be taken to be the eigenvalue that makes the (1, 1)-element of $A - \lambda I$ the largest in magnitude (why?).]
3. Let $A \in \mathbb{R}^{n\times n}$ be quasi-triangular. Write an INFL program to reduce A to a quasi-triangular form in which the 2×2 blocks have only complex eigenvalues.
4. Let $A \in \mathbb{R}^{n\times n}$ be quasi-triangular and let the ith diagonal element belong to a 1×1 block (that is, α_{ii} is an eigenvalue). Show how an eigenvector corresponding to α_{ii} may be computed, and give INFL code implementing the algorithm. [*Hint*: The computation proceeds as in Exercise 3.4, except that each 2×2 block generates a system of two equations for two components of the eigenvector.]
5. Let $A \in \mathbb{R}^{n\times n}$ be quasi-triangular, and suppose the 2×2 block with diagonal elements $\alpha_{i-1,i-1}$ and α_{ii} has a pair of complex conjugate eigenvalues λ and $\bar{\lambda}$. Show how to compute the corresponding eigenvectors and give INFL code implementing the computation. [*Hint*: It is only necessary to determine one eigenvector x, since the other is $\bar{x}$. Set $\xi_i = 1$ and find an equation for ξ_{i-1}. Thereafter the reduction proceeds as in Exercise 4, except that the equations are complex.]

NOTES AND REFERENCES

The QR algorithm has its origin in Rutishauser's *qd* algorithm. Rutishauser (1958) observed that the operations of the progressive form of the *qd* algorithm could be interpreted as factoring a tridiagonal matrix A into a product LR of lower and upper triangular matrices and forming the product RL in reverse order. He called the natural generalization of this procedure to full matrices the LR algorithm and gave the first proof of the analog of our Theorem 3.2. Since the LR algorithm is based on the LU decomposition, the practical algorithm requires pivoting for stability, which complicates the theory considerably.

The QR algorithm, which is simply an orthogonal variant of the LR algorithm, was discovered independently by Francis (1961, 1962) and Kublanovskaya (1961). Francis' paper, a model of practical numerical analysis, presents the complete method, including the double shift strategy and the economization techniques. In addition he describes an important technique for taking advantage of two consecutive subdiagonal elements whose *product* is negligible. A complete treatment of both the LR and QR algorithms is given by Wilkinson (AEP, Chapter 8).

The relation between the QR method and the power method as exhibited in Theorem 3.1 is the basis for proofs of Theorem 3.2 and its variants. Early proofs of Theorem 3.2 were based on determinantal arguments [see, e.g., Householder (1964)]; however, Wilkinson (1965c; AEP, Chapter 9) has given a proof cast entirely in terms of matrices (cf. Exercise 3.9). The first published discussion of the relation between the QR algorithm and the Rayleigh quotient iteration is by Kahan and Parlett (1968), who exploit it to establish the convergence properties of the shifted algorithm. Wilkinson (1968) also discusses the convergence of the shifted algorithm for symmetric matrices.

The QR algorithm for symmetric tridiagonal matrices has proved to be a tinkerer's playground. The algebraic nature of the formulas invites rearrangement and tempts one to add improvements or remedy shortcomings, real and imagined. Variants have been published by Ortega and Kaiser (1963) [this one is unstable; see Welsch (1967)], Reinsch and Bauer (1968, HACLA/II/6), Stewart (1970a), and Boothroyd (1970).

A program for the doubly shifted QR algorithm for real Hessenberg matrices has been published by Martin, Peters, and Wilkinson (1970, HACLA/II/14); and LR program for complex matrices by Martin and Wilkinson (1968, HACLA/II/16). Peters and Wilkinson have modified these programs to find eigenvectors (1970a, HACLA/II/15). For symmetric tridiagonal matrices an explicitly shifted QR program has been published by Bowdler, Martin, Reinsch, and Wilkinson (1968, HACLA/II/3) and an implicitly shifted QR program by Dubrulle, Martin, and Wilkinson (1968, HACLA/II/4). Stewart (1970b) gives a FORTRAN program for the complete solution of the real symmetric eigenvalue problem.

5. COMPUTING SINGULAR VALUES AND VECTORS

In this section and the next we shall apply the QR algorithm to the solution of two important problems. In this section we shall describe an algorithm for computing the singular values and vectors of a matrix A. In the next section we shall consider the problem of finding "generalized eigenvalues" λ for which the system of order n

$$Ax = \lambda Bx \tag{5.1}$$

has a nontrivial solution. Since our chief purpose is to illustrate the power of QR techniques, we shall confine ourselves to sketches of the algorithms. The reader who is interested in the details may consult the literature, where detailed descriptions and program listings may be found.

Both the singular value problem and the generalized eigenvalue problem may be reduced to an ordinary eigenvalue problem. The squares of the singular values of A are the eigenvalues of the real symmetric matrix $A^{\mathrm{T}}A$. Similarly if B is nonsingular, the generalized eigenvalues of (5.1) are the eigenvalues of the matrix $B^{-1}A$. This suggests that one compute, say, the singular values of A by finding the eigenvalues of $A^{\mathrm{T}}A$. However, this is inadvisable for the following reason. If A is rounded to t figures, it follows from Theorem 6.6.6 that a singular value of A of size $10^{-p} \| A \|_2$ will be affected in at worst its $t - p$ digit. On the other hand, this same singular value corresponds to an eigenvalue of $A^{\mathrm{T}}A$ of size $10^{-2p} \| A \|_2^2$. If $A^{\mathrm{T}}A$ is rounded to t figures, this eigenvalue will be determined to $t - 2p$ figures. In particular, if $2p > t$, we can expect no accuracy in the computed eigenvalue of $A^{\mathrm{T}}A$, even though the corresponding singular value may have several accurate figures.

EXAMPLE 5.1. The matrix

$$A = \begin{pmatrix} 1.005 & 0.995 \\ 0.995 & 1.005 \end{pmatrix}$$

has singular values 2 and 0.01. On the other hand the matrix $A^{\mathrm{T}}A$, rounded to four figures, is

$$\begin{pmatrix} 2.000 & 2.000 \\ 2.000 & 2.000 \end{pmatrix}$$

whose eigenvalues are 4 and 0. Obviously all information about the smallest singular value of A has been lost in passing to the rounded $A^{\mathrm{T}}A$.

Likewise the attempt to solve the generalized eigenvalue problem by finding the eigenvalues of $B^{-1}A$ may fail when B is nearly singular.

To circumvent these difficulties we shall take the following approach to both problems.

1. Find a class of unitary transformations that leave the answers to the problems unchanged.
2. Use these transformations to reduce the problem to a compact form.

3. By considering the implicitly shifted QR algorithm for the associated eigenvalue problem (i.e., $A^{\mathrm{T}}A$ or $B^{-1}A$), derive an iteration that uses the transformations of part 1 to reduce the results of part 2 to a simple form from which the answers may be read off.

Since at no stage of this process do we actually work with the associated eigenvalue problem, we may hope to obtain accurate answers even when the associated eigenvalue problem would give inaccurate answers.

We shall now show how this process may be applied to the computation of the singular values of a matrix A. For definiteness we shall assume that $A \in \mathbb{R}^{m \times n}$, with $m > n$. The first step is to find a class of transformations which leave the singular values of A unaltered. Such a class is described in the following theorem, whose proof is obvious from the definition of singular value and singular vector.

THEOREM 5.2. Let $A \in \mathbb{R}^{m \times n}$ and let $P \in \mathbb{R}^{m \times m}$ and $Q \in \mathbb{R}^{m \times n}$ be unitary. Then the singular values of A and PAQ are the same. If u is a right (left) singular vector of A, then $Q^{\mathrm{H}}u$ (Pu) is a right (left) singular vector of PAQ.

Two matrices A and B related by the equation $B = PAQ$, where P and Q are nonsingular, are said to be *equivalent* matrices. If P and Q are unitary, then A and B are *unitarily equivalent*. In this terminology, Theorem 5.2 states that unitarily equivalent matrices have the same singular values.

The second step in our program is to use unitary equivalence to reduce A to a condensed form. For the singular value decomposition we shall reduce A to an upper bidiagonal form illustrated below for $m = 6$ and $n = 5$:

$$\begin{matrix} x & x & 0 & 0 & 0 \\ 0 & x & x & 0 & 0 \\ 0 & 0 & x & x & 0 \\ 0 & 0 & 0 & x & x \\ 0 & 0 & 0 & 0 & x \\ 0 & 0 & 0 & 0 & 0 \end{matrix}.$$

The reduction proceeds as follows. First an elementary reflector P_1 is chosen to introduce zeros into the positions distinguished in the illustration below:

$$\begin{matrix} x & x & x & x & x \\ \textcircled{x} & x & x & x & x \\ \textcircled{x} & x & x & x & x \\ \textcircled{x} & x & x & x & x \\ \textcircled{x} & x & x & x & x \\ \textcircled{x} & x & x & x & x \end{matrix} .$$

Then P_1A has the form

$$\begin{matrix} x & x & \textcircled{x} & \textcircled{x} & \textcircled{x} \\ 0 & x & x & x & x \\ 0 & x & x & x & x \\ 0 & x & x & x & x \\ 0 & x & x & x & x \\ 0 & x & x & x & x \end{matrix} . \tag{5.2}$$

Next an elementary reflector Q_1 is chosen so that P_1AQ_1 has zeros in the positions distinguished in (5.2). The matrix Q_1 can be chosen so that it does not change the first column of any matrix it postmultiplies. Then P_1AQ_1 has the form

$$\left(\begin{array}{c|cccc} x & x & 0 & 0 & 0 \\ \hline 0 & x & x & x & x \\ 0 & x & x & x & x \\ 0 & x & x & x & x \\ 0 & x & x & x & x \\ 0 & x & x & x & x \end{array}\right) . \tag{5.3}$$

The process is continued by working with the trailing submatrix in (5.3). For $m > n$, the reduction terminates with a matrix

$$P_nP_{n-1} \cdots P_1AQ_1 \cdots Q_{n-3}Q_{n-2}$$

in upper bidiagonal form. By construction, the matrix $Q_1Q_2 \cdots Q_{n-2}$ does not alter the first column of any matrix it postmultiplies.

The third step in our program is to determine an iterative method for reducing the bidiagonal matrix A to diagonal form. Since the bidiagonal form is zero below the nth row, we may assume that A is square of order n. To derive the iteration, suppose that one step of the QR method with shift x

applied to the tridiagonal matrix A^TA yields the orthogonal transformation Q. Consider the following algorithm.

1) Find a matrix Q_0 whose first column is the same as Q
2) Use the technique described above to reduce AQ_0 to bidiagonal form, yielding a matrix A'

Now $Q' = Q_0Q_1 \cdots Q_{n-2}$, where $Q_1, Q_2, \ldots, Q_n$ are the matrices obtained from the reduction of AQ_0 to bidiagonal form. Then Q' has the same first column as Q. Moreover

$$A'^TA' = Q'^TA^T(P_{n-1} \cdots P_1)^T(P_{n-1} \cdots P_1)AQ'' = Q'^T(A^TA)Q'.$$

Since both A'^TA' and Q^TA^TAQ are tridiagonal, it follows from Theorem 4.2 that $Q = Q'$ and $A'^TA' = Q^TA^TAQ$. In other words A'^TA' is the tridiagonal matrix that would result from applying one QR step with shift $\varkappa$ to the tridiagonal matrix A^TA; however, in the algorithm suggested above we need never form A^TA.

Of course this tentative algorithm will be complete only when we specify how to compute Q_0 and how to simplify the reduction to bidiagonal form. To compute Q_0, recall from the discussion of the implicit shift technique for symmetric tridiagonal matrices that the first column of A is the same as the first column of R_{12}^T, where R_{12} is the rotation in the (1, 2)-plane that annihilates the (2, 1)-element of $A^TA - \varkappa I$. But the first column of $A^TA - \varkappa I$ is given by

$$(\alpha_{11}^2 - \varkappa, \alpha_{11}\alpha_{12}, 0, \ldots, 0)^T.$$

Hence the numbers γ and σ determining $Q_0 = R_{12}^T$ are given by the INFL statement

$$(\gamma, \sigma, \nu) = \text{rot}(\alpha_1{}^2 - \varkappa, \alpha_{11}\alpha_{12}).$$

The reduction of AQ_0 to bidiagonal form can be simplified considerably. Because A is bidiagonal and $Q_0{}^T$ is a rotation in the (1, 2)-plane, AQ_0 has the form

$$\begin{matrix} x & x & 0 & 0 & 0 \\ x & x & x & 0 & 0 \\ 0 & 0 & x & x & 0 \\ 0 & 0 & 0 & x & x \\ 0 & 0 & 0 & 0 & 0 \end{matrix}.$$

Let the matrix P_1 be the rotation in the (1, 2)-plane that annihilates the element distinguished above. Then P_1AQ_0 then has the form

$$\begin{matrix} x & x & \textcircled{x} & 0 & 0 \\ 0 & x & x & 0 & 0 \\ 0 & 0 & x & x & 0 \\ 0 & 0 & 0 & x & x \\ 0 & 0 & 0 & 0 & x \end{matrix} . \tag{5.4}$$

If Q_1 is chosen to be a rotation in the (2, 3)-plane that annihilates the element distinguished in (5.4), then $P_1AQ_0Q_1$ has the form

$$\begin{array}{c|cccc} x & x & 0 & 0 & 0 \\ \hline 0 & x & x & 0 & 0 \\ 0 & x & x & x & 0 \\ 0 & 0 & 0 & x & x \\ 0 & 0 & 0 & 0 & x \end{array} ,$$

whose trailing submatrix has the same form as AQ_0.

More generally, the reduction proceeds by using plane rotations to annihilate elements in the order shown below.

$$\begin{matrix} x & x & z^2 & 0 & 0 & 0 \\ z^1 & x & x & z^4 & 0 & 0 \\ 0 & z^3 & x & x & z^6 & 0 \\ 0 & 0 & z^5 & x & x & z^8 \\ 0 & 0 & 0 & z^7 & x & x \\ 0 & 0 & 0 & 0 & z^9 & x \end{matrix} .$$

The annihilation of one element results in the introduction of the next. The odd numbered elements are annihilated by a premultiplication, the even by a postmultiplication.

The above considerations may be combined to give a general algorithm for computing the singular values of a matrix A. First unitary matrices P_0 and P_0 are determined so that

$$A_1 = P_0AQ_0$$

is bidiagonal. Then, with shifts $\varkappa_1, \varkappa_2, \varkappa_3, \ldots,$ the bidiagonal matrices

$A_1, A_2, \ldots$ are determined as described above so that $A_{\nu+1}^{\mathrm{T}} A_{\nu+1}$ is the symmetric tridiagonal matrix that would be obtained by applying one step of the QR algorithm with shift $\varkappa_\nu$ to the matrix $A_\nu{}^{\mathrm{T}} A_\nu$. If the shifts are appropriately chosen, $A_\nu{}^{\mathrm{T}} A_\nu$ will converge to a diagonal matrix; and hence so will A_ν (except perhaps for the signs of the diagonal elements, which may change). The diagonal elements of the limit will be the singular values of A.

The above description leaves much unsaid. In practice one must compute the shifts $\varkappa_\nu$ and take advantage of the emergence of effectively zero elements in the matrix A_ν. Singular vectors may be found by accumulating the products of the transformations P_ν and Q_ν. These details, along with examples of various applications, are discussed in the literature.

NOTES AND REFERENCES

The reduction to bidiagonal form is due to Golub and Kahan (1965). The variant of the QR algorithm to compute singular values is due to Golub. An ALGOL program has been published by Golub and Reinsch (1970, HACLA/I/10), a FORTRAN program by Businger and Golub (1969).

6. THE GENERALIZED EIGENVALUE PROBLEM $A - \lambda B$

Let A and B be matrices of order n. The generalized eigenvalue problem $A - \lambda B$ is the problem of determining all values of λ for which the equation

$$Ax = \lambda Bx \tag{6.1}$$

has a nontrivial solution. Such numbers λ are called eigenvalues of the problem $A - \lambda B$. The solution x of (6.1) is an eigenvector of the problem.

Since (6.1) has a solution if and only if $A - \lambda B$ is singular, the eigenvalues of $A - \lambda B$ are the roots of the generalized characteristic equation

$$\det(A - \lambda B) = 0. \tag{6.2}$$

When B is the identity matrix, the generalized eigenvalue problem $A - \lambda B$ reduces to the ordinary eigenvalue problem in A. If B is nonsingular, Equation (6.1) is equivalent to the equation

$$B^{-1}Ax = \lambda x;$$

hence the eigenvalues of $A - \lambda B$ are the usual eigenvalues of $B^{-1}A$. They are also the eigenvalues of $AB^{-1} = B(B^{-1}A)B^{-1}$.

The algebraic and analytic theory of the generalized eigenvalue problem is considerably more complicated then the corresponding theory of the ordinary eigenvalue problem, and we shall not develop it here. However, two unusual features of the generalized eigenvalue problem should be mentioned. First, the characteristic equation (6.2) may vanish identically. For example, this happens whenever A and B have a common null vector. In this case, any number λ is an eigenvalue. We shall exclude this pathological case from our treatment.

When B is singular, the characteristic equation (6.2) has a degree less than n; that is the problem $A - \lambda B$ has fewer than n eigenvalues. This should not be regarded as a pathological case. If A is nonsingular, then (6.1) is equivalent to the equation

$$A^{-1}Bx = \lambda^{-1}x \equiv \mu x$$

and the missing eigenvalues correspond to zero eigenvalues of the reciprocal problem $B - \mu A$. If we adopt the convention that the reciprocal of zero is infinity, it is natural to refer to the missing eigenvalues of $A - \lambda B$ as "infinite eigenvalues." There is more than just convention in this. If B is perturbed slightly so that it is no longer singular, there will appear some very large eigenvalues that grow unboundedly as the perturbation is reduced to zero. The other eigenvalues, if they are well conditioned, will not be much affected by such perturbations.

We turn now to the numerical solution of the generalized eigenvalue problem. One technique that naturally suggests itself is to use the techniques of Sections 3 and 4 to solve the equivalent eigenvalue problem for the matrix $B^{-1}A$. When B is not ill conditioned with respect to inversion this method has much to recommend it. However, when B is nearly singular, $B^{-1}A$ cannot be computed accurately. And even if it could, the elements of $B^{-1}A$ will be large compared with those of A and B. This means that the smaller eigenvalues of $B^{-1}A$, which may be well conditioned in the original problem, cannot be computed accurately. Of course, if A is well conditioned with respect to inversion, we can work with the reciprocal problem $A^{-1}B$; however, if A is also nearly singular, this is not possible.

In an important special case one customarily forms $B^{-1}A$, after a fashion. In many applications, A turns out to be symmetric and B positive definite. In such cases it is important to reduce the amount of computation by taking advantage of the symmetry of the problem. This may be done as follows. Since B is positive definite, it may be written in the form LL^{T}, where L is a lower triangular matrix, which may be computed by Algorithm 3.3.9. If we set

$$y = L^{\mathrm{T}}x, \tag{6.3}$$

then Equation (6.1) is equivalent to the equation

$$L^{-1}AL^{-\mathrm{T}}y = \lambda y.$$

Thus the eigenvalue problem $A = \lambda B$ is equivalent to the symmetric eigenvalue problem for the matrix $L^{-1}AL^{-\mathrm{T}}$, which may be efficiently solved by reduction to tridiagonal form and the QR algorithm. The eigenvectors of the two problems are related by (6.3). Advantage may be taken of the symmetry of A in forming $L^{-1}AL^{-\mathrm{T}}$. When B, and hence L, is nearly singular, the objections registered above to the formation of $B^{-1}A$ also apply to this algorithm.

The alternative to forming $B^{-1}A$ or $A^{-1}B$ in order to solve the generalized eigenvalue problem is to attempt to reduce the matrices A and B to simple forms from which the solution of the problem may be readily computed. We shall now describe an algorithm for reducing A and B to upper triangular form (actually A will be quasi-triangular, which makes no real difference). The eigenvalues of the problem $A - \lambda B$ will then be given by the ratios of the diagonal elements of A and $B : \lambda_i = \alpha_{ii}/\beta_{ii}$. Naturally, in performing the reduction we must use only transformations that do not change the eigenvalues of the problem. A description of such a class is contained in the following theorem.

THEOREM 6.1. Let $A, B, U, V \in \mathbb{C}^{n\times n}$ with U and V nonsingular. Then the eigenvalues of the problems $A - \lambda B$ and $UAV - \lambda UBV$ are the same. If x is an eigenvector of $A - \lambda B$, then $V^{-1}x$ is an eigenvector of $UAV - \lambda UBV$.

PROOF. Since

$$\det(UAV - \lambda UBV) = \det[U(A - \lambda B)V] = \det(U)\det(V)\det(A - \lambda B)$$

and since $\det(U) \neq 0$ and $\det(V) \neq 0$, the characteristic equations of the two problems are nonzero multiples of one another and have the same roots. The assertion about the eigenvectors is proved by a trivial computation. ■

In the terminology of Section 5, if A and B are transformed by the same equivalence, the eigenvalues of the problem $A - \lambda B$ are unchanged. We say that two problems obtained in this way are equivalent. The device of forming $B^{-1}A$ to solve the generalized problem is the same as working with the equivalent problem $UAV - \lambda UBV$ in which $U = B^{-1}$ and $V = I$. This indicates that nothing will be gained by using equivalence transformations unless we restrict the matrices U and V to some class of well-conditioned matrices. In what follows we shall take U and V to be unitary.

The reduction of A and B to triangular form follows the pattern described in Section 5. First A and B are reduced to condensed forms; in this case A is reduced to upper Hessenberg form while B is reduced to upper triangular form. Then the effect of one shifted QR step on AB^{-1} is simulated by unitary equivalences on the problem $A - \lambda B$. This algorithm applied iteratively will reduce A to triangular or quasi-triangular form.

The reduction of A to Hessenberg form and B to triangular form may be accomplished as follows. First B is reduced to upper triangular form by means of Algorithm 5.3.8 and the transformations applied to A. Then A and B have the form illustrated below for $n = 5$ (A is on the left, B on the right):

$$\begin{matrix} x & x & x & x & x \\ x & x & x & x & x \\ x & x & x & x & x \\ x & x & x & x & x \\ \circledx & x & x & x & x \end{matrix}, \qquad \begin{matrix} x & x & x & x & x \\ 0 & x & x & x & x \\ 0 & 0 & x & x & x \\ 0 & 0 & 0 & x & x \\ 0 & 0 & 0 & 0 & x \end{matrix}.$$

Choose a plane rotation Q_{45} in the (4, 5)-plane so that $Q_{45}A$ has a zero in the position indicated. Then $Q_{45}B$ has the form

$$\begin{matrix} x & x & x & x & x \\ 0 & x & x & x & x \\ 0 & 0 & x & x & x \\ 0 & 0 & 0 & x & x \\ 0 & 0 & 0 & \circledx & x \end{matrix}.$$

Let Z_{45} be chosen as the plane rotation in the (4, 5)-plane for which $Q_{45}BZ_{45}$

is upper triangular. Then $Q_{45}AZ_{45}$ has the form

$$\begin{matrix} x & x & x & x & x \\ x & x & x & x & x \\ x & x & x & x & x \\ \textcircled{x} & x & x & x & x \\ 0 & x & x & x & x \end{matrix}\,.$$

Now choose a rotation Q_{34} in the (3, 4)-plane so that $Q_{34}Q_{45}AZ_{45}$ has a zero in the position indicated. Then $Q_{34}Q_{45}BZ_{45}$ has the form

$$\begin{matrix} x & x & x & x & x \\ 0 & x & x & x & x \\ 0 & 0 & x & x & x \\ 0 & 0 & \textcircled{x} & x & x \\ 0 & 0 & 0 & 0 & x \end{matrix}\,.$$

A rotation Z_{34} in the (3, 4)-plane may then be chosen so that $Q_{34}Q_{45}BZ_{45}Z_{34}$ is again upper triangular.

In general, zeros are introduced into the array A in the order shown below by premultiplying by plane rotations $Q_{i,i+1}$:

$$\begin{matrix} x & x & x & x & x \\ x & x & x & x & x \\ x^3 & x & x & x & x \\ x^2 & x^5 & x & x & x \\ x^1 & x^4 & x^6 & x & x \end{matrix}\,.$$

Each premultiplication of B by a rotation $Q_{i,i+1}$ introduces a nonzero subdiagonal element in the $(i + 1, i)$-position, which is immediately annihilated by postmultiplying by an appropriate plane rotation $Z_{i,i+1}$.

To develop an analog of the QR algorithm for the generalized eigenvalue problem, assume that A and B have been reduced as described above and that B is nonsingular. Let $C = AB^{-1}$. Since A is upper Hessenberg and B is upper triangular, C is upper Hessenberg. Let one step of the doubly shifted QR algorithm be applied to C to yield the upper Hessenberg matrix $C' = QCQ^{\mathrm{H}}$. Now suppose that we can determine unitary matrices Q' and Z with the following properties:

1. Q' has the same first row as Q,
2. $A' = Q'AZ$ is upper Hessenberg,
3. $B' = Q'BZ$ is upper triangular.

Then $A'B'^{-1} = Q'AZZ^{-1}B^{-1}Q'^{\mathrm{H}} = Q'CQ'^{\mathrm{H}}$ is upper Hessenberg. Hence by Theorem 4.2 $Q' = Q$ and $C' = A'B'^{-1}$.

Ignoring for the moment the practical problem of computing Q and Z, suppose that we apply the procedure outlined above to yield sequences of matrices $A_1, A_2, A_3, \ldots$ and $B_1, B_2, B_3, \ldots$. If we set $C_\nu = A_\nu B_\nu^{-1}$, then the sequence $C_1, C_2, C_3, \ldots$ is precisely the sequence of matrices that would result from the doubly shifted QR algorithm. The shifts may be chosen so that the C_ν approach a quasi-triangular form, and since B_ν is upper triangular it follows that $A_\nu = C_\nu B_\nu$ must approach a quasi-triangular form. Of course, we never form the matrices C_ν; instead we work directly with A_ν and B_ν.

So far as the practical algorithm is concerned, we must show how to find the first row of Q and then how to compute A' and B'. From Section 4 we know that the first row of Q is the same as the first row of an elementary reflector Q_0 that introduces zeros into the (2, 1)- and (3, 1)-positions of the matrix $(C - \varkappa_1 I)(C - \varkappa_2 I)$, where $\varkappa_1$ and $\varkappa_2$ are the shifts. This reflector depends only on the first two columns of C, which, since B is upper triangular, are given by

$$(c_1, c_2) = (a_1, a_2)\begin{pmatrix} \beta_{11} & \beta_{12} \\ 0 & \beta_{22} \end{pmatrix}^{-1}.$$

Thus to compute Q_0, we need only invert the 2×2 leading principal submatrix of B.

To find A' and B' we form Q_0A and Q_0B. If we can find unitary matrices U and V such that UQ_0AV is upper Hessenberg and UQ_0BV is upper triangular and such that UQ_0 has the same first row as Q_0, then $Q' = UQ_0$ has the same first row as Q; that is, UQ_0AV and UQ_0BV are the required matrices A' and B'. This reduction may be accomplished as follows.

The matrix Q_0 has the form

$$Q_0 = \begin{pmatrix} R_0 & 0 \\ 0 & I_{n-3} \end{pmatrix},$$

where R_0 is an elementary reflector of order 3. Hence Q_0A and Q_0B have the forms illustrated below for $n = 6$:

$$
\begin{matrix}
x & x & x & x & x & x \\
x & x & x & x & x & x \\
x & x & x & x & x & x \\
0 & 0 & x & x & x & x \\
0 & 0 & 0 & x & x & x \\
0 & 0 & 0 & 0 & x & x
\end{matrix}
, \qquad
\begin{matrix}
x & x & x & x & x & x \\
x^2 & x & x & x & x & x \\
x^1 & x^1 & x & x & x & x \\
0 & 0 & 0 & x & x & x \\
0 & 0 & 0 & 0 & x & x \\
0 & 0 & 0 & 0 & 0 & x
\end{matrix}
.
$$

First B is reduced to upper triangular form as follows. Choose an elementary reflector Z_0 to introduce zeros into the positions distinguished by the superscript unity, and then choose a rotation Z_0' in the (1, 2)-plane to introduce a zero into the position distinguished by the superscript two. Then $Q_0AZ_0Z_0'$ and $Q_0BZ_0Z_0'$ will have the forms

$$
\begin{matrix}
x & x & x & x & x & x \\
x & x & x & x & x & x \\
\textcircled{x} & x & x & x & x & x \\
\textcircled{x} & x & x & x & x & x \\
0 & 0 & 0 & x & x & x \\
0 & 0 & 0 & 0 & x & x
\end{matrix}
, \qquad
\begin{matrix}
x & x & x & x & x & x \\
0 & x & x & x & x & x \\
0 & 0 & x & x & x & x \\
0 & 0 & 0 & x & x & x \\
0 & 0 & 0 & 0 & x & x \\
0 & 0 & 0 & 0 & 0 & x
\end{matrix}
.
$$

Next choose an elementary reflector Q_1 to introduce zeros into the positions indicated above. Then $Q_1Q_0AZ_0Z_0'$ and $Q_1Q_0AZ_0Z_0'$ have the forms

$$
\left[\begin{array}{c:ccccc}
x & x & x & x & x & x \\ \hdashline
x & x & x & x & x & x \\
0 & x & x & x & x & x \\
0 & x & x & x & x & x \\
0 & 0 & 0 & x & x & x \\
0 & 0 & 0 & 0 & x & x
\end{array}\right]
, \qquad
\left[\begin{array}{c:ccccc}
x & x & x & x & x & x \\ \hdashline
0 & x & x & x & x & x \\
0 & x & x & x & x & x \\
0 & x & x & x & x & x \\
0 & 0 & 0 & 0 & x & x \\
0 & 0 & 0 & 0 & 0 & x
\end{array}\right]
.
$$

The reduction is continued by working with the submatrices indicated above, which have the same form as the original matrices. At the last a special step analogous to statements 3–7 in Algorithm 4.5 must be performed to complete the reduction. At no stage is the first row affected by a premultiplication by a Q_i (except of course Q_0); hence the final matrices are the desired A' and B'.

This brief sketch of what is called the QZ algorithm for the generalized eigenvalue problem has necessarily left a number of important questions untreated. Provided only that the leading principal submatrix $B^{[2]}$ is nonsingular, the QZ algorithm may be applied to problems where both A and B are singular. As with the QR algorithm, considerable economy can be achieved by working with submatrices whenever a subdiagonal element of A becomes effectively zero. Eigenvectors may be found by finding the eigenvectors of the triangular problem and transforming back to the original problem. For more details and program listings, the reader should consult the literature.

NOTES AND REFERENCES

The QZ algorithm is due to Moler and Stewart (1971). Although the method is applicable to the case where A is symmetric and B positive definite, the transformations do not preserve symmetry and the method is computationally just as expensive as for the general problem. When B is well conditioned one can of course apply the Cholesky QR method described in the text. When B is ill conditioned, Wilkinson (AEP, page 345) has observed that there may be some advantage in decomposing B in the form $B = XDX^{\mathrm{H}}$, where X is unitary and D is diagonal and positive definite, and solving the eigenvalue problem for $D^{-1/2}X^{\mathrm{H}}AXD^{-1/2}$. In particular, if the diagonal elements of D are arranged in ascending order, $D^{-1/2}X^{\mathrm{H}}AXD^{-1/2}$ will tend to have the kind of graded structure that the techniques described in this chapter work well on.

A perturbation theory in the spirit of Section 6.4 for the generalized eigenvalue problem has been developed by Stewart (1971).

An excellent survey of other computational techniques has been given by Peters and Wilkinson (1970c).

APPENDIX 1 THE GREEK ALPHABET AND LATIN NOTATIONAL CORRESPONDENTS

Alpha	A	α	a	Nu	N	ν	n, v
Beta	B	β	b	Xi	Ξ	ξ	x
Gamma	Γ	γ	c, g	Omicron	O	o	o
Delta	Δ	δ	d	Pi	Π	π	p
Epsilon	E	ε	e	Rho	P	ϱ	r
Zeta	Z	ζ	z	Sigma	Σ	σ	s
Eta	H	η	h, y	Tau	T	τ	t
Theta	Θ	θ		Upsilon	Y	υ	u
Iota	I	ι	i	Phi	Φ	φ	f
Kappa	K	$\varkappa$	k	Chi	X	χ	
Lambda	Λ	λ	l, l	Psi	Ψ	ψ	
Mu	M	μ	m, u	Omega	Ω	ω	w

APPENDIX 2 DETERMINANTS

The determinant is a function from $\mathbb{R}^{n\times n}$ (or $\mathbb{C}^{n\times n}$) into $\mathbb{R}$ (or $\mathbb{C}$) satisfying the conditions

$$\det(AB) = \det(A)\det(B)$$

and

$$\det(I) = 1.$$

Historically the theory of determinants preceded matrix theory, and many familiar results of matrix theory were originally formulated in terms of determinants. However, as linear transformations began to be studied systematically, the theory of determinants fell increasingly into eclipse, until today it is relegate to the backwaters of undergraduate linear algebra courses, usually to be skipped over.

This situation is unfortunate, for no one has been able to dispense entirely with determinants. For example, the definition of the multiplicity of an eigenvalue is most naturally cast in terms of determinants (see Section 6.2). Moreover, a determinantal argument will often establish easily a result whose proof would otherwise be laborious. Accordingly, in this section we shall develop some of the elementary theory of determinants that arises in applied linear algebra. For definiteness we shall confine ourselves to real matrices of order n, although analogous results will obviously hold for complex matrices. In most cases we shall only sketch the proofs, leaving the details as exercises.

The classical definition of the determinant requires some elementary facts about permutations, which we collect here. A permutation on the integers $1, 2, \ldots, n$ is a function $\sigma : \{1, 2, \ldots, n\} \to \{1, 2, \ldots, n\}$ that is 1–1 and onto. We shall denote the set of permutations on $1, 2, \ldots, n$ by $\mathcal{S}_n$.

The product of two permutations σ and τ in $\mathcal{S}_n$ is the composition function $\sigma\tau$ defined by

$$\sigma\tau(i) = \sigma[\tau(i)] \qquad (i = 1, 2, \ldots, n).$$

The identity function ι on $\{1, 2, \ldots, n\}$ obviously satisfies

$$\iota\sigma = \sigma\iota = \sigma$$

for all $\sigma \in \mathcal{S}_n$. Moreover, if $\sigma \in \mathcal{S}_n$, then the function σ has a unique inverse σ^{-1} satisfying

$$\sigma\sigma^{-1} = \sigma^{-1}\sigma = \iota. \tag{1}$$

These considerations may be summed up by saying that with the product defined above, the elements of $\mathcal{S}_n$ form a group.

A member $\tau \in \mathcal{S}_n$ is a *transposition* if there are distinct integers k and l in $\{1, 2, \ldots, n\}$ such that

$$\begin{aligned} \tau(k) &= l, \\ \tau(l) &= k, \\ \tau(i) &= i \qquad (i \neq k, l). \end{aligned}$$

If $\sigma \in \mathcal{S}_n$, we may decompose σ into a product of transpositions:

$$\sigma = \tau_1\tau_2 \cdots \tau_m. \tag{2}$$

The decomposition is not unique; however, if m is even, then it must be even for any other decomposition, and if m is odd, it must likewise be odd for any other decomposition. This justifies defining the *signum* of σ by the equation

$$\operatorname{sign}(\sigma) = (-1)^m,$$

where m is the number of terms in a decomposition of form (2).

It is easy to verify that

$$\operatorname{sign}(\sigma\tau) = \operatorname{sign}(\sigma)\operatorname{sign}(\tau). \tag{3}$$

In particular,

$$\operatorname{sign}(\iota) = 1,$$

which with (3) implies

$$\operatorname{sign}(\sigma^{-1}) = \operatorname{sign}(\sigma).$$

Let $A \in \mathbb{R}^{n\times n}$. The *determinant* of A is the number

$$\det(A) = \sum_{\sigma\in\mathfrak{S}_n} \operatorname{sign}(\sigma)\alpha_{\sigma(1),1}\alpha_{\sigma(2),2}\cdots\alpha_{\sigma(n),n}. \tag{4}$$

The determinant has a number of elementary properties that follow immediately from its definition. The first property is that the determinant of a matrix is a continuous function of its elements; that is

A2.1. If $\lim A_\nu = A$, then $\lim \det(A_\nu) = \det(A)$.

Certain simple transformations on the matrix A leave its determinant unchanged or change it in a simple manner. Some of these transformations are considered below.

A2.2. $\det(A^{\mathrm{T}}) = \det(A)$.

PROOF.

$$\begin{aligned}\det(A^{\mathrm{T}}) &= \sum_{\sigma\in\mathfrak{S}_n} \operatorname{sign}(\sigma)\alpha_{1\sigma(1)}\cdots\alpha_{n\sigma(n)}\\ &= \sum_{\sigma\in\mathfrak{S}_n} \operatorname{sign}(\sigma)\alpha_{\sigma^{-1}(1),1}\cdots\alpha_{\sigma^{-1}(n),n}.\end{aligned}$$

But if $\tau = \sigma^{-1}$, then $\operatorname{sign}(\tau) = \operatorname{sign}(\sigma)$, and as σ ranges over $\mathfrak{S}_n$ so does τ. Hence

$$\det(A^{\mathrm{T}}) = \sum_{\tau\in\mathfrak{S}_n} \operatorname{sign}(\tau)\alpha_{\tau(1),1}\cdots\alpha_{\tau(n),n} = \det(A). \quad \blacksquare$$

A2.3. $\det[(\beta_1 a_1, \ldots, \beta_n a_n)] = \beta_1\cdots\beta_n \det[(a_1, \ldots, a_n)]$.

A2.4. If $\tau \in \mathfrak{S}_n$, then

$$\det[(a_{\tau(1)}, \ldots, a_{\tau(n)})] = \operatorname{sign}(\tau)\det[(a_1, \ldots, a_n)].$$

PROOF.

$$\begin{aligned}\det[(a_{\tau(1)}, \ldots, a_{\tau(n)})] &= \sum_{\sigma \in \mathfrak{S}_n} \operatorname{sign}(\sigma)\alpha_{\sigma(1)\tau(1)} \cdots \alpha_{\sigma(n)\tau(n)} \\ &= \sum_{\sigma \in \mathfrak{S}_n} \operatorname{sign}(\tau) \operatorname{sign}(\sigma\tau^{-1})\alpha_{\sigma\tau^{-1}(1),1} \cdots \alpha_{\sigma\tau^{-1}(n),n} \\ &= \operatorname{sign}(\tau) \det(a_1, \ldots, a_n). \quad \blacksquare\end{aligned}$$

A2.5. $\det(a_1, \ldots, a_k + a_k', \ldots, a_n) = \det(a_1, \ldots, a_k, \ldots, a_n) + \det(a_1, \ldots, a_k', \ldots, a_n)$.

A2.6. If two columns of A are linearly dependent, then $\det(A) = 0$.

PROOF. Suppose that, say, a_1 and a_2 are linearly independent. Then one of a_1 or a_2 is a multiple of the other, say $a_2 = \beta a_1$. By A2.3

$$\begin{aligned}\det[(a_1, a_2, \ldots, a_n)] &= \det[(a_1, \beta a_1, \ldots, a_n)] \\ &= \beta \det[(a_1, a_1, \ldots, a_n)] \\ &= \det[(\beta a_1, a_1, \ldots, a_n)] = \det[(a_2, a_1, \ldots, a_n)].\end{aligned}$$

But by A2.4, $\det[(a_1, a_2, \ldots, a_n)] = -\det[(a_2, a_1, \ldots, a_n)]$. Hence $\det[(a_1, a_2, \ldots, a_n)] = 0$.

The properties A2.3–6 are cast in terms of the columns of a matrix. For example, A2.3 says that the effect of multiplying a column of a matrix by a scalar is to multiply its determinant by the same scalar. Likewise, it follows from A2.4 that interchanging two columns of a matrix changes its determinant. It is important to note that because of A2.2 the same results hold for the rows of a matrix.

We are now in a position to prove that the determinant of a product is the product of the determinants.

A2.7. Let $A, B \in \mathbb{R}^{n \times n}$. Then

$$\det(AB) = \det(A)\det(B).$$

PROOF. With A partitioned by columns we have

$$\det(AB) = \det[(\beta_{11}a_1 + \cdots + \beta_{n1}a_n, \ldots, \beta_{1n}a_1 + \cdots + \beta_{nn}a_n)]. \quad (5)$$

By applying A2.5 and A2.3 repeatedly to the right-hand side of (5), we obtain

$$\det(AB) = \sum_{i_1,\ldots,i_n=1}^{n} \beta_{i_1,1} \cdots \beta_{i_n,n} \det(a_{i_1}, \ldots, a_{i_n}). \tag{6}$$

But by A2.6, if any of the indices in the sum in (6) are equal, $\det(a_{i_1}, \ldots, a_{i_n}) = 0$. Hence

$$\det(AB) = \sum_{\sigma \in \mathcal{S}_n} \beta_{\sigma(1),1} \cdots \beta_{\sigma(n),n} \det(a_{\sigma(1)}, \ldots, a_{\sigma(n)}),$$

and by A2.4

$$\begin{aligned}\det(AB) &= \sum_{\sigma \in \mathcal{S}_n} \operatorname{sign}(\sigma)\beta_{\sigma(1),1} \cdots \beta_{\sigma(n),n} \det(a_1, \ldots, a_n) \\ &= \det(A)\det(B). \quad \blacksquare\end{aligned}$$

The result A2.7 has a number of important consequences. In the first place by calculating the determinants of the matrices in A2.1, we may determine the effects of the elementary row operations on the determinant of a matrix A.

1. Multiplying a row of A by λ multiplies the determinant by λ (this is implicit in A2.2 and A2.3).
2. Interchanging two rows of A changes the sign of the determinant (A2.2 and A2.4).
3. Adding a multiple of one row of A to a different row leaves the determinant unchanged.

Of course similar results hold for the columns of A. More generally, since an elementary lower triangular matrix has determinant unity, premultiplying or postmultiplying a matrix by an elementary lower triangular matrix does not change its determinant.

Another consequence of A2.7 is the following classical characterization of nonsingularity.

A2.8. Let $A \in \mathbb{R}^{n\times n}$. Then A is nonsingular if and only if $\det(A) \neq 0$.

PROOF. Suppose A is nonsingular. Then from the equation $AA^{-1} = I$, it follows that $\det(A)\det(A^{-1}) = 1$ and hence that $\det(A) \neq 0$.

Conversely suppose A is singular. Then there is a nonzero vector x such that $Ax = 0$. Let U be a nonsingular matrix whose first column is x. Then the first column of AU is zero, whence

$$0 = \det(AU) = \det(A)\det(U).$$

But since U is nonsingular, $\det(U) \neq 0$. Hence $\det(A) = 0$. ∎

An immediate corollary of the proof of A2.8 is the fact that if A is nonsingular,

$$\det(A^{-1}) = \frac{1}{\det(A)}. \tag{7}$$

Since the vanishing of the determinant of a matrix implies that the matrix is singular, it might be thought that a small determinant would be an indication of near singularity. However, this is fallacious, for the size of the determinant depends strongly on how the matrix is scaled. In fact it follows from A2.3 that if A is of order n,

$$\det(\beta A) = \beta^n \det(A).$$

Thus if A is of order 50, the determinant of A can be reduced by a factor of 10^{-50} simply by dividing the elements of A by 10. The reader is warned that the misconception that a small determinant indicates a nearly singular matrix is widespread, and many otherwise unobjectionable programs give an error return when a determinant becomes "too small."

The material presented above contains all that is required in the body of the text. However, the reader will encounter many other results from determinant theory in the literature. Three of the more important of these are Cramer's rule, the adjugate, and expansion by minors, which we shall now present.

The description of Cramer's rule can be simplified by introducing some notation. For any matrix $A \in \mathbb{R}^{n \times n}$ and vector $b \in \mathbb{R}^n$, let $A_j(b)$ denote the matrix obtained by replacing the jth column of A by b.

A2.9. (Cramer's rule). Let $A \in \mathbb{R}^{n \times n}$, $b \in \mathbb{R}^n$, and $Ax = b$. Then

$$\det[A_j(b)] = \det(A)\xi_j \qquad (j = 1, 2, \ldots, n). \tag{8}$$

PROOF. First assume that A is nonsingular. Then $A^{-1}A_j(b) = I_j(x)$. However $\det[I_j(x)] = \xi_j$. Hence

$$\xi_j = \det(A^{-1})\det[A_j(b)],$$

which in view of (7) is equivalent to (8). If A is singular and $Ax = b$, then b lies in the column space of A, and $\operatorname{rank}[A_j(b)] \leq \operatorname{rank}(A) < n$. Hence $\det[A_j(b)] = \det(A) = 0$, which implies (8). ■

When A is nonsingular, Equation (8) gives explicit determinantal expressions for the components of the solution of the equation $Ax = b$. Although Cramer's rule cannot be recommended as a numerical technique for solving linear systems, it has numerous theoretical applications, especially to systems having special structure.

When A is nonsingular, the ith column x_i of the inverse of A satisfies the equation

$$Ax_i = e_i.$$

Hence by Cramer's rule the (j, i)-element of A^{-1} is given by

$$\xi_{ji} = \frac{\det[A_j(e_i)]}{\det(A)}. \tag{9}$$

This result is usually expressed as follows. Let A_{ij} denote the submatrix of A obtained by deleting the ith row and jth column of A. Then it is easily verified that

$$\det[A_j(e_i)] = (-1)^{i+j}\det(A_{ij}).$$

[The determinant $\det(A_{ij})$ is called the *cofactor* of α_{ij}.] Hence if we define the *adjugate* (sometimes called the *adjoint* of A) as the matrix $\operatorname{adj}(A)$ whose (i, j)-element is $(-1)^{i+j}\det(A_{ji})$, it follows from (9) that

$$A^{-1} = \frac{\operatorname{adj}(A)}{\det(A)}. \tag{10}$$

More generally we have for any matrix:

A2.10.

$$A \operatorname{adj}(A) = \operatorname{adj}(A)A = \det(A)I. \tag{11}$$

PROOF. For A nonsingular, the result is equivalent to (10). If A is singular, then $A + \varepsilon I$ is nonsingular for all sufficiently small ε (Exercise 6.3.22). Hence

$$(A + \varepsilon I)\operatorname{adj}(A + \varepsilon I) = \det(A + \varepsilon I)I$$

for all sufficiently small ε. In view of A2.1, we may take the limit as ε approaches zero to obtain the result. ∎

The algebraically inclined reader will rightly feel uneasy about the appeal to continuity in the proof of A2.10, since it does not suggest any way of generalizing the result to fields where the notion of limit does not exist. In fact there is a purely algebraic proof of A2.10. Our proof was chosen to illustrate a valuable technique for dealing with singular matrices.

One consequence of A2.10 is to give a determinantal expression for a nontrivial solution of the homogeneous equation $Ax = 0$ when $\text{rank}(A) = n - 1$. For in this case at least one submatrix of A of order $n - 1$ must be nonsingular, and hence some column, say the ith, of $\text{adj}(A)$ is nonzero. It follows from the equation $A\,\text{adj}(A) = 0$ that

$$x = (-1)^{i+1}\big(\det(A_{i1}), -\det(A_{i2}), \ldots, (-1)^{n-1}\det(A_{in})\big)^{\mathrm{T}}$$

is a nontrivial solution of $Ax = 0$.

If we evaluate the (i, i)-elements of the left- and right-hand sides of (11), we obtain the following important expression for the determinant:

A2.11. (*Expansion by cofactors*)

$$\det(A) = \sum_{j=1}^{n} (-1)^{i+j}\alpha_{ij}\det(A_{ij}). \tag{12}$$

Equation (12) expresses $\det(A)$ in terms of the elements of a single row and their cofactors. It is used most often to simplify a determinant one of whose rows consists mostly of zeros. A similar expression in terms of columns may be obtained by evaluating the middle term in (11).

We conclude this appendix with a brief comment on the calculation of determinants. Expression (4) by which the determinant was defined contains $n!$ terms and obviously cannot be used to evaluate $\det(A)$, even when n is fairly small.

The evaluation of the determinant of a general matrix is best accomplished by reducing it to triangular form. If, for example, Gaussian elimination with partial pivoting is used to reduce A to triangular form, the result is the matrix

$$A_n = M_{n-1}P_{n-1}\cdots M_1P_1A,$$

where each M_k is an elementary lower triangular matrix and each P_k is an elementary permutation. Hence

$$\det(A) = \frac{\det(A_n)}{\det(M_{n-1})\det(P_{n-1}) \cdots \det(M_1)\det(P_1)}.$$

Now since M_k is an elementary lower triangular matrix, $\det(M_k) = 1$. Moreover, $\det(P_k) = -1$ if an interchange was made at the kth stage of the reduction; otherwise $\det(P_k) = 1$. Since A_n is upper triangular, its determinant is the product of its diagonal elements. Hence

$$\det(A) = (-1)^l \alpha_{11}^{(1)} \alpha_{22}^{(2)} \cdots \alpha_{nn}^{(n)},$$

where $\alpha_{kk}^{(k)}$ is the kth pivot element and l is the number of interchanges required in the reduction. Of course the determinant can be calculated in a similar manner from any of the standard reductions to triangular form.

We have seen that the determinant of a matrix can be very large or very small; so much so, in fact, that its computation can easily cause an overflow or underflow. Consequently, elimination routines that also compute the determinant often return it scaled by a factor that is specified in logarithmic form.

APPENDIX 3 ROUNDING-ERROR ANALYSIS OF SOLUTION OF TRIANGULAR SYSTEMS AND OF GAUSSIAN ELIMINATION

In this appendix we shall establish the rounding-error results cited in Section 3.5. Throughout the appendix we shall assume that all computations are done in floating-point arithmetic without underflows or overflows. The arithmetic operations are assumed to satisfy the bounds of Section 2.1; namely

$$\mathrm{fl}(\alpha \circ \beta) = \alpha \circ \beta(1 + \varrho),$$

and

$$\mathrm{fl}(\alpha \pm \beta) = \alpha(1 + \varrho) \pm \beta(1 + \sigma),$$

where " $\circ$ " denotes multiplication or division and

$$|\varrho|, |\sigma| \leq \mu \cdot 10^{-t}.$$

We begin by extending the "fl" notation of Section 2.2. If e is an arithmetic expression, $\mathrm{fl}(e)$ will denote the result of evaluating e in floating-point arithmetic. When there is more than one possible way of evaluating the expression e, it will be assumed that one of them has been fixed upon.

For example,

$$\mathrm{fl}(\alpha + \beta + \gamma)$$

may denote either

$$\mathrm{fl}[\mathrm{fl}(\alpha + \beta) + \gamma] \tag{1}$$

or

$$\mathrm{fl}[\alpha + \mathrm{fl}(\beta + \gamma)].$$

Of course the expression (1) represents the natural way of summing α, β, and γ in order.

One of the most frequently occurring computations in numerical linear algebra is the calculation of an inner product $y^{\mathrm{T}}x$. Since the analysis is sufficiently complicated to illustrate many of the features of more extensive analyses and since it is required for the analysis of the solution of triangular systems, we shall present it here in detail.

The usual algorithm for computing the inner product $y^{\mathrm{T}}x$ $(x, y \in \mathbb{R}^n)$ may be described as follows.

$$\begin{array}{ll} 1) & \sigma_1 = \eta_1\xi_1 \\ 2) & \left| \begin{array}{ll} \text{For} & i = 2, 3, \ldots, n \\ 1) & \sigma_i = \eta_i\xi_i + \sigma_{i-1} \end{array} \right. \end{array} \tag{2}$$

If exact calculations are performed, σ_n is the inner product $y^{\mathrm{T}}x$.

Now suppose the calculations in (2) are done in floating-point arithmetic, and let σ_i denote the *computed* values (the use of variables appearing in an algorithm to denote computed values is common in rounding-error analyses, since we are often uninterested in the exact values). Then

$$\sigma_1 = \mathrm{fl}(\eta_1\xi_1) = \eta_1\xi_1(1 + \pi_1) \tag{3}$$

and

$$\begin{aligned} \sigma_i &= \mathrm{fl}[\mathrm{fl}(\eta_i\xi_i) + \sigma_{i-1}] \\ &= \mathrm{fl}[\eta_i\xi_i(1 + \pi_i) + \sigma_{i-1}] \\ &= \eta_i\xi_i(1 + \pi_i)(1 + \varrho_i) + \sigma_{i-1}(1 + \tau_i), \end{aligned} \tag{4}$$

where

$$|\pi_i|, |\varrho_i|, |\tau_i| \leq \mu \cdot 10^{-t}. \tag{5}$$

If expressions (3) and (4) are combined, the result is

$$\begin{aligned}\sigma_n = \eta_1\xi_1(1+\pi_1)(1+\tau_2)(1+\tau_3)\cdots(1+\tau_n)\\ +\eta_2\xi_2(1+\pi_2)(1+\varrho_2)(1+\tau_3)\cdots(1+\tau_n)\\ +\eta_3\xi_3(1+\pi_3)(1+\varrho_3)(1+\tau_4)\cdots(1+\tau_n)\\ \cdots+\eta_n\xi_n(1+\pi_n)(1+\varrho_n).\end{aligned} \tag{6}$$

Expression (6) is somewhat complicated to work with. One simplification can be achieved by noting that the error terms in (6) all satisfy the same inequality (5). Hence what is important in (6) is not the individual error terms but the number of them. This suggests that we introduce the symbol

$$\langle k\rangle$$

as a generic symbol for a quotient of the form

$$\frac{(1+\pi_1)(1+\pi_2)\cdots(1+\pi_m)}{(1+\varrho_1)(1+\varrho_2)\cdots(1+\varrho_n)}, \tag{7}$$

where $m+n=k$ and $|\pi_i|, |\varrho_i| \leq \mu \cdot 10^{-t}$ (the reason for allowing divisions will become clear later). In this notation expression (6) becomes

$$\sigma_n = \eta_1\xi_1\langle n\rangle + \eta_2\xi_2\langle n\rangle + \eta_3\xi_2\langle n-1\rangle + \cdots + \eta_n\xi_n\langle 2\rangle.$$

Incidentally, because the $\langle k\rangle$ symbols satisfy the relations

$$\langle k\rangle\langle l\rangle = \langle k+l\rangle \tag{8}$$

and

$$\frac{\langle k\rangle}{\langle l\rangle} = \langle k+l\rangle, \tag{9}$$

they may be used to simplify the derivation of expressions such as (6). For example,

$$\sigma_1 = \text{fl}(\eta_1\xi_1) = \eta_1\xi_1\langle 1\rangle$$

and

$$\begin{aligned}\sigma_2 &= \text{fl}[\text{fl}(\eta_2\xi_2)+\sigma_1]\\ &= \text{fl}(\eta_2\xi_2\langle 1\rangle+\sigma_1)\\ &= \sigma_2\langle 1\rangle + \eta_2\xi_2\langle 1\rangle\langle 1\rangle\\ &= \eta_1\xi_1\langle 2\rangle + \eta_2\xi_2\langle 2\rangle.\end{aligned}$$

The expression for σ_3 may be computed similarly, after which it is not difficult to guess the general result.

Although the $\langle k \rangle$ symbols are a useful way of presenting intermediate results, they contain no more information than expressions such as (7). What is needed in the final result is a simple bound on quantities of the form (7). The following theorem (Exercise 2.1.4), gives such a bound.

A3.1. If $k\mu \cdot 10^{-t} < .1$, then

$$\langle k \rangle = 1 + \varrho,$$

where

$$|\varrho| \leq 1.06k\mu \cdot 10^{-t}.$$

In the sequel, we shall always assume that $k\mu \cdot 10^{-t} < .1$, and we shall set

$$\mu' = 1.06\mu.$$

Our result for the computation of an inner product may be summarized by saying that

$$\mathrm{fl}(y^{\mathrm{T}}x) = \eta_1\xi_1(1+\gamma_1) + \eta_2\xi_2(1+\gamma_2) + \cdots + \eta_n\xi_n(1+\gamma_n)$$

where

$$|\gamma_1| \leq n\mu' \cdot 10^{-t}$$

and

$$|\gamma_i| \leq (n-i+2)\mu' \cdot 10^{-t} \qquad (i = 2, 3, \ldots, n).$$

If we set

$$e = \xi_i\gamma_i,$$

then we have

$$\mathrm{fl}(y^{\mathrm{T}}x) = y^{\mathrm{T}}(x+e),$$

where the components of e are small compared to the corresponding components of x. In the language of Section 2.1, algorithm (2) is a stable way of computing the inner product.

A rounding-error analysis of the solution of triangular systems follows easily from the above results on the inner product. For definiteness suppose we are to solve the lower triangular system

$$Lx = b,$$

by the algorithm

$$
1)\left|\begin{array}{ll}
\text{For} & i = 1, 2, \ldots, n \\
1) & \sigma = \sum_{j=1}^{i-1} \lambda_{ij}\xi_j \\
2) & \xi_i = \dfrac{(\beta_i - \sigma)}{\lambda_{ii}}
\end{array}\right. \tag{10}
$$

As above, let the variables in (10) denote the computed values. Then

$$\xi_i = \mathrm{fl}\left[\frac{\beta_i - \sigma}{\lambda_{ii}}\right] = \left[\frac{\beta_i\langle 1\rangle - \sigma\langle 1\rangle}{\lambda_{ii}}\right]\langle 1\rangle.$$

Hence

$$\frac{\xi_i\lambda_{ii}}{\langle 1\rangle} = \beta_i\langle 1\rangle - \sigma\langle 1\rangle,$$

or

$$\frac{\sigma\langle 1\rangle}{\langle 1\rangle} + \frac{\xi_i\lambda_{ii}}{\langle 1\rangle\langle 1\rangle} = \beta_i.$$

From relations (8) and (9) it follows that

$$\sigma\langle 2\rangle + \xi_i\lambda_{ii}\langle 2\rangle = \beta_i.$$

But from the result on inner products

$$\sigma = \xi_1\lambda_{i1}\langle i-1\rangle + \xi_2\lambda_{i2}\langle i-1\rangle + \cdots + \xi_{i-1}\lambda_{i,i-1}\langle 2\rangle.$$

Hence

$$\begin{aligned}\xi_1\lambda_{i1}\langle i+1\rangle + \xi_2\lambda_{i2}\langle i+1\rangle + \xi_3\lambda_{i3}\langle i\rangle + \cdots + \xi_{i-1}\lambda_{i,i-1}\langle 4\rangle \\ + \xi_i\lambda_{ii}\langle 2\rangle = \beta_i.\end{aligned} \tag{11}$$

It follows from A3.1 that

$$\sum_{j=1}^{i} \xi_j\lambda_{ij}(1 + \varrho_{ij}) = \beta_i \qquad (i = 1, 2, \ldots, n), \tag{12}$$

where certainly

$$|\varrho_{ij}| \leq (n+1)\mu' \cdot 10^{-t}.$$

If we set

$$\varepsilon_{ij} = \lambda_{ij}\varrho_{ij},$$

then (12) asserts that the computed solution x satisfies

$$(L + E)x = b,$$

where

$$|\varepsilon_{ij}| \leq (n + 1)|\lambda_{ij}|\mu' \cdot 10^{-t}.$$

This is essentially Theorem 3.5.1 for lower triangular matrices. The consequences of this result are discussed in Section 3.5. Note that the factor $n + 1$, which was obtained by taking an upper bound for the arguments of the $\langle k \rangle$ symbols in (11) is obviously an overestimate for most of the ε_{ij}.

We conclude this appendix with an error analysis of Gaussian elimination with pivoting. By Theorem 3.2.9 we may assume that the interchanges have been made before the start of the reduction and analyze the simpler Algorithm 3.2.4 applied to a matrix $A_1 \in \mathbb{R}^{n \times n}$. This algorithm generates elementary lower triangular matrices $M_1, M_2, \ldots, M_{n-1}$ and matrices $A_2, A_3, \ldots, A_n$ satisfying

$$A_{k+1} = M_k A_k \qquad (k = 1, 2, \ldots, n - 1)$$

according to the formulas

$$\begin{aligned} &1) \quad \mu_{ik} = \frac{\alpha_{ik}^{(k)}}{\alpha_{kk}^{(k)}} \qquad (i = k + 1, k + 2, \ldots, n) \\ &2) \quad \alpha_{ik}^{(k+1)} = 0 \qquad (i = k + 1, k + 2, \ldots, n) \\ &3) \quad \alpha_{ij}^{(k+1)} = \alpha_{ij}^{(k)} - \mu_{ik}\alpha_{kj}^{(k)} \qquad (i, j = k + 1, k + 2, \ldots, n) \end{aligned} \tag{13}$$

The matrix A_n is upper triangular, and if we set

$$L = M_1^{-1} M_2^{-1} \cdots M_{n-1}^{-1}, \tag{14}$$

then L is unit lower triangular and

$$A_1 = L A_n.$$

If we assume that either partial or complete pivoting is used, then

$$|\mu_{ik}| \leq 1 \qquad (k = 1, 2, \ldots, n - 1; \quad i = k + 1, k + 2, \ldots, n). \tag{15}$$

Now suppose that (13) is performed in floating-point arithmetic and let the variables in (13) stand for the computed values. Let

$$\beta_k = \max\{|\alpha_{ij}^{(k)}| : i, j = 1, 2, \ldots, n\}. \tag{16}$$

By an elementary application of the rounding-error bounds, we have

$$\mu_{ik} = \left(\frac{\alpha_{ik}^{(k)}}{\alpha_{kk}^{(k)}}\right)(1 + \pi_{ik}^{(k)}) \qquad (i = k+1, k+2, \ldots, n)$$

and

$$\alpha_{ij}^{(k+1)} = \alpha_{ij}^{(k)}(1 + \pi_{ij}^{(k)}) - \mu_{ik}\alpha_{kj}^{(k)}(1 + \varrho_{ij}^{(k)}) \qquad (i, j = k+1, k+2, \ldots, n),$$

where

$$|\pi_{ij}^{(k)}| \leq \mu' \cdot 10^{-t}, \qquad |\varrho_{ij}^{(k)}| \leq 2\mu' \cdot 10^{-t}. \tag{17}$$

Recalling that $\alpha_{ik}^{(k+1)} = 0$ and setting $\varrho_{ik}^{(k)} = 0$ $(i = k+1, k+2, \ldots, n)$, we may rewrite these equations in the form

$$\alpha_{ij}^{(k)} - \mu_{ik}\alpha_{kj}^{(k)} = \alpha_{ij}^{(k+1)} - \alpha_{ij}^{(k)}\pi_{ij}^{(k)} + \mu_{ik}\alpha_{kj}^{(k)}\varrho_{ij}^{(k)}$$
$$(i = k+1, k+2, \ldots, n; \ \ j = k, k+1, \ldots, n). \tag{18}$$

Hence if we define E_{k+1} as the matrix whose first $k - 1$ columns and first k rows are zero and whose other elements are given by

$$\varepsilon_{ij}^{(k+1)} = \alpha_{ij}^{(k)}\pi_{ij}^{(k)} - \mu_{ik}\alpha_{kj}^{(k)}\varrho_{ij}^{(k)} \qquad (i = k+1, \ldots, n; \ \ j = k, \ldots, n),$$

then (18) becomes

$$M_k A_k = A_{k+1} - E_{k+1}. \tag{19}$$

To complete the analysis, we need two properties of the matrices E_k. First, since the first k rows of E_{k+1} are zero and M_l is an elementary lower triangular matrix of index l, it follows that

$$M_l^{-1}E_{k+1} = E_{k+1} \qquad (l = 1, 2, \ldots, k). \tag{20}$$

Second we need a bound on the elements of E_{k+1}. This is easily obtained from (15)–(17).

$$\begin{aligned} |\varepsilon_{ij}^{(k+1)}| &\leq |\alpha_{ij}^{(k)}|\,|\pi_{ij}^{(k)}| + |\mu_{ik}|\,|\alpha_{kj}^{(k)}|\,|\varrho_{ij}^{(k)}| \\ &\leq \beta_k(\mu' \cdot 10^{-t}) + 1 \cdot \beta_k(2\mu' \cdot 10^{-t}) \\ &= 3\beta_k\mu' \cdot 10^{-t} \leq 3\beta\mu' \cdot 10^{-t}, \end{aligned} \tag{21}$$

where

$$\beta = \max\{\beta_1, \beta_2, \ldots, \beta_n\}.$$

Now from (19) and (20) we have the equations

$$A_{n-1} = M_{n-1}^{-1}A_n - E_n,$$
$$A_{n-2} = M_{n-2}^{-1}A_{n-1} - E_{n-1},$$
$$\vdots$$
$$A_1 = M_1^{-1}A_2 - E_2.$$

From the first two equations and the fact that $M_{n-2}^{-1}E_n = E_n$, we have

$$A_{n-2} = M_{n-2}^{-1}M_{n-1}^{-1}A_n - E_{n-1} - E_n.$$

Continuing this substitution, we obtain

$$\begin{aligned} A_1 &= M_1^{-1}M_2^{-1} \cdots M_{n-1}^{-1}A_n - E_2 - E_3 - \cdots - E_n \\ &\equiv LA_n - E, \end{aligned}$$

where L is defined by (14) and

$$E = E_2 + E_3 + \cdots + E_n.$$

From (21) it follows that the computed L and A_n satisfy

$$LA_n = A_1 + E,$$

where

$$|\,\varepsilon_{ij}\,| \leq 3(n-1)\beta\mu' \cdot 10^{-t}. \tag{22}$$

This result is essentially Theorem 3.5.2 and its implications are discussed there. Two points suggested by the above analysis bear mentioning here. First, the factor $n - 1$ in (22) is clearly an overestimate for most of the ε_{ij}, since the matrices E_k may have many of their elements equal to zero. Second, the result appears to depend on the fact that the pivoting strategy has been chosen so that $|\,\mu_{ik}\,| \leq 1$. This is in fact not true. The crux of the matter is the derivation of the bound (21) for $|\,\varepsilon_{ij}^{(k+1)}\,|$. When $|\,\alpha_{ij}^{(k)}\,| \geq \frac{1}{2}\,|\,\mu_{ik}\alpha_{kj}^{(k)}\,|$ the same bound holds with a factor of 5 instead of 3. When $|\,\alpha_{ij}^{(k)}\,| < \frac{1}{2}\,|\,\mu_{ik}\alpha_{kj}^{(k)}\,|$, a reasonable bound may be obtained in terms of β_{k+1} instead of β_k. The fact that the error analysis does not depend on the size of the μ_{ik} is important for positive definite matrices, where μ_{ik} can be very large indeed, but the numbers $\beta_1, \beta_2, \ldots, \beta_n$ cannot grow (Exercise 3.5.2).

APPENDIX 4 OF THINGS NOT TREATED

An introductory textbook in numerical linear algebra must needs exclude a large amount of important material. In this appendix we shall mention some of the subjects not covered in the text, giving references for the reader who is interested in pursuing a particular subject in greater depth.

SOLUTION OF SYMMETRIC SYSTEMS THAT ARE NOT POSITIVE DEFINITE. It is easily shown that if Gaussian elimination without pivoting is applied to a symmetric matrix, the sequence of trailing submatrices so produced are symmetric. Unfortunately the only pivoting strategy that preserves this symmetry is one in which two rows and the same two columns are interchanged. This limits the choice of pivots to the diagonal elements of the matrix, which may all be zero, even in well-conditioned cases. Bunch and Parlett (1972) have devised a block Gaussian elimination scheme which circumvents this difficulty. Since the algorithm takes advantage of the symmetry of the matrix, it is faster than Gaussian elimination. It is about as stable as Gaussian elimination with complete pivoting.

MATRIX INVERSION BY THE METHOD OF MODIFICATION. Assuming the necessary inverses exist, one can write for $A \in \mathbb{C}^{n\times n}$, $U, V \in \mathbb{C}^{n\times r}$, and $s \in \mathbb{C}^{r\times r}$

$$(A - USV^{\mathrm{H}})^{-1} = A^{-1} - A^{-1}UTV^{\mathrm{H}}A^{-1},$$

where

$$T = (V^{\mathrm{H}}A^{-1}U - S^{-1})^{-1}.$$

This formula, due to Sherman and Morrison for $r = 1$ and generalized by Woodbury [see Householder (1964, page 141)] expresses the inverse of A modified by a matrix of rank r in terms of the inverse of A and inverses of matrices of order r. By a judicious choice of rank 1 modifications it is possible to build up the inverse of a given matrix starting from the inverse of, say, its diagonal. While this procedure is not often used to invert matrices, it finds application in iterative methods for solving nonlinear equations which require the inverse of a matrix that from step to step is modified by a matrix of low rank [see Ortega and Rheinboldt (1970, Chapter 7)].

THE CONJUGATE GRADIENT METHOD. The conjugate gradient method is a method proposed by Hestenes and Stiefel (1952) for solving the linear system $Ax = b$, where A is positive definite of order n. Starting with an approximate solution x_0, the method generates a sequence $x_0, x_1, x_2, \ldots$ of approximate solutions in such a way that x_n is the exact solution of the system. The method requires only that one be able to form the product Ax's; hence it is well suited for problems involving large sparse matrices with special structure. Reid (1971b) has observed that the method may at an early stage produce a sufficiently accurate solution. A program for the conjugate gradient method is given by Ginsberg (HACLA/I/5).

SOLUTION OF LARGE SPARSE LINEAR SYSTEMS. Many physical problems require the solution of very large systems of linear equations whose matrices consist mostly of zero elements (sparse matrices). For example, the attempt to solve partial differential equations by difference or finite element methods leads to large systems whose matrices have special structure reflecting the nature of the differential operator and the geometry of the problem. These problems may be solved by iterative methods, such as successive over relaxation (SOR), whose analysis depends on the special properties of the matrix [for detailed treatments see the books by Varga (1962), Householder (1964), and Young (1971)]. More recently efficient direct methods for solving such systems have been devised [see the surveys by Dorr (1970) and Widlund (1972) and papers by Martin and Rose (1972) and George (1972)].

Other problems, such as network problems, give rise to large sparse systems whose matrices are less structured. For developments in this area as well as extensive bibliographies the reader is referred to the proceedings of three recent sparse matrix conferences [Willoughby (1968), Reid (1970), and Rose and Willoughby (1972)].

SOLUTION OF THE MATRIX EQUATION $AX + XB = C$. This equation is important in control theory. Bartels and Stewart (1972) give a program in which the QR algorithm is used to transform the equation into an equivalent equation in which A and B are quasi-triangular. The solution may then be calculated by a back substitution process.

THE JACOBI METHOD AND ITS EXTENSIONS. From the early 1950's to the discovery of the QR algorithm, the principal method for solving the real symmetric eigenvalue problem was by Jacobi's method. In this iteration plane rotations are used to annihilate two symmetrically placed off-diagonal elements. With a suitable strategy for choosing the off-diagonal elements, the matrix can be made to converge to a diagonal matrix, ultimately quadratically. Eigenvectors may be found by accumulating the transformations. For a brief historical discussion, further references, and an elegant program see Rutishauser (1966, HACLA/II/1).

A Jacobi-like method for general matrices has been proposed by Eberlein (1962). Ruhe (1968) has shown that such methods can converge quadratically. Programs have been published by Eberlein and Boothroyd (1968, HACLA/II/12) for real matrices and by Eberlein (1970, HACLA/II/17) for complex matrices.

EIGENVALUE TECHNIQUES BASED ON FINDING ROOTS OF THE CHARACTERISTIC EQUATION $\det(A - \lambda I) = 0$. When only a few eigenvalues and eigenvectors of a matrix are required, it may be inefficient to reduce the matrix to quasi-triangular form via the QR algorithm. An alternative is to find the eigenvalues by iterating to find roots of the characteristic equation. The corresponding eigenvectors may then be found by the inverse power method.

The eigenvalues of a real symmetric matrix may be found by first reducing the matrix to tridiagonal form and then applying the method of Sturm sequence first suggested by Givens (1954). The method has the advantage of allowing the user to specify which eigenvalues he wants.

For general matrices the procedure is to reduce the matrix to Hessenberg form, after which $\det(A - \lambda I)$ and its derivatives may be evaluated efficiently by the method of Hyman (AEP, Chapter 7). A standard iterative technique, such as Newton's method or Laguerre's method, may then be used to find the eigenvalues. There are deflation techniques that insure that an eigenvalue once found will not be found again. This technique is particularly useful when one has a sequence of eigenvalue problems, each differing only slightly from its predecessor. In this case the eigenvalues of one matrix may be used as good starting approximations for the eigenvalues of the next.

Methods based on the characteristic equation are discussed in detail by Wilkinson (AEP, Chapters 5, 7). Parlett (1964) has treated the use of Laguerre's method to find eigenvalues.

THE GENERALIZED EIGENVALUE PROBLEM $Ax = \lambda Bx$. Peters and Wilkinson have proposed a technique for nearly singular B in which the ill-conditioned part of B is deflated from the problem, after which the problem may be solved in the form $B^{-1}A$. The method is quite efficient; however, it requires that the user decide on the amount of ill-conditioning he is willing to tolerate in B. In general the method works best when the singular values of B divide sharply into two classes: those that are acceptably large and those that are negligible. A similar method for A symmetric and B positive definite has been proposed by Fix and Heiburger (1972).

THE GENERALIZED EIGENVALUE PROBLEM $\det[A(\lambda)] = 0$. Here the elements of A are polynomials in λ. This problem arises in a number of applications [see, e.g., Lancaster (1966)]. Peters and Wilkinson (1970c) discuss the computational aspects of the problem. The usual practice is to use iterative techniques to find the zeros of the function $\det[A(\lambda)]$; however, this subject has scarcely been touched and better techniques may be in the offing.

CONSTRAINED PROBLEMS. This area, which includes linear programming, has a vast literature. It has not until recently attracted the attention it deserves from numerical analysts, especially those concerned with devising efficient stable algorithms. Bartels (1971) has investigated the stability of the simplex method, and a program based on his results has been published by Bartels, Stoer, and Zenger (HACLA/I/11). Golub and Sanders (1969) and Stoer (1971) have described algorithms for the constrained least squares problem.

BIBLIOGRAPHY

Two abbreviations have been used throughout the notes and references section and in this bibliography. Wilkinson's book "The Algebraic Eigenvalue Problem" is referred to as AEP. The collection of linear algebra algorithms in Volume II of "The Handbook for Automatic Computation" edited by Wilkinson and Reinsch is referred to HACLA. The notation HACLA/I/3 refers to the third contribution in part one of the handbook.

Bartels, R. H. (1971). A stabilization of the simplex method. *Numer. Math.* **16**, 414–434.

Bartels, R. H., and Stewart, G. W. (1972). Algorithm, 432, solution of the matrix equation $AX + XB = C$. *ACM* **15**, 820–826.

Bartels, R. H., Stoer, J., and Zenger, C. (HACLA/I/11). A realization of the simplex method based on triangular decomposition.

Bauer, F. L. (1957). Das Verfahren der Treppeniteration und verwandte Verfahren zur Lösung algebraischer Eigenwertprobleme. *Z. Angew. Math. Phys.* **8**, 214–235.

Bauer, F. L. (1959). Sequential reduction to tridiagonal form. *SIAM* **7**, 107–113.

Bauer, F. L. (1963). Optimally scaled matrices. *Numer. Math.* **5**, 73–87.

Bauer, F. L. (1967). Theory of Norms. Tech. Rep. No. CS75. Comput. Sci. Dept., Stanford Univ., Stanford, California.

Bauer, F. L., and Fike, C. T. (1960). Norms and exclusion theorems. *Numer. Math.* **2**, 137–141.

Bauer, F. L., and Reinsch, C. (HACLA/I/3). Inversion of positive definitive matrices by the Gauss–Jordan method.

Björk, Å. (1967a). Solving linear least squares problems by Gram–Schmidt orthogonalization. *BIT* **7**, 1–21.

Björk, Å. (1967b). Iterative refinement of linear least squares solution I. *BIT* **7**, 251–278.

Björk, Å. (1968). Iterative refinement of linear least squares solution II. *BIT* **8**, 8–30.

Björk, Å. and Golub, G. H. (1967). Iterative refinement of linear least squares solution by Householder transformation. *BIT* **7**, 322–337.

Boothroyd, J. (1970). The *QR* algorithm for symmetric tridiagonal matrices using a semi-implicit shift of origin. *Austral. Comput. J.* **2**, 55–60.

Boullion, T. L. and Odell, P. L. (1971). "Generalized Inverse Matrices." Wiley, New York.

Bowdler, H. J., Martin, R. S., Peters, G., and Wilkinson, J. H. (1966, HACLA/I/7). Solution of real and complex systems of linear equations. *Numer. Math.* **8**, 217–234.

Bowdler, H., Martin, R. S., Reinsch, C., and Wilkinson, J. H. (1968, HACLA/II/3). The QR and QL algorithms for symmetric matrices. *Numer. Math.* **11**, 293–306.

Boyer, C. B. (1968). "A History of Mathematics." Wiley, New York.

Bunch, J. R., and Parlett, B. N. (1972). Direct methods for solving symmetric indefinite systems of linear equations. *SIAM J. Numer. Anal.* **8**, 639–655.

Businger, P. A., and Golub, G. H. (1965, HACLA/I/8). Linear least squares solutions by Householder transformations. *Numer. Math.* **7**, 269–276.

Businger, P. A., and Golub, G. H. (1969). Algorithm 358, singular value decomposition of a complex matrix. *Comm. ACM* **12**, 564–565.

Clint, M., and Jennings, A. (1970). The evaluation of eigenvalues and eigenvectors of real symmetric matrices by simultaneous iteration. *Comput. J.* **13**, 76–80.

Cryer, C. W. (1968). Pivot size in Gaussian elimination. *Numer. Math.* **12**, 335–345.

Davis, C. and Kahan, W. M. (1970). The rotation of eigenvectors by a perturbation. III. *Siam J. Numer. Anal.* **7**, 1–46.

Dorr, F. W. (1970). The direct solution of the discrete Poisson equation on a rectangle. *SIAM Rev.* **12**, 248–263.

Dubrulle, A., Martin, R. S., and Wilkinson, J. H. (1968, HACLA/II/4). The implicit QL algorithm. *Numer. Math.* **12**, 377–383.

Eberlein, P. J. (1962). A Jacobi-like method for the automatic computation of eigenvalues and eigenvectors of an arbitrary matrix. *SIAM J. Appl. Math.* **10**, 74–88.

Eberlein, P. J. (1970, HACLA/II/17). Solution to the complex eigenproblem by a norm reducing Jacobi type method. *Numer. Math.* **14**, 232–245.

Eberlein, P. J., and Boothroyd, J. (1968, HACLA/12). Solution to the eigenproblem by a norm reducing Jacobi type method. *Numer. Math.* **11**, 1–12.

Eckart, C., and Young, G. (1936). The approximation of one matrix by another of lower rank. *Psychometrika* **1**, 211–218.

Eckart, C., and Young, G. (1939). A principal axis transformation for non-Hermitian matrices. *Bull. Amer. Math. Soc.* **45**, 118–121.

Faddeev, D. K., and Faddeeva, V. N. (1960). "Computational Methods of Linear Algebra," Russian ed., Freeman, San Francisco, California.

Faddeev, D. K., and Faddeeva, V. N. (1963). "Computational Methods of Linear Algebra." Freeman, San Francisco, California.

Feller, W., and Forsythe, G. E. (1951). New matrix transformations for obtaining characteristic vectors. *Quart. Appl. Math.* **8**, 325–331.

Fix, G., and Heiberger, R. (1972). An algorithm for the ill-conditioned generalized eigenvalue problem. *SIAM J. Numer. Anal.* **9**, 78–88.

Forsythe, G. E., and Strauss, E. G. (1955). On best conditioned matrices. *Proc. Amer. Math. Soc.* **6**, 340–345.

Forsythe, G., and Moler, C. B. (1967). "Computer Solution of Linear Algebraic Systems." Prentice-Hall, Englewood Cliffs, New Jersey.

Francis, J. G. F. (1961, 1962). The *QR* transformation I, II. *Computer J.* **4**, 265–271, 332–345.

George, J. A. (1972). Block eliminations on finite element systems of equations. *In* "Sparse Matrices and their Applications" (D. J. Rose and R. A. Willough by, eds.), 101–114.

Ginsberg, T. (HACLA/I/5). The conjugate gradient method.

Givens, J. W. (1954). Numerical computation of the characteristic values of a real symmetric matrix. Rep. ORNL-1574. Oak Ridge Nat. Lab., Oak Ridge, Tennessee.

Gohberg, I. C., and Krein, M. G. (1969). "Introduction to the Theory of Linear Nonself-adjoint Operators." Amer. Math. Soc., Providence, Rhode Island.

Golub, G. H. (1965). Numerical methods for solving linear least squares problems. *Numer. Math.* **7**, 206–216.

Golub, G. H. (1969). Matrix decompositions and statistical calculations. Tech. Rep. No. CS124. Comput. Sci. Dept., Stanford Univ., Stanford, California.

Golub, G. H., and Kahan, W. (1965). Calculating the singular values and pseudo-inverse of a matrix. *SIAM J. Numer. Anal.* **2**, 202–224.

Golub, G. H., and Reinsch, C. (1970, HACLA/I/10). Singular value decomposition and least squares solutions. *Numer. Math.* **14**, 403–420.

Golub, G. H., and Sanders, M. A. (1969). Linear least squares and quadratic programming. Tech. Rept. No. CS134. Comput. Sci. Dept., Stanford Univ., Stanford, California.

Golub, G. H., and Wilkinson, J. H. (1966). Note on iterative refinement of least squares solution. *Numer. Math.* **9**, 139–148.

Halmos, P. R. (1958). "Finite Dimensional Vector Spaces," 2nd ed. Van Nostrand-Reinhold, Princeton, New Jersey.

Hestenes, M. R., and Stiefel, E. (1952). Methods of conjugate gradients for solving linear systems. *J. Res. Nat. Bur. Standards* **49**, 409–436.

Hoffman, A. J., and Wielandt, H. W. (1953). The variation of the spectrum of a normal matrix. *Duke Math. J.* **20**, 37–39.

Householder, A. S. (1958a). The approximate solution of matrix problems. *J. Assoc. Comput. Mach.* **5**, 204–243.

Householder, A. S. (1958b). Unitary triangularization of a nonsymmetric matrix. *J. Assoc. Comput. Mach.* **5**, 339–342.

Householder, A. S. (1964). "The Theory of Matrices in Numerical Analysis." Ginn (Blaisdell), Boston, Massachusetts.

Householder, A. S., and Bauer, F. L. (1958). On certain methods for expanding the characteristic polynomial. *Numer. Math.* **1**, 29–37.

Householder, A. S., and Young, G. (1938). Matrix approximation and latent roots. *Amer. Math. Monthly* **45**, 165–171.

Issacson, E., and Keller, H. B. (1966). "Analysis of Numerical Methods." Wiley, New York.

Kahan, W. (1966). Numerical linear algebra. *Canad. Math. Bull.* **9**, 756–801.

Kahan, W., and Parlett, B. N. (1968). On the convergence of a practical QR algorithm. *Proc. IFIP Congr.* A25-A30.

Kublanovskaya, V. N. (1961). On some algorithms for the solution of the complete eigenvalue problem. *Z. Vyčisl. Mat. i Mat. Fiz.* **1**, 555–570; *USSR Comput. Math. Math. Phys.* **3**, 637–657.

Lancaster, P. (1966). "Lambda-matrices and Vibrating Systems." Pergamon, Oxford.

Lanczos, C. (1950). An iteration method for the solution of the eigenvalue problem of linear differential and integral operators. *J. Res. Nat. Bur. Standards* **45**, 255–282.

Martin, M. S. and Rose, D. J. (1972). Complexity bounds for regular finite difference and finite element grids. Univ. of Denver, Report MS-R-7221, Denver, Colorado.

Martin, R. S., and Wilkinson, J. H. (1965, HACLA/I/4). Symmetric decomposition of positive definite band matrices. *Numer. Math.* **7**, 335–361.

Martin, R. S., and Wilkinson, J. H. (1967, HACLA/I/6). Solution of symmetric and unsymmetric band equations and the calculation of eigenvectors of band matrices. *Numer. Math.* **9**, 279–301.

Martin, R. S., and Wilkinson, J. H. (1968a, HACLA/II/13). Similarly reduction of a general matrix to Hessenberg form. *Numer. Math.* **12**, 369–368.

Martin, R. S., and Wilkinson, J. H. (1968b, HACLA/II/16). The modified *LR* algorithm for complex Hessenberg matrices. *Numer. Math.* **12**, 379–376.

Martin, R. S., Peters, G., and Wilkinson, J. H. (1965, HACLA/I/1). Symmetric decomposition of a positive definite matrix. *Numer. Math.* **7**, 363–383.

Martin, R. S., Peters, G., and Wilkinson, J. H. (1966, HACLA/I/2). Iterative refinement of the solution of a positive definite system of equations. *Numer. Math.* **8**, 203–216.

Martin, R. S., Reinsch, C., and Wilkinson, J. H. (1968, HACLA/II/2). Householder tridiagonalization of a real symmetric matrix. *Numer. Math.* **11**, 181–195.

Martin, R. S., Peters, G., and Wilkinson, J. H. (1970, HACLA/II/14). The *QR* algorithm for real Hessenberg matrices. *Numer. Math.* **14**, 219–231.

Moler, C. B. (1967). Iterative refinement in floating point. *J. Assoc. Comput. Mach.* **14**, 316–321.

Moler, C. B., and Stewart, G. W. (1971). An algorithm for the generalized matrix eigenvalue problem $Ax = \lambda Bx$. Rept. CNA-32. Center for Numer. Anal., Univ. of Texas, Austin, Texas. *SIAM J. Numer. Anal.* (to appear).

Moore, E. H. (1919–1920). On the reciprocal of the general algebraic matrix, Abstract. *Bull. Amer. Math. Soc.* **26**, 394–394.

Noble, B. (1969). "Applied Linear Algebra." Prentice-Hall, Englewood Cliffs, New Jersey.

Ortega, J. M., and Kaiser, H. F. (1963). The LL^T and *QR* methods for symmetric tridiagonal matrices. *Comput. J.* **6**, 99–101.

Ortega, J. M., and Rheinboldt, W. C. (1970). "Iterative Solution of Nonlinear Equations in Several Variables." Academic Press, New York.

Ostrowski, A. (1958–1959). On the convergence of the Rayleigh quotient iteration for the computation of characteristic roots and vectors I, II, III, IV, V, VI, *Arch. Rational Mech. Anal.* **1**, 233–241; **2**, 423–428; **3**, 325–340, 341–347, 472–481; **4**, 153–165.

Ostrowski, A. M. (1960). "Solution of Equations and Systems of Equations." Academic Press, New York.

Ostrowski, A. M. (1966). "Solution of Equations and Systems of Equations," 2nd ed. Academic Press, New York.

Parlett, B. N. (1964). Laguerre's method applied to the matrix eigenvalue problem. *Math. Comp.* **18**, 464–485.

Penrose, R. (1955). A generalized inverse for matrices. *Proc. Cambridge Philos Soc.* **51**, 406–413.

Penrose, R. (1956). On best approximate solution of linear matrix equations. *Proc. Cambridge Philos. Soc.* **52**, 17–19.

Peters, G., and Wilkinson, J. H. (1970a, HACLA/II/15). Eigenvectors of real and complex matrices by *LR* and *QR* triangularizations. *Numer. Math.* **16**, 181–204.

Peters, G., and Wilkinson, J. H. (1970b). The least squares problem and pseudoinverses. *Comput. J.* **13**, 309–316.

Peters, G., and Wilkinson, J. H. (1970c). $Ax = \lambda Bx$ and the generalized eigenproblem. *SIAM J. Numer. Anal.* **7**, 479–492.

Rao, C. R. and Mitra, S. K. (1971). "Generalized Inverse of Matrices and Its Applications." Wiley, New York.

Reid, J. K., ed. (1971a). "Large Sparse Sets of Linear Equations." Academic Press, New York.

Reid, J. K. (1971b). On the method of conjugate gradients for the solution of large sparse systems of linear equations. *In* "Large Sparse Sets of Linear Equations" (J. K. Reid, ed.), pp. 231–254. Academic Press, New York.

Reinsch, C., and Bauer, F. L. (1968, HACLA/II/6). Rational *QR* transformation with Newton shift for symmetric tridiagonal matrices. *Numer. Math.* **11**, 264–272.

Rice, J. R. (1966). Experiments on Gram–Schmidt orthogonalization. *Math. Comp.* **20**, 325–328.

Rose, D. J. and Willoughby, R. A., eds. (1972). "Sparse Matrices and Their Applications." Plenum Press, New York.

Ruhe, A. (1968). On the quadratic convergence of a generalization of the Jacobi method to arbitrary matrices. *BIT* **8**, 210–231.

Ruhe, A. (1970a). An algorithm for numerical determination of the structure of a general matrix. *BIT* **10**, 196–216.

Ruhe, A. (1970b). Perturbation bounds for means of eigenvalues and invariant subspaces. *BIT* **10**, 343–354.

Rutishauser, H. (1958). Solution of eigenvalue problems with the LR-transformation. *Nat. Bur. Standards Appl. Math. Ser.* **49**, 47–81.

Rutishauser, H. (1966, HACLA/II/1). The Jacobi method for real symmetric matrices. *Numer. Math.* **9**, 1–10.

Rutishauser, H. (1969). Computational aspects of F. L. Bauer's simultaneous iteration method. *Numer. Math.* **13**, 4–13.

Rutishauser, H. (1970, HACLA/II/9). Simultaneous iteration method for symmetric matrices. *Numer. Math.* **16**, 205–223.

Stewart, G. W. (1969a). On the continuity of the generalized inverse. *SIAM J. Appl. Math.* **17**, 33–45.

Stewart, G. W. (1969b). Accelerating the orthogonal iteration for the eigenvalues of a Hermitian matrix. *Numer. Math.* **13**, 362–376.

Stewart, G. W. (1970a). Incorporating origin shifts into the *QR* algorithm for symmetric tridiagonal matrices. *Comm. ACM* **13**, 365–367.

Stewart, G. W. (1970b). Algorithm 384, eigenvalues and eigenvectors of a real symmetric matrix. *Comm. ACM* **13**, 369–371; Errata **13**, 750.

Stewart, G. W. (1972). On the sensitivity of the eigenvalue problem. *SIAM J. Numer. Anal.* **9**, 669–686.

Stewart, G. W. (1972). Error bounds for invariant subspaces of closed operators. *SIAM J. Numer. Anal.* **8**, 796–808.

Stoer, J. (1971). On the numerical solution of constrained least-squares problems. *SIAM J. Numer. Anal.* **8**, 382–411.

Sylvester, J. J. (1889). *Messenger of Math.* **19**, 42. Cited by Eckart and Young (1939).

Taussky, O. (1949). A recurring theorem on determinants. *Amer. Math. Monthly* **56**, 672–676.

Turing, A. M. (1948). Rounding-off errors in matrix processes. *Quart. J. Mech. Appl. Math.* **1**, 287–308.

Varah, J. M. (1967). The computation of bounds for the invariant subspaces of a general matrix operator. Tech. Rep. CS66. Comput. Sci. Dept., Stanford Univ., Stanford, California.

Varah, J. M. (1970). Computing invariant subspaces of a general matrix when the eigensystem is poorly conditioned. *Math. Comp.* **24**, 137–149.

Varga, R. S. (1962). "Matrix Iterative Analysis." Prentice-Hall, Englewood Cliffs, New Jersey.

Varga, R. S. (1965). Minimal Gerschgorin sets. *Pacific J. Math.* **15**, 719–729.

von Neumann, J., and Goldstine, H. H. (1947). Numerical inverting of matrices of high order. *Bull. Amer. Math. Soc.* **53**, 1021–1099.

Welsch, J. H. (1967). Certification of algorithm 253, eigenvalues of a real symmetric matrix by the *QR* method. *Comm. ACM* **10**, 376.

Wendroff, B. (1969). "First Principles of Numerical Analysis." Addison-Wesley, Reading, Massachusetts.

Widlund, O. B. (1972). On the use of fast methods for separable finite difference equations for the solution of general elliptic problems. In "Sparse Matrices and Their Applications" (D. J. Rose and R. A. Willoughby, eds.), 121–134.

Wielandt, H. (1944). Das Iterationsverfahren bei nicht selbstadjungierten linearen Eigenwertaufgaben. *Math. Z.* **50**, 93–143.

Wilkinson, J. H. (1959). The evaluation of the zeros of ill-conditioned polynomials. *Numer. Math.* **1**, 150–180.

Wilkinson, J. H. (1960). Error analysis of floating-point computation. *Numer. Math.* **2**, 319–340.

Wilkinson, J. H. (1961). Error analysis of direct methods of matrix inversion. *J. Assoc. Comput. Mach.* **8**, 281–330.

Wilkinson, J. H. (1963). "Rounding Errors in Algebraic Processes." Prentice-Hall, Englewood Cliffs, New Jersey.

Wilkinson, J. H. (1965a). "The Algebraic Eigenvalue Problem." Oxford Univ. Press (Clarendon), London and New York.

Wilkinson, J. H. (1965b). Error analysis of transformations based on the use of matrices of the form $I\text{-}2ww^{\mathrm{H}}$. In "Error in Digital Computation" (L. B. Rall, ed.), Vol. 2, pp. 77–101. Wiley, New York.

Wilkinson, J. H. (1965c). Convergence of the *LR*, *QR*, and related algorithms. *Comput. J.* **8**, 77–84.

Wilkinson, J. H. (1968). Global convergence of tridiagonal *QR* algorithm with origin shifts. *Lin. Alg. and Appl.* **1**, 409–420.

Wilkinson, J. H. (1971). Modern error analysis. *SIAM Rev.* **13**, 548–568.

Wilkinson, J. M., and Reinsch, C., eds. (1971). "Handbook for Automatic Computation, Vol. II, Linear Algebra." Springer-Verlag, Berlin and New York.

Willoughby, R. A., ed. (1968). "Sparse Matrix Proceedings." IBM Research Report RAI, Yorktown Heights, New York.

Young, D. M. (1971). "Iterative Solution of Large Linear Systems." Academic Press, New York.

INDEX OF NOTATION

INDEX OF ALGORITHMS

INDEX

M

R

S

Computer Science and Applied Mathematics

A SERIES OF MONOGRAPHS AND TEXTBOOKS

Editor

Werner Rheinboldt

University of Maryland

HANS P. KÜNZI, H. G. TZSCHACH, and C. A. ZEHNDER. Numerical Methods of Mathematical Optimization: With ALGOL and FORTRAN Programs, Corrected and Augmented Edition

AZRIEL ROSENFELD. Picture Processing by Computer

JAMES ORTEGA AND WERNER RHEINBOLDT. Iterative Solution of Nonlinear Equations in Several Variables

AZARIA PAZ. Introduction to Probabilistic Automata

DAVID YOUNG. Iterative Solution of Large Linear Systems

ANN YASUHARA. Recursive Function Theory and Logic

JAMES M. ORTEGA. Numerical Analysis: A Second Course

G. W. STEWART. Introduction to Matrix Computations

CHIN-LIANG CHANG AND RICHARD CHAR-TUNG LEE. Symbolic Logic and Mechanical Theorem Proving

C. C. GOTLIEB AND A. BORODIN. Social Issues in Computing

ERWIN ENGELER. Introduction to the Theory of Computation

F. W. J. OLVER. Asymptotics and Special Functions

DIONYSIOS C. TSICHRITZIS AND PHILIP A. BERNSTEIN. Operating Systems

ROBERT R. KORFHAGE. Discrete Computational Structures

PHILIP J. DAVIS AND PHILIP RABINOWITZ. Methods of Numerical Integration

A. T. BERZTISS. Data Structures: Theory and Practice, Second Edition

In preparation

ALBERT NIJENHUIS AND H. S. WILF. Combinatorial Algorithms

5
6
C 7
D 8
E 9
F 0
G 1
H 2
I 3
J 4